# Microbial Gas Metabolism

**Special Publications of the Society for General Microbiology**

Publications Officer: Colin Ratledge, 62 London Road, Reading, UK.

This book is based on a Symposium of the SGM held at Reading, 5 January 1984.

# Microbial Gas Metabolism

## Mechanistic, Metabolic and Biotechnological Aspects

Edited by

### Robert K. Poole

*Department of Microbiology*
*Queen Elizabeth College*
*University of London*
*London, United Kingdom*

and

### Crawford S. Dow

*Department of Biological Sciences*
*University of Warwick*
*Coventry, United Kingdom*

### 1985

Published for the
**Society for General Microbiology**
by
**ACADEMIC PRESS**
*(Harcourt Brace Jovanovich, Publishers)*
London   Orlando   San Diego   New York
Toronto   Montreal   Sydney   Tokyo

ACADEMIC PRESS INC. (LONDON) LTD.
24–28 Oval Road
LONDON NW1 7DX


*United States Edition published by*
ACADEMIC PRESS, INC.
Orlando, Florida 32887


British Library Cataloguing in Publication Data

Microbial gas metabolism : mechanistic,
    metabolic and biotechnological aspects.—
    (Special publications of the Society for
    General Microbiology; no. 14)
    1. Microbial metabolism    2. Gases
    I. Poole, Robert    II. Dow, Crawford S.
    III. Series
    576'.1133        QR88


Library of Congress Cataloging in Publication Data
Main entry under title:

Microbial gas metabolism.

    (Special publications of the Society for General
Microbiology ; 14)
    Includes index.
    1. Microbial metabolism.    2. Gas—Metabolism.
I. Poole, Robert K.    II. Dow, Crawford S.    III. Series.
QR88.M525    1985      576'.11      84-21546
ISBN 0—12—561480—2 (alk. paper)


PRINTED IN THE UNITED STATES OF AMERICA

85 86 87 88      9 8 7 6 5 4 3 2 1

# Contents

# Contributors

P. R. ALEFOUNDER   *Department of Biochemistry, University of Birmingham, Birmingham B15 2TT, United Kingdom*

WILLIAM J. ASTON   *Biotechnology Centre, Cranfield Institute of Technology, Cranfield, Bedfordshire MK43 0AL, United Kingdom*

BALDEV S. BAINES   *Department of Microbiology, Queen Elizabeth College, University of London, London W8 7AH, United Kingdom (Present address: Biotechnology Department, Microbiology Division, Glaxo Group Research Ltd., Greenford Road, Greenford, Middlesex UB6 0HE, United Kingdom)*

STUART P. BALLANTINE   *Department of Biochemistry, Medical Sciences Institute, Dundee University, Dundee, Tayside DD1 4HN, United Kingdom*

JUDITH M. BELL   *Biology Department, Sunderland Polytechnic, Sunderland SR1 3SD, United Kingdom*

DAVID H. BOXER   *Department of Biochemistry, Medical Sciences Institute, Dundee University, Dundee, Tayside DD1 4HN, United Kingdom*

RICHARD CAMMACK   *Department of Plant Sciences, King's College London, University of London, London SE24 9JF, United Kingdom*

JOHN COLBY   *Biology Department, Sunderland Polytechnic, Sunderland SR1 3SD, United Kingdom*

R. P. COX   *Institute of Biochemistry, Odense University, 5230 Odense M, Denmark*

HOWARD DALTON   *Department of Biological Sciences, University of Warwick, Coventry CV4 7AL, United Kingdom*

GRAHAM DAVIS   *Biotechnology Centre, Cranfield Institute of Technology, Cranfield, Bedfordshire MK43 0AL, United Kingdom*

GABRIELE DIEKERT   *Mikrobiologie, Fachbereich Biologie, Philipps-Universität Marburg, D-3550 Marburg, Federal Republic of Germany*

S. J. FERGUSON   *Department of Biochemistry, University of Birmingham, Birmingham B15 2TT, United Kingdom*

GEORG FUCHS   *Universität Ulm, Abteilung Angewandte Mikrobiologie, D-7900 Ulm, Federal Republic of Germany*

A. J. GREENFIELD   *Department of Biochemistry, University of Birmingham, Birmingham B15 2TT, United Kingdom*

DAVID O. HALL   *Department of Plant Sciences, King's College London, University of London, London SE24 9JF, United Kingdom*

I. JOHN HIGGINS   *Biotechnology Centre, Cranfield Institute of Technology, Cranfield, Bedfordshire MK43 0AL, United Kingdom*

H. ALLEN O. HILL   *Inorganic Chemistry Laboratory, University of Oxford, Oxford OX1 3QR, United Kingdom*

K. HILLMAN   *Department of Microbiology, University College, Cardiff CF2 1TA, Wales, United Kingdom*

JULIA A. M. HUBBARD   *Departments of Chemistry and Microbiology, Queen Elizabeth College, University of London, London W8 7AH, United Kingdom*

MARTIN N. HUGHES   *Department of Chemistry, Queen Elizabeth College, University of London, London W8 7AH, United Kingdom*

B. B. JENSEN   *Institute of Biochemistry, Odense University, 5230 Odense M, Denmark*

LARS JOERGENSEN  *Institute of Biochemistry, Odense University, 5230 Odense M, Denmark*

DAVID J. LEAK  *Department of Biological Sciences, University of Warwick, Coventry CV4 7AL, United Kingdom (Present address: Centre for Biotechnology, Imperial College of Science and Technology, London SW7 2AZ, United Kingdom)*

DAVID LLOYD  *Department of Microbiology, University College, Cardiff CF2 1TA, Wales, United Kingdom*

MARIE-ANDREÉ MANDRAND-BERTHELOT  *Laboratoire de Microbiologie, INSA, 69621 Villeurbanne, France*

J. E. G. McCARTHY  *Department of Biochemistry, University of Birmingham, Birmingham B15 2TT, United Kingdom*

ORTWIN MEYER  *Institut für Mikrobiologie der Georg-August-Universität Göttingen, D-3400 Göttingen, Federal Republic of Germany*

ROBERT K. POOLE  *Department of Microbiology, Queen Elizabeth College, University of London, London W8 7AH, United Kingdom*

K. KRISHNA RAO  *Department of Plant Sciences, King's College London, University of London, London SE24 9JF, United Kingdom*

ROBERT I. SCOTT  *Department of Microbiology, University College, Cardiff CF2 1TA, Wales, United Kingdom (Present address: The Polytechnic of Central London, School of Engineering and Science, 115 New Cavendish Street, London W1M 8JS, United Kingdom)*

STEPHEN H. STANLEY  *Department of Biological Sciences, University of Warwick, Coventry CV4 7AL, United Kingdom*

RUDOLF K. THAUER  *Mikrobiologie, Fachbereich Biologie, Philipps-Universität Marburg, D-3550 Marburg, Federal Republic of Germany*

ANTHONY P. F. TURNER  *Biotechnology Centre, Cranfield Institute of Technology, Cranfield, Bedfordshire MK43 0AL, United Kingdom*

ROBERT WAUGH  *Department of Biochemistry, Medical Sciences Institute, Dundee University, Dundee, Tayside DD1 4HN, United Kingdom*

TIMOTHY N. WHITMORE  *Department of Microbiology, University College, Cardiff CF2 1TA, Wales, United Kingdom*

A. G. WILLIAMS  *Department of Animal Nutrition and Production, Hannah Research Institute, Ayr KA6 5HL, Scotland, United Kingdom*

EDWIN WILLIAMS  *Microbiology Department, The University, Newcastle-upon-Tyne NE1 7RU, United Kingdom*

HUW D. WILLIAMS  *Department of Microbiology, Queen Elizabeth College, University of London, London W8 7AH, United Kingdom*

T. NORMAN WILLIAMS  *Department of Microbiology, University College, Cardiff CF2 1TA, Wales, United Kingdom*

MARC P. WOODLAND  *Department of Biological Sciences, University of Warwick, Coventry CV4 7AL, United Kingdom (Present address: Centre for Biotechnology, Imperial College of Science and Technology, London SW7 2AZ, United Kingdom)*

# Introduction

On January 5, 1984, the Cell Biology Group of the Society for General Microbiology held a Symposium, at the 99th Ordinary Meeting in Reading, on Microbial Gas Metabolism.

Of course, the subject area is not new, for microbiologists have been interested in gases since Pasteur's time. The concept of such a symposium is not new either. A much larger meeting (though confined to the production and utilization of hydrogen, methane and carbon monoxide) was held in Göttingen in 1975 and the proceedings published in a valuable book (Schlegel *et al.*, 1976). Not even the main title is novel (see Cole, 1976). What *is* new, or at least growing rapidly, is the detail in which some of the underlying metabolic pathways and biochemical mechanisms can now be described and the intense commercial and biotechnological interest expressed in an increasing number of these processes, especially photosynthesis, nitrogen fixation and the metabolism of methane and carbon monoxide. Thus arose the sub-title of the proceedings: ''Mechanistic, Metabolic and Biotechnological Aspects''.

In his review, Cole (1976) identified four gases that provide the biosphere with reservoirs of potentially useful carbon, nitrogen and free energy, namely carbon dioxide, nitrogen, oxygen and hydrogen. The last three were subjects of invited papers in the Reading symposium (albeit bearing their *newer* names: dinitrogen, dioxygen and dihydrogen). In addition, there were papers on carbon monoxide, methane, powerful analytical techniques only recently applied to microbiological systems and, significantly, the inorganic chemistry of the metalloenzymes that are largely responsible for gas metabolism.

Figure 1 attempts to emphasize the close interrelationships between these four gases as well as two others which have more recently acquired great importance, carbon monoxide and methane. Some of these processes are uniquely microbial; all can be carried out by prokaryotes. The cycles are so inextricably linked that Fig. 1 is a gross oversimplification. Especially significant are the mutual inhibitory interactions between some of these reactions, such as the inhibition of nitrogen fixation by oxygen and of oxygen binding and reduction by carbon monoxide.

This volume presents the contents of the seven invited lectures, together with extended abstracts based on many of the pertinent posters displayed at the meeting. These abstracts represent a departure from previous publications in this series; the editors hope that they increase the breadth and interest of the book. The eloquence and pellucidity of the question-and-answer sessions following each main talk have only a little to do with the fact that written forms of the

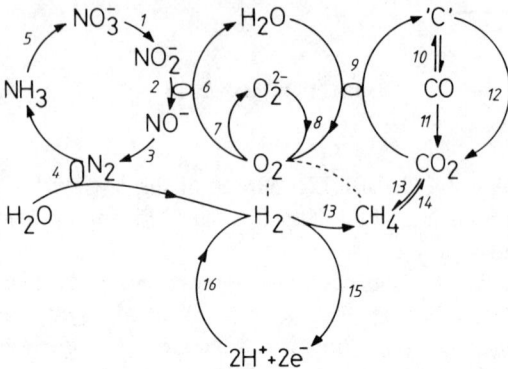

**Fig. 1.** A simplified scheme of the interactions between the natural cycles of nitrogen, oxygen, carbon dioxide and hydrogen. (1) Nitrate reduction; (2) nitrite reduction; (1–3) denitrification; (4) nitrogen fixation and concomitant hydrogen evolution; (5) nitrification; (6) $O_2$ reduction to water may be catalysed by the same enzyme as (2), e.g. in *Pseudomonas aeruginosa;* (7) $O_2$ reduction to peroxide; (8) catalase; (9) photosynthesis; (10) CO assimilation by carboxydobacteria: (11) carbon monoxide oxidation (e.g. by cytochrome $c$ oxidase); (12) "respiratory" carbon dioxide evolution; (13) methanogenesis; (14) methane utilization; (15) hydrogen-assimilating hydrogenase; (16) hydrogen-evolving hydrogenase. Dashed lines indicate interactions (with oxygen), rather than transformations.

questions were solicited from participants and then passed to the speakers for rumination at leisure. The chairmen have given their assurances, however, that these dialogues are a true representation of the symposium proceedings!

The organizer and editors take this opportunity of thanking all who spoke, presented posters and submitted written material, as well as the meetings staff of the Society and the excellent local organization at Reading.

*Robert K. Poole*
*Crawford S. Dow*

REFERENCES

Cole, J. A. (1976). Microbial gas metabolism. *In* "Advances in Microbial Physiology", Vol. 14 (Eds. A. H. Rose and D. W. Tempest), pp. 1–92. Academic Press, London and New York.
Schlegel, H. G., Gottschalk, G. and Pfenning, N. (Eds.) (1976). "Microbial Production and Utilization of Gases ($H_2$, $CH_4$, CO)". Akademie der Wissenschaften, Göttingen.

# Microbial Gas Metabolism

# Part I
## Inorganic Aspects

# 1

# The Inorganic Chemistry of Microbial Gas Metabolism

MARTIN N. HUGHES

*Department of Chemistry, Queen Elizabeth College, University of London, London, United Kingdom*

## Introduction

Much current work in inorganic chemistry is relevant, directly or indirectly, to the topic of microbial gas metabolism. Inorganic chemists are interested in the activation of these gases by metal complexes, in kinetic and structural aspects of model compounds for the active sites of the appropriate metalloenzymes and in the application of instrumental techniques in the study of these enzymes. In addition, there is a great deal of interest in the catalysis of the reactions of carbon monoxide, dihydrogen and methane by simple metal complexes, metal clusters and metal surfaces in order to find new routes to chemical feedstocks as petroleum reserves dwindle. These particular reactions have no biological flavour about them, and, in some cases, have no biological analogues. Nevertheless they may provide useful parallels with biological catalysis involving both isolated and multimetal centres.

The gases of microbiological interest are shown in Table 1, together with some indication of the processes in which they are involved. There is considerable overlap between some of these topics. Thus hydrogenases are involved in the chemolithotrophic, methanogenic, acetogenic and dinitrogen-fixing bacteria. The possibility of inhibitory or destructive interactions must also be borne in mind. Thus it is well known that nitrogenase is most sensitive to dioxygen. It is also remarkable that aerobic carbon monoxide-utilising bacteria can withstand the toxic effects of carbon monoxide. This leads to interesting questions about competition between dioxygen and carbon monoxide for binding sites on heme proteins.

The importance of transition metals in microbial gas metabolism is demonstrated readily on inspection of Table 1. Reactions of gases such as dioxygen,

3

MICROBIAL GAS METABOLISM:
MECHANISTIC, METABOLIC
AND BIOTECHNOLOGICAL ASPECTS

**Table 1.**  *Some gases of microbiological interest*

| Gas | Processes involved |
| --- | --- |
| Dihydrogen | Anaerobic production or oxidation of $H_2$ |
|  | Aerobic hydrogen-oxidising bacteria |
| Methane | Methanogenesis |
|  | Decomposition, $CH_4 \rightarrow CH_3OH$ |
| Carbon monoxide | Oxidation to $CO_2$ by a variety of processes |
|  | Reduction to methane |
| Dinitrogen | Reduction to ammonia |
| Nitrogen monoxide<br>Dinitrogen monoxide | $\Big\}$ Intermediates in denitrification (?) |
| Dioxygen | Cytochrome oxidase, oxygenases and monooxygenases; parallels with carriers |
|  | $O_2$ evolution by cyanobacteria |
| Carbon dioxide | Photosynthesis |
|  | Reduction to methane (*Methanobacterium thermoautotrophicum*) |

dinitrogen and other gases in the nitrogen cycle involve their coordination to a transition metal centre. It seems entirely reasonable also that carbon monoxide will be activated by binding to a nickel centre. It is noteworthy too that methane has recently been shown to undergo activation by metal complexes (Watson, 1983), and to add oxidatively to an organoiridium complex (Hoyano *et al.*, 1983). The reactions of these gases subsequent to coordination involve redox pathways and electron transfer, usually by membrane-bound transition metalloproteins.

The transition metals iron, nickel, copper and molybdenum are of particular interest in this symposium. Attention must be drawn to nickel, as only comparatively recently has its importance been recognised. Nevertheless, nickel is probably quite widespread, for example, in membrane-bound hydrogenases. The requirement for nickel was probably overlooked in the past because nickel was present as an impurity to an extent sufficient to prevent it from becoming growth-limiting. Molybdenum is the only second row transition element known to have a biological role, and is present in nitrogenase, and in carbon monoxide oxidase in the aerobic carboxydobacteria.

## Transition Metals in Microbial Gas Metabolism

The transition metals have special properties, arising from the presence of only partly filled *d* orbitals, that make them well suited for the binding of these gases and for the subsequent control of their reactions. The electronic and chemical properties of the transition metals and their coordination complexes are well described in standard texts (such as Cotton and Wilkinson, 1980; Purcell and Kotz,1977), which will give detailed information on preferred oxidation states, ligands, coordination numbers and geometry for each metal, together with data on redox potentials. Transition metal centres bound by proteins or other biological molecules may have rather different, indeed unique, properties reflecting the influence of the biological ligand (Hughes, 1981). Thus, metals may be bound in unusual or distorted geometries, or with open sites, with resulting enhanced reactivity. Again, less usual oxidation states may become more important.

The metal will usually be bound either by a tetrapyrrole macrocycle (porphyrin or occasionally a hydroporphyrin) or will form part of a dimer or cluster in which the metal ions are bridged by sulphide or oxo groups. Both cluster and macrocycle complex are easily polarised, and are therefore subject to fine control by interaction with the protein, which will also provide additional ligands for the metal. Nickel appears to be bound in some cases by a protein (where it may be part of a cluster) and in others by a tetrapyrrole. Factor 430, an example of the latter class, has been studied by NMR spectroscopy (Pfaltz et al., 1982), but a definitive answer awaits an X-ray diffraction study. The NMR evidence suggests that the nickel is bound by a hydroporphyrin, but in other cases the ligand may well be a corrinoid group.

It is therefore of interest to compare the reactivities of porphyrin- and corrin-bound metal ions (Fig. 1). These may be related to the size of the cavity. Cobalt in vitamin $B_{12}$ coenzymes fits tightly into the smaller cavity provided by the corrin. Thus it does not move relative to the corrin plane, and remains coplanar with the four N donor atoms. The cobalt remains low-spin, with the $d$ electrons paired up as much as possible. Reactivity is then controlled by loss of axial ligands, giving 5- and 4-coordinate complexes. The formal oxidation state changes from Co(III) to Co(II) and Co(I) as this takes place, in accord with the free-radical behaviour associated with these coenzymes.

The cavity in the porphyrin ligand is larger. Fe(II) and Fe(III) are too small to

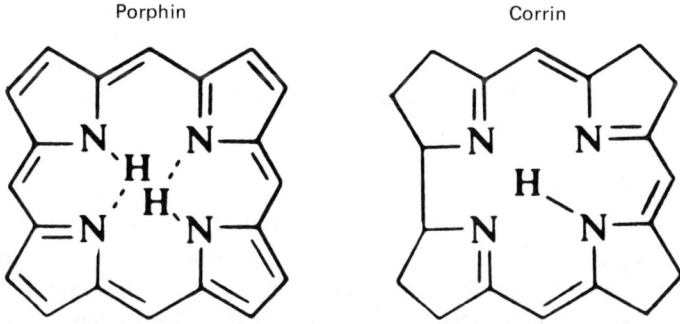

Porphin                    Corrin

**Fig. 1.** Porphin and corrin ring structures.

fit tightly into the cavity, and the ring must be buckled to allow good interaction between the N donor atoms and the metal. This misfitting allows the iron centre to move in and out of the plane in response to various factors, with resulting changes from low spin to high spin from heme to heme. This will have important implications for the control of redox potential and rates of electron transfer reactions, as the occupancy of orbitals by electrons will be affected (Moore and Williams, 1977; Scheidt and Reed, 1981).

The protein will control the physiological function of the heme. It can give a hidden or exposed heme. It may make the heme 5- or 6-coordinate, and so, in the former case, able to bind small molecules. It may control directly the electronic properties of the heme by providing axial ligands of a particular type, such as good $\sigma$ donors and $\pi$ acceptors, which in turn control the spin state and redox properties, and the extent of electron transfer to reducible substrates (Buchler *et al.*, 1978). The redox potential of a cytochrome may also be affected by steric effects and the electrostatic charge on the protein.

The magnetic properties of the metal centre may also change on binding of the substrate at the vacant coordination site. The strong field ligand carbon monoxide will tend to give low-spin complexes, while the binding of dioxygen to iron(II) in hemoglobin produces a change from high spin to low spin because the iron moves into the plane of the porphyrin and becomes subject fully to the ligand field of the prophyrin.

Iron in iron hemes normally cycles between Fe(II) and Fe(III). However, formal high oxidation states are also known, namely Fe(IV), Fe(V) and Fe(VI). Thus compound 1 formed during peroxidase activity is formally Fe(V), with two oxidising equivalents above Fe(III). Compound 1 is actually an Fe(IV)–oxo complex, with an extra oxidising equivalent present on the porphyrin, which is then a cation radical species $O\!=\!Fe^{IV}P^+$. In other cases, additional oxidising equivalents may be associated with an oxidised amino acid residue.

### Dimers and Clusters

These involve oxide ($O^{2-}$) or sulphide ($S^{2-}$) bridging ligands. The former ligand is correctly termed an "oxo" group, but there is a lack of consistency in naming the latter ligand. Both bridges allow antiferromagnetic interaction between metal centres.

*Iron–sulphur proteins.* These are involved in electron transport and in certain complex oxidation–reduction enzymes. Bacterial iron–sulphur proteins have been reviewed by Yoch and Carithers (1979). The familiar structures of the two-iron and four-iron sulphur centres are shown in Fig. 2. In each case, the iron is surrounded by a tetrahedral array of sulphur donors, either cysteine residues or, in the case of bridged systems, sulphide groups. Each structural unit can transfer one electron. Thus, the oxidised form of the two-iron protein involves two

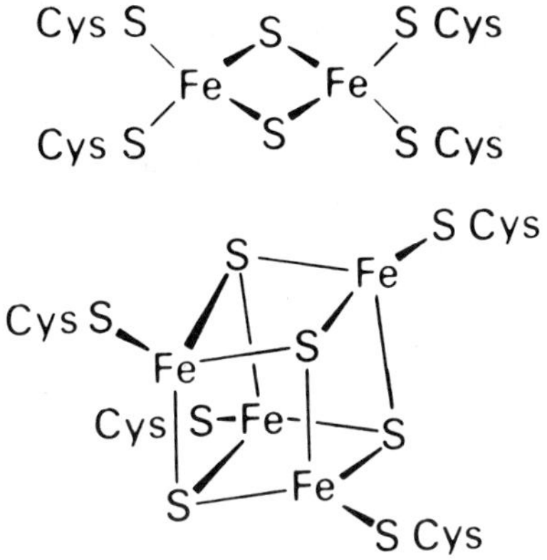

**Fig. 2.** The two-iron and four-iron centres in iron–sulphur proteins.

Fe(III), $d^5$ centres, and is diamagnetic through antiferromagnetic interaction between the two $d^5$ centres. The reduced form contains formally Fe(II) and Fe(III) and is paramagnetic, as one electron has been added to the diamagnetic oxidised protein.

The redox potentials of these proteins are probably controlled in a fine sense by the protein through small structural changes which affect the detailed geometry of the cluster and control the hydrophobic character of the protein environment around the cluster. Thus the synthetic cluster $[Fe_4S_4(SCH_2Ph)_4]^{2-/3-}$ has a redox potential of $-1.25$ V in dimethyl formamide, compared to values for 4Fe ferredoxin of $-0.93$ V in dimethyl sulphoxide and $-0.67$ V in aqueous solution (for *Clostridium pasteurianum*). Peptide analogues have more positive potentials than the simple cube, but a difference of about 0.12 V between the native ferredoxins and the synthetic peptide analogues still exists (Hill *et al.*, 1977). However, a further positive shift in redox potential of the cluster $[Fe_4S_4(Z\text{-}Cys\text{-}Gly\text{-}Ala\text{-}OMe)_4]^{2-}$, used as a model for *Peptococcus aerogenes* ferredoxin, has been attributed by Ueyama *et al.* (1983) to hydrogen bonding between the alanine NH group and the cysteine sulphur, which is a ligand to iron in the cube. A related 4Fe–4S dipeptide complex, which was unable to form this intramolecular NH··S bond, showed no positive shift in potential.

The protein also exerts a gross control over the redox potential of the 4Fe–4S cluster. This cluster offers three alternative overall oxidation states, as shown below.

| | | |
|---|---|---|
| $[Fe_4S_4]^{1-}$ | III, III, III, II | oxidised HiPIP |
| $\updownarrow$ 1e | | |
| | | reduced HiPIP |
| $[Fe_4S_4]^{2-}$ | III, III, II, II | oxidised Fd |
| $\updownarrow$ 1e | | (diamagnetic) |
| $[Fe_4S_4]^{3-}$ | III, II, II, II | reduced Fd |

Thus ferredoxin and high potential iron protein (HiPIP) have similar 4Fe–4S clusters but quite different redox potentials, as they cycle between different oxidation states of the cluster. Oxidised Fd and reduced HiPIP have the same electronic configuration, but the protein prevents reduced HiPIP being reduced further and oxidised Fd being oxidised further.

The 3Fe protein is still subject to uncertainty with respect to both structure and function (Xavier et al., 1981). In some cases, the presence of the cluster is thought to result from damage to a 4Fe–4S cluster during the isolation procedure. Thus Ruzicka and Beinert (1978) have shown that the enzyme aconitase contains an iron–sulphur cluster, which in the aerobically isolated, inactive enzyme is a 3Fe centre. However, this inactive enzyme may be activated by reduction, followed by the addition of Fe(II), or on long standing. The role of the additional iron has been the subject of much controversy, but one possibility is that the 3Fe protein is formed by degradation of a 4Fe cluster during extraction, and that activation involved the reformation of the 4Fe cluster, either by addition of Fe(II) or by cannibalisation of the 3Fe cluster. The use of resonance Raman spectroscopy (Johnson et al., 1983) has now confirmed this proposal. As activation occurs, so the characteristic Raman spectrum of the iron–sulphur cluster changes from that of a 3Fe cluster to that of a 4Fe–4S centre. Similarly, the conversion of 3Fe–3S clusters into 4Fe–4S clusters in a Desulfovibrio gigas ferredoxin and the reverse process in Bacillus stearothermophilus have been demonstrated by electron paramagnetic resonance (EPR) and Mössbauer spectroscopy (Kent et al., 1982; Bell et al., 1982).

The structures of the 3Fe clusters are still in doubt; that of the Azotobacter vinelandii ferredoxin I, determined by Ghosh et al. (1981), is shown in Fig. 3. However the Fe–Fe distance in this structure (4.2 Å) is different from that obtained by extended X-ray absorption fine structure (EXAFS) measurements on D. gigas ferredoxin II (Antonio et al., 1982a) So either the two 3Fe centres have substantially different core structures or one of the structures is in error.

A possible explanation of this problem is that the 3Fe cluster may exist either as a 3Fe–3S cluster or as a 3Fe–4S cluster. The former has a relatively flat structure, while the latter would maintain the essential features of the 4Fe–4S cluster (Beinert et al., 1983). This would explain the difference in Fe–Fe distance between A. vinelandii ferredoxin I and aconitase. Figure 4 shows a possible structure for the 3Fe–4S cluster and its relationship to the 4Fe–4S cube. This structure accounts for the Mössbauer evidence, which indicates that there are two types of tetrahedrally coordinated iron, and that 2Fe–2S clusters are extruded from aconitase.

One problem in working with iron–sulphur proteins is the question of rapid and unambiguous differentiation between 2Fe, 3Fe and 4Fe proteins. The application of EPR has allowed some distinctions to be made: thus EPR-active species are oxidised 3Fe, oxidised high potential iron proteins, reduced 2Fe and

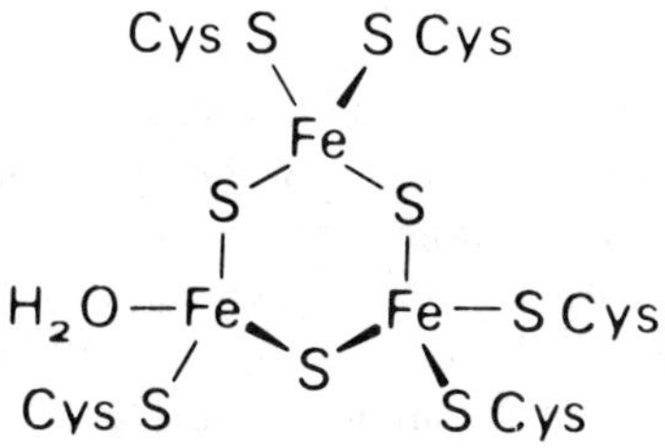

Cys S
S Cys
Fe
S
S
$H_2O$—Fe
Fe—S Cys
S
Cys S
S Cys

**Fig. 3.** The 3Fe–3S cluster in ferredoxin I (*Azotobacter vinelandii*).

reduced 4Fe (Fd). Further distinction may now be made by consideration of the dependence on magnetic fields of the shift in *g* value induced by an externally applied electric field (LEFE). This effect is similar for 3Fe and 4Fe (Fd) clusters, but is considerably different for both 2Fe and HiPIPs, which are also different from each other. Thus differentiation between all four classes of iron–sulphur protein may now be made on the basis of both EPR and LEFE properties, but not by the use of each technique individually (Peisach *et al.*, 1983).

*Molybdenum-containing clusters.* Mixed metal clusters are also important in biology. Thus the iron- and molybdenum-containing cofactor from nitrogenase probably involves a cluster in which a molybdenum has replaced an iron. The exact structure of this cofactor has not yet been determined, but a number of iron–molybdenum–sulphide clusters have been prepared and comparisons made (Holm, 1981; Teo *et al.*, 1983). Information is now accumulating on the reactivities of substitutionally labile Mo and Fe atoms in a common cluster, having the same spin state ($S$=3/2) as, and coordination sites resembling those in, the Fe–Mo cofactor of nitrogenase (Palermo and Holm, 1983).

RS
SR
Fe
$S_4$
$S_3$
Fe
$S_2$
Fe
SR
Fe
$S_1$
RS
SR
RS
Fe
$S_3$
$S_4$
$S_2$
Fe
Fe
RS
$S_1$
SR

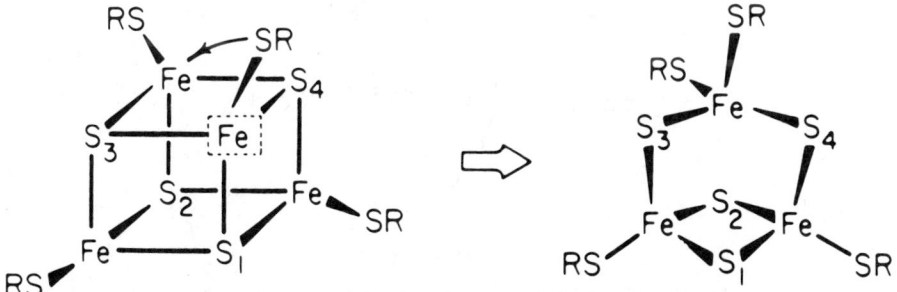

**Fig. 4.** The conversion of a 4Fe–4S cluster into a 3Fe–4S cluster.

## Dimers with Oxo Bridges

The importance of another type of iron binuclear structure is currently receiving recognition. This structure involves two oxo-bridged iron atoms, and is similar to that found in the dioxygen carrier hemerythrin (Fig. 5). It is interesting that this structural feature is present in several systems, such as acid phosphatase and ribonucleotide reductase (Theil *et al.*, 1983), and is also a component of methane monooxygenase (Woodland and Cammack, Chapter 13, this volume). It is probable that this grouping is distributed quite widely, but has not been recognised as such. Indeed it may have been assigned as a 2Fe–2S centre.

Oxo bridges linking two copper centres are found in hemocyanin. Such a structure may well be characteristic of the binuclear type 3 copper site, found, for example, in laccase.

## Concerted Electron Transfer

The redox reactions of compounds involving main group elements (for example dioxygen, or nitrogen compounds) usually involve two-electron changes. The electron transfer proteins catalyse the transfer of one electron. Accordingly there is a need for concerted electron transfer from two one-electron centres to couple with a two-electron process. Thus the heme $a_3$ and copper site for dioxygen in cytochrome oxidase can convert the dioxygen into peroxide. In bacterial cytochrome $c$ peroxidase, two hemes are coupled together so that hydrogen peroxide undergoes a two-electron reduction to water [Eq. (1)].

$$H_2O_2 + 2cyt\ Fe^{2+} + 2H^+ \rightarrow 2cyt\ Fe^{3+} + 2H_2O \tag{1}$$

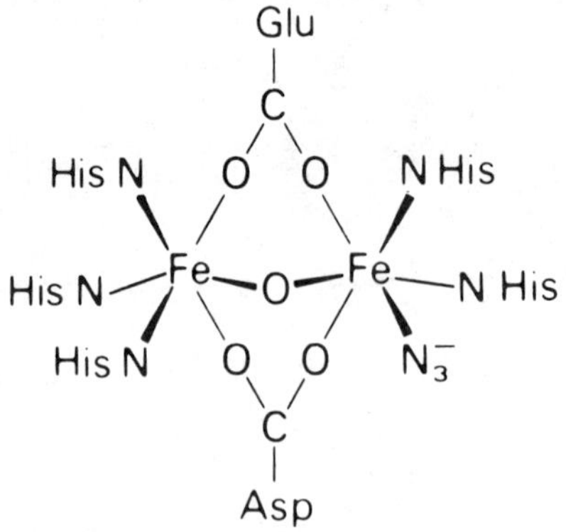

**Fig. 5.** The oxo-bridged binuclear iron centre from hemerythrin.

A particularly interesting example involves the NADPH-sulphite reductase of *Escherichia coli,* which also catalyses the six-electron reduction of nitrite to ammonia in the presence of a suitable electron donor. The two metal centres in this enzyme, siroheme (an iron isobacteriochlorin) and a 4Fe–4S cluster, are covalently linked, and can transfer two electrons to the substrate (Christner *et al.,* 1983).

*Nickel*

The biochemistry of nickel seems to be concerned particularly with gas reactions. Thus it is present in carbon monoxide dehydrogenase in acetogenic bacteria [Eq. (2)], in methanogenic bacteria [Eq. (3)], and in hydrogenases from several different types of bacteria [Eq. (4)], including those which carry out the Knallgas reaction [Eq. (5)]. A feature of some of these reactions is the apparent involvement of nickel(III). Although complexes of Ni(III) are sometimes considered to be novel, they are in fact quite well known, and a large number have been synthesised during the last 20 years (Haines and McAuley, 1981). In many cases, they have been prepared with the Ni(III) bound by macrocyclic ligands that provide four good donor atoms (usually nitrogen) in a plane. Complexes with peptides are also well known; these usually provide deprotonated peptide groups as binding sites. In all cases, the Ni(II) complex is usually first prepared and then oxidised chemically or electrochemically to the Ni(III) species.

$$CO + H_2O \rightarrow CO_2 + H_2 \qquad 2CO_2 + 4H_2 \rightarrow CH_3COOH + 2H_2O \qquad (2)$$

$$CO_2 + 4H_2 \rightarrow CH_4 + 2H_2O \qquad (3)$$

$$H_2 + 2Ni(III) \rightarrow 2H^+ + 2Ni(II) \qquad (4)$$

$$2H_2 + O_2 \rightarrow 2H_2O \qquad (5)$$

Nickel is required for the growth of methanogenic bacteria. It is present in factor 430, a low-molecular-weight compound, which is the prosthetic group of methyl coenzyme M reductase. This catalyses the conversion of methyl cofactor M into methane (Ellefson *et al.,* 1982) [Eq. (6)].

$$CH_3SCH_2CH_2SO_3^- \rightarrow CH_4 + HSCH_2CH_2SO_3^- \qquad (6)$$

Several lines of argument have led to the proposal that F430 is a tetrapyrrole. Pfaltz *et al.* (1982) have suggested, mainly on NMR evidence, that the product of methanolysis of F430 (F430M) has the tetrahydroporphyrin structure shown in Fig. 6. This structure has similarities both to siroheme and to corrin in some respects. As noted earlier, nickel can only interact well with the porphyrin donor atoms if the porphyrin ring is buckled. The NMR studies confirm that buckling of the ring occurs to a substantial extent.

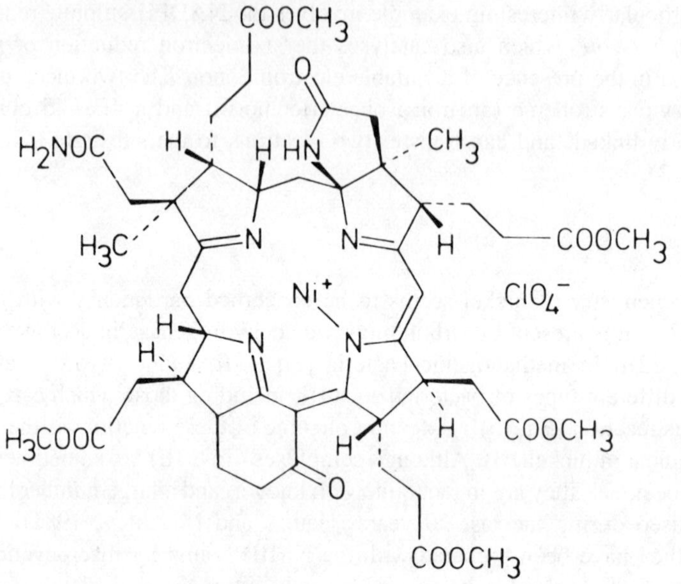

**Fig. 6.** The structure of F430M.

Nickel is also found in the carbon monoxide dehydrogenase of acetogenic bacteria, for example *Acetobacterium woodii* (Diekert and Ritter, 1982), *Clostridium thermoaceticum* (Drake *et al.*, 1980) and *Clostridium pasteurianum* (Drake, 1982). The nickel in carbon monoxide dehydrogenase is also bound by a tetrapyrrole, but this appears to be different from F430 (Ragsdale *et al.*, 1982). It is noteworthy that the EPR evidence (Ragsdale *et al.*, 1982) indicates that the carbon monoxide is bound to the nickel during reaction. The spectroscopic data have been interpreted in terms of a nickel(III) complex with a bound radical species derived either from CO or from $CO_2$. This conclusion is based upon analogies with cobalt in $B_{12}$ coenzymes. It is possible that nickel is also bound by a corrinoid ring structure in carbon monoxide dehydrogenase. This would be consistent with the observation of radical intermediates during reaction.

In contrast to the examples discussed above, nickel in hydrogenase is bound to protein. It is suggested to cycle between Ni(III) and Ni(II) during reaction. The Ni(III), a $d^7$ species, is paramagnetic, and is detected by characteristic EPR signals at *g* factors 2.3, 2.2 and 2.0 (Lancaster, 1982; Moura *et al.*, 1982; Cammack *et al.*, 1982). The Ni(III) in membrane-bound extracts from *Methanobacterium bryantii* (Lancaster, 1982) is suggested to be present in a rhombically distorted octahedral coordination site, and to be low-spin. As noted earlier, Ni(III) is quite well known in model compounds, but it is a surprising species to be found in a hydrogenase. The midpoint potential of model Ni(III) peptide complexes (about 0.9 V) is much greater than that found for the Ni(III)/ Ni(II) couple of the *D. gigas* hydrogenase ($-0.145$ V). Cammack *et al.* (1982) have therefore suggested that nickel in the hydrogenase is bound by very different amino acid residues from those involved in the peptide model complexes. It seems likely that the nickel in hydrogenase could be bound by sulphur ligands, and clusters should be considered. Cammack *et al.* (1982) have also pointed out that the potential is higher than the $2H^+/H_2$ couple by about 0.28 V, so that there are mechanistic difficulties in postulating a role for the Ni(III)/Ni(II) couple in the activation of dihydrogen, although there are claims for an EPR-active intermediate during the activation of dihydrogen, which may be a hydride species (Moura *et al.*, 1982). It is noteworthy that about half of the chemically detectable nickel is present as Ni(III). Thus, many questions remain to be answered, particularly on the structure of these nickel proteins. It is possible that the nickel centre involves a Ni(II)–Ni(III) dimer or some more complicated cluster.

## The Activation of Small Molecules

The binding of gaseous molecules to transition metal centres may result in their activation and subsequent reaction.

### Carbon Monoxide and Dinitrogen

These may usefully be considered together, as they are isoelectronic molecules. Both are particularly stable triply bonded molecules, with one sigma and two pi bonds, but both may be reduced microbially. Their stability with respect to oxidation is somewhat different, and reflects differences in their highest occupied molecular orbitals. Thus, for dinitrogen this is a sigma orbital, while for carbon monoxide it is a $\pi_{2p}$ orbital.

These molecules are not basic, as is demonstrated by their failure to undergo protonation. They form complexes with transition metals because of the unfilled *d* orbitals of the transition metal. This is due to the synergic effect, which is shown in Fig. 7. The formation of the metal–ligand bond reflects the presence of two bonding components. There is a ligand-to-metal sigma effect and a "back-bonding" metal-to-ligand pi effect. The sigma donation to the metal can only take place because electron density is transferred from the metal through the *d* orbital into the antibonding $\pi$ orbitals of the ligand. Thus the metal–ligand bond is created at the expense of the N–N or C–O bond. This bond weakening is

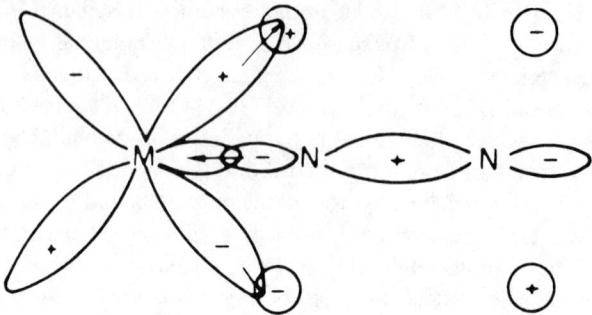

**Fig. 7.** The synergic effect in a metal–dinitrogen complex.

shown in metal carbonyls and metal dinitrogen complexes by the drop in the $\nu_{N_2}$ or $\nu_{CO}$ frequencies compared to those of the free molecules. This bond-weakening effect and enhanced polarisation resulting from metal–ligand bond formation results in enhanced activity. This will be elaborated upon later.

In Fig. 7, the dinitrogen molecule is shown bound end-on to the metal. However, metal complexes with bridging or side-on dinitrogen are known. The situation for dinitrogen bound to nitrogenase is not known.

The relative magnitudes of the sigma and pi components in the synergic effect are not necessarily similar. Thus, in many carbonyl complexes the back-bonding $\pi$ component is dominant. This explains why carbon monoxide complexes of many cytochromes undergo photolysis with loss of the carbon monoxide, whereas dioxygen cannot be removed in this way.

*Dioxygen*

This is used as a substrate by a vast number of enzymes (Malmström, 1982). The oxygen molecule is a good oxidising agent, but in general requires activation by binding Fe(II) or Cu(I). This allows control mechanisms to be established. Electron transfer then takes place from the metal to the antibonding molecular orbitals of $O_2$. The extent to which this occurs is crucial in determining whether a heme protein acts as a carrier of dioxygen, an oxygenase or an oxidase. This is controlled by variation in the axial ligand supplied by the protein, and in particular its $\sigma$-donor and $\pi$-acceptor properties.

The oxygen molecule is a paramagnetic diradical. Addition of electrons [Eq. (7)] to the antibonding orbitals gives sequentially superoxide (paramagnetic, bond order 1½) and peroxide (diamagnetic, bond order 1). Addition of a third electron results in cleavage of the O–O bond to give oxide and hydroxyl radical, while complete reduction to water occurs on addition of a fourth electron. These

$$O_2 \xrightarrow{\epsilon} O_2^- \xrightarrow{\epsilon} O_2^{2-} \xrightarrow{\epsilon} O_2^{3-} \xrightarrow[2H+]{\epsilon} O^- + H_2O \xrightarrow[2H^+]{\epsilon} 2H_2O \tag{7}$$

various reduction products of dioxygen are good oxidants, and they do not suffer the kinetic limitation that makes $O_2$ inert. Thus free superoxide and peroxide are toxic species (Hill, 1981).

Redox potentials show that it is more favourable to reduce $O_2$ to peroxide than superoxide. Hence oxidases have binuclear centres to allow the synchronous transfer of two electrons to dioxygen. Heme-bound intermediate species are utilised in certain enzymes. Thus monooxygenases use bound superoxide (heme $Fe^{III}O_2H$), while cytochrome P450 involves a bound peroxo group in its hydroxylation reactions (White and Coon, 1980). Oxo complexes are important in the high-oxidation-state intermediates formed in the reactions of catalase and peroxidase. There is much interest at present in the reactivity of such high-oxidation-state oxo iron porphyrins. They may be prepared by oxidising Fe(III) porphyrins with various oxidising agents to give Fe(V) oxo complexes. These compounds can transfer their oxo ligand to a variety of organic substrates, and can, for example, convert hydrocarbons into alcohols (Groves and Nemo, 1983).

The bacterial cytochrome oxidases appear to offer a particularly interesting opportunity to study the reduction of dioxygen. Alternative cytochromes are synthesised under different environmental conditions. These have remarkably varied properties, for example, in terms of their affinity for dioxygen and the stability of partly reduced oxygen species. The bacterial cytochrome oxidases have been reviewed by Poole (1983).

*Dihydrogen*

The role of metal surfaces and metal complexes in the activation of dihydrogen is well known. Homogeneous catalysis in solution usually involves the generation of coordinatively unsaturated transition metal species, for example, by photochemical dissociation of carbon monoxide from a carbonyl complex, or the thermal cleavage of a metal–metal bond in a dimeric complex. This allows the oxidative addition of dihydrogen to give a metal hydride. An example is the role of cobalt carbonyl complexes in the oxo process for the formation of aldehydes by the reaction of olefins with carbon monoxide and dihydrogen. The first stages are shown in Eqs. (8)–(10).

$$Co_2(CO)_8 \rightleftharpoons Co_2(CO)_7 + CO \tag{8}$$

$$Co_2(CO)_7 + H_2 \rightleftharpoons HCo(CO)_4 + HCo(CO)_3 \tag{9}$$

$$HCo(CO)_3 + CHR{=}CH_2 \rightarrow \overset{CH_2}{\underset{RCH \ |}{\|}}{\to}Co(CO)_3 \overset{CO}{\to} RCH_2CH_2Co(CO)_4 \tag{10}$$
$$H$$

## Some Examples of Microbial Gas Reactions

This discussion will in some cases be very speculative, due to the absence of detailed information on the structure and electronic properties of the biological centre that appears to bind the gaseous substrate. An attempt will be made to compare these reactions with appropriate examples of homogeneous catalysis taken from inorganic and industrial contexts, which involve the use of metal complexes and organometallic compounds as catalysts. The activity of these catalysts is very dependent on subtle combinations of steric and electronic effects in the ligand (Kochi, 1979; Nakamura, 1979; Muetterties and Krause, 1983).

### Reactions of Carbon Monoxide

Carbon monoxide is an extremely important industrial chemical, and the catalysis of its reactions has been well studied. The water gas shift reaction (11) usually involves heterogeneous catalysis, but homogeneous catalysis by simple

$$CO + H_2O \rightleftharpoons CO_2 + H_2 \tag{11}$$

metal complexes and metal clusters is well known. Other important processes involve conversion to methanol or methane, while a wide range of saturated and unsaturated compounds are obtained from carbon monoxide and hydrogen by Fischer–Tropsch syntheses.

Microbial reactions involving carbon monoxide are also of considerable importance. Oxidation to $CO_2$ takes place in aerobic and anaerobic bacteria, while reduction to methane occurs under anaerobic conditions by certain methanogens. The oxidation of carbon monoxide by microorganisms is recognised to be a major factor in the removal of carbon monoxide from the atmosphere (Spratt and Hubbard, 1981; Uffen, 1981; Hegeman, 1980). These reactions will be summarised.

*Aerobic organisms.* Certain methylotrophic bacteria catalyse the oxidation of carbon monoxide via their methane monooxygenase [Eq. (12)]. The process appears to be non-productive, CO being oxidised advantitiously in place of the normal substrate (Ferenci *et al.*, 1975).

$$CO + O_2 + NAD(P)H + H^+ \rightarrow CO_2 + H_2O + NAD(P)^+ \tag{12}$$

Much more important are the carboxydobacteria, which are aerobic and use carbon monoxide as the sole carbon and energy source, with $O_2$ as the final electron acceptor (Meyer and Schlegel, 1983) [Eq. (13)].

$$CO + H_2O \rightarrow CO_2 + \text{reduced carrier} \tag{13}$$

The carbon monoxide oxidase from several sources has been purified, notably that from *Pseudomonas carboxydovorans* (Meyer and Schlegel, 1980) which has an apparent $K_m$ for binding CO of 53 $\mu M$. The enzyme contains two Mo, two FAD and four 2 Fe–2S clusters per mole (a dimer of molecular weight 300,000), and is similar to xanthine oxidase and other molybdo proteins. The molybdenum is bound to a pterin group, as found for other molybdo proteins (Johnson, 1980). There is indirect evidence to implicate molybdenum in the catalytic site for the oxidation of CO. The remarkable lack of toxicity of carbon monoxide to the organism is attributed to the presence of a novel cytochrome *o* (cyt $b_{563}$) as terminal oxidase, which is suggested to be insentitive to carbon monoxide (Cypionka and Meyer, 1983).

An interesting parallel reaction is the cytochrome oxidase-catalysed oxidation of CO. Young *et al.* (1979) suggest that cytochrome oxidase acts as an oxygenase, and that 2 moles of CO bound in turn to the "invisible" copper are oxidised by 1 mole of $O_2$ bound to cytochrome $a_3$ via oxygen atom transfer reactions.

*Anaerobic organisms.* A few methanogens grow slowly using carbon monoxide as the sole carbon and energy source, for example *Methanobacterium thermoautotrophicum* [Eq. (14)]. *Rhodopseudomonas gelatinosa* uses carbon monoxide for growth in the dark, with the production of dihydrogen via a formic acid intermediate.

$$CO + H_2O \rightarrow (CO_2 + H_2) \rightarrow CH_4 \qquad (14)$$

Carbon monoxide is oxidised to carbon dioxide by several other bacteria, including *C. pasteurianum* and *C. thermoaceticum*. These contain the enzyme carbon monoxide dehydrogenase, which contains nickel bound in a tetrapyrrole cofactor different from F430 from methanogens. It appears that carbon monoxide dehydrogenase is involved in the synthesis of acetate by acetogenic bacteria, such as *A. woodii* (Diekert and Ritter, 1982).

*Some mechanistic considerations.* The enhanced activity of coordinated carbon monoxide has been noted. However, the magnitudes of the $\sigma$ and $\pi$ components of the synergic effect may vary. In cationic and neutral metal carbonyls, the back donation must be small to avoid the build-up of positive charge on the metal. Therefore, the carbon of the CO becomes negatively charged and subject to attack by nucleophiles. The situation is different in negatively charged or electron-rich carbonyls, for here $\pi$ donation may be considerable. This results in the oxygen atom being negatively charged and subject to electrophilic attack from a proton or a metal ion. Attack by a second metal centre gives a bridging carbonyl group. Addition of an electrophile to the oxygen of the CO group will lead to further weakening of the CO bond, to give a bond order similar to that in $CO_2$ (and possibly to C–O bond cleavage). In addition, nucleophilic attack upon the carbon of the bound carbon monoxide will be further enhanced (Shriver, 1983).

There is no evidence that simple C–O bond cleavage of this type occurs in any of these microbial reactions. The oxygenase mechanism for oxidation via cytochrome oxidase does provide a binuclear site, which seems appropriate for the binding and cleavage of CO. However, present information on this reaction (Young et al., 1979; Nicholls and Chanady, 1981) seems to exclude this approach, although there are many questions unanswered.

The carbon monoxide oxidase-catalysed reaction in carboxydobacteria utilises water rather than $O_2$ as the source of oxygen for $CO_2$ formation. Of particular interest is the lack of toxicity of CO towards the terminal oxidase, which is postulated to be an unusual cytochrome o. Recently there has been much interest in the binding of carbon monoxide and dioxygen by heme proteins and various model porphyrins, in the hope of pinpointing factors that could result in selectivity towards these gases. Normally carbon monoxide gives a linear Fe–CO group, while dioxygen gives a bent Fe–$O_2$ structure. It is well known that the Fe–CO group is forced to adopt a non-linear geometry in hemoproteins through interaction with various distal groups (often histidine residues). This is supposed to lower the affinity for CO and hence provide protection against it. Attempts have been made to synthesise model compounds with "pockets" to accommodate bent Fe–$O_2$ units, and which provide steric hindrance against the preferred linear binding of Fe–CO (Collman et al., 1983a,b). Such studies have led to the conclusion that there is substantially lowered affinity for CO in the sterically hindered porphyrins. Kinetic studies have been carried out on the binding of $O_2$ and CO by cyclophane hemes (Traylor et al., 1981b) and for cyclophane hemes and cofacial diporphyrins (Ward et al., 1981). Both studies have shown that distal steric hindrance affects the ligand association rate constants and has no effect on the dissociation rates. The association rates showed a differentiation factor against binding of CO in the range 3 to 8. However, these are not very large factors in view of substantial differences between these models and natural systems. Traylor et al. (1981a) have shown that changes in the electronic nature of the heme group have little effect on CO affinity. At this stage, therefore, it can only be said that some measure of protection against carbon monoxide toxicity could be provided by distal steric effects. This uncertain answer serves to emphasise the remarkable resistance of the carboxydobacteria to the toxic effects of carbon monoxide.

The stoicheiometry of oxidation of CO by the anaerobic bacteria is similar to that of the water gas shift reaction. The latter reaction is catalysed by simple carbonyl complexes, including $Ni(CO)_4$, although their activity falls with time due to the formation of clusters. The mechanism of the water gas shift reaction involves nucleophilic attack of hydroxide on the carbon atom of the CO [Eq. (15); other ligands are not shown].

$$\overset{\delta-\ \ \delta+}{\text{Ni---C}\!\equiv\!\text{O}} \overset{\text{OH}^-}{\longrightarrow} \text{Ni---}\overset{\overset{\displaystyle O^-}{\displaystyle \parallel}}{\underset{\displaystyle OH}{\text{C}}} \overset{\text{H}_2\text{O}}{\longrightarrow} \text{Ni} + \text{CO}_2 + \text{H}_2 + \text{OH}^- \qquad (15)$$

An effective, novel catalyst is a water-soluble ruthenium porphyrin, which can allow the binding of both hydroxide and carbonyl groups on one side of the porphyrin plane (Pawlik *et al.*, 1983).

The mechanism of carbon monoxide dehydrogenase from *C. thermoaceticum* (Ragsdale *et al.*, 1982) is suggested to involve the formation of a Ni(III) centre with a bound radical species derived from CO or $CO_2$. Any mechanism must be speculative, and it is unusual to postulate a carbonyl complex with metals in oxidation states as high as II or III. The presence of a bound radical suggests that carbon monoxide interacts initially with Ni(II), but substantial electron transfer occurs to the carbon monoxide, generating a Ni(III) radical species. The formation of the carbon monoxide anion radical would suggest the possibility of electophilic attack (by a proton, for example) on the oxygen atom. This would favour attack of hydroxide ion upon the carbon with rearrangement to give a formate radical species, which could either decompose back to Ni(II) plus $CO_2$ and $H_2$, or be incorporated into acetate synthesis.

*Denitrification*

In this process, nitrate is reduced to nitrite, and then to nitrous oxide and dinitrogen. It is catalysed by denitrifying bacteria, and by some non-denitrifiers. The pathway in denitrification is shown in Eq. (16). Nitrous oxide is well established as an intermediate. The pathway of conversion of nitrite to nitrous

$$\text{NO}_3^- \rightarrow \text{NO}_2^- \rightarrow ? \rightarrow \text{N}_2\text{O} \rightarrow \text{N}_2 \qquad (16)$$

oxide (a four-electron jump, overall) is still confused. It is often suggested that nitric oxide is an intermediate, as (1) NO has been detected, (2) NO may be reduced, (3) labelled NO exchanges with the nitrogen pool, and (4) there is good evidence for the formation of iron(II) heme nitrosyls. None of this evidence means that nitric oxide is an intermediate. Rather the intermediate is the nitroxyl ion $NO^-$, formed in a two-electron reduction of nitrite. Coordinated nitric oxide does not correspond to uncharged NO, but rather to the species $NO^-$, nitroxyl ion. In a minority of cases, a nitrosyl group may be present as $NO^+$, which is chemically equivalent to nitrite; such metal nitrosyls with $NO^+$ will carry out nitrosation reactions. In some cases nitro complexes and nitrosyl ($NO^+$) complexes exist in a pH-dependent equilibrium. Thus coordinated nitrite may readily be reduced via a two-electron reduction of the $NO^+$ form to give $NO^-$. This requires a change in geometry from linear $M-NO^+$ to bent $M-NO^-$ (where M is the metal).

The next step in denitrification involves the formation of nitrous oxide from $NO^-$, probably through the formation of a N–N bonded intermediate. This stage has been considered by Averill and Tiedje (1982), who point out that the dimerisation of bound $NO^-$ to give hyponitrite seems unlikely as the heme centres in nitrite reductase appear to be isolated from each other. The formation of hyponitrite would require substantial reorganisation of the protein, although it would undergo decomposition to give $N_2O$. Hollocher et al. (1980) have assessed the possibility that hyponitrite is a precursor of nitrous oxide in *Paracoccus denitrificans* by attempting to trap labelled hyponitrite, produced by denitrification of added nitrite, into a pool of added hyponitrite. They concluded from these experiments that hyponitrite was not an intermediate, and also showed that cell lysates did not catalyse the reduction of hyponitrite. These experiments, however, while showing that hyponitrite is unlikely to be a free intermediate, do not rule out absolutely the possibility that enzyme-bound hyponitrite is an intermediate. Added hyponitrite may not obtain access to this site.

Another precursor to nitrous oxide could be trioxodinitrate. The intermediacy of this species has been suggested by Averill and Tiedje (1982), and its possible formation attributed to reaction of free nitrite with bound $NO^+$ to give bound $N_2O_3$, which could undergo a two-electron reduction to give bound trioxodinitrate. Alternatively, trioxodinitrate could be formed by reaction of free nitrite with bound $NO^-$, a pathway rejected incorrectly by Averill and Tiedje on chemical grounds. Hughes and Wimbledon (1977) have shown previously that the stabilisation of trioxodinitrate by added nitrite results from the recombination of nitrite and nitroxyl ion. This has been confirmed isotopically by Akhtar et al. (1979).

Coordinated trioxodinitrate could then undergo a two-electron reduction to nitrous oxide. These reactions are summarised in Eqs. (17) and (18), which show only the uncoordinated nitrogen species. The hypothesis may be

$$NO^+ + NO_2^- \rightarrow N_2O_3 \xrightarrow{2\epsilon} N_2O_3^{2-} \xrightarrow[4H^+]{2\epsilon} N_2O + 2H_2O \qquad (17)$$

$$NO^- + NO_2^- \rightarrow N_2O_3^{2-} \xrightarrow[4H^+]{2\epsilon} N_2O + 2H_2O \qquad (18)$$

tested by checking if added trioxodinitrate is reduced directly to nitrous oxide. However, the reaction of trioxodinitrate with a nitrite-reducing system from *E. coli* resulted in the formation of a nitrosyl heme complex, apparently by reaction with the $NO^-$ group formed by self-decomposition of trioxodinitrate (Hubbard et al., Chapter 16, this volume) [Eq. (19)]. Recent isotopic work

$$HN_2O_3^- \rightleftharpoons HNO + NO_2^- \qquad (19)$$

by Garber *et al.* (1983) on the effects of denitrifying bacteria on the conversion of trioxodinitrate to nitrous oxide also suggests that the N–N bond of trioxodinitrate is not conserved and that nitroxyl is an intermediate in the production of nitrous oxide. It appears, therefore, that trioxodinitrate is probably not an intermediate in denitrification.

## Fixation of Dinitrogen

The chemical fixation of dinitrogen to ammonia under the extreme conditions of, for example, the Haber process is well established. Only recently have inorganic chemists achieved fixation under conditions that approximate those of biological fixation. As noted earlier, dinitrogen forms complexes with a range of transition metals. In most of these, the ligand is bound end-on to the metal, but examples where the dinitrogen is side-on to the metal or bridging are known. The reduction of the dinitrogen to ammonia occurs readily in certain of these complexes, notably for some dinitrogen complexes of molybdenum or tungsten phosphines. In these the dinitrogen undergoes protonation, with electron transfer from the metal, to give a sequence of coordinated intermediates. These are shown in the case of the tungsten complex in Eq. (20) and (21) (Chatt *et al.*, 1978).

$$cis\text{-}[W(N_2)_2(\text{phosphine})_4] \xrightarrow[\text{(H}_2\text{SO}_4)]{\text{CH}_3\text{OH}} 2NH_3 + N_2 + W^{VI} \text{ species} \qquad (20)$$

$$W^0{\cdots}N{\equiv}N \rightarrow W{\cdots}N{\equiv}NH \rightarrow W{\equiv}N{-}NH_2 \rightarrow W{\cdots}NH{-}NH_2$$

$$\qquad\qquad\qquad\qquad\qquad\qquad\qquad\qquad\qquad H^+ \Big\downarrow \epsilon \qquad\qquad (21)$$

$$W^{VI} + NH_3 \xleftarrow[2\epsilon]{2H^+} W{\cdots}NH + NH_3 \longleftarrow W{\cdots}NH{-}NH_3^+$$

The reaction is slightly different for the molybdenum analogue, which diverges from the above scheme at the (N–NH$_2$) stage. This may result from release of the ligand from the metal and its subsequent disproportionation into dinitrogen and ammonia. In the reaction shown in Eq. (20), the tungsten undergoes a change in formal oxidation state of six. The phosphine ligands are strongly electron-releasing, so making the dinitrogen basic and readily protonated. The mechanism may be applied to nitrogenase-catalysed reduction of dinitrogen. Electrons may be supplied via the iron–sulphur centres to the dinitrogen site, so that the molybdenum would not have to undergo such a wide change in oxidation state. The problem described for the molybdenum complex above, where an intermediate was lost prematurely from the metal, could be overcome by the ready availability of electrons and, perhaps, binding constrictions.

Nitrogenase, the nitrogen-fixing enzyme, has been isolated from a number of sources and always has similar properties, irrespective of its origin. It is made up of two proteins, which are both very sensitive to dioxygen. The roles of these two proteins are known. The iron protein (Fe protein) binds MgATP and transfers electrons from the electron donor to the second protein. This is the molybdenum–iron protein, which binds and reduces dinitrogen. The Fe protein contains an $Fe_4S_4$ cluster. The MoFe protein probably contains four $Fe_4S_4$ clusters and two complex clusters containing both molybdenum and iron. An iron–molybdenum cofactor has been isolated from this protein which contains six to eight Fe and four to six $S^{2-}$ per molybdenum. It is clear, therefore, that electron transfer in nitrogenase is a complex process because of the number of redox-active metal clusters available to receive and transfer electrons. An indication of the complexity of these processes has been obtained from EPR and Mössbauer spectroscopy.

The iron–molybdenum cofactor (Fe–Mo–co) has not yet been shown to catalyse the reduction of dinitrogen to ammonia. Nevertheless, it has been much studied by a range of techniques, and compared with various model Fe–Mo clusters (Coucouvanis, 1981; Holm, 1981). At present, it appears that no synthetic model represents all the key features of the Fe–Mo cofactor, although specific properties of the cofactor may be represented well by particular models. The ultimate test of a synthetic cluster would be its incorporation into the Fe–Mo protein. Some examples of recent work are given. The use of $^{95}$Mo electron nuclear double resonance (ENDOR) spectroscopy has indicated that the molybdenum is electronically integrated into the Fe–Mo–co electronic system, and that it is best viewed as a nominally tetrahedral Mo(II) or Mo(VI) ion (Hoffman *et al.*, 1982). Iron EXAFS measurements on the Fe–Mo cofactor (Antonio *et al.*, 1982b) have been interpreted in terms of structural similarities with two model compounds. One involves two $Fe_3S_3$ units bridged by a molybdenum atom. The other is a $[L_3MoFe_7S_6(SR)_7]^{2-}$ cluster containing a $MoFe_7S_6$ core, having the metal atoms situated at the corners of a cube with quadruply bridging sulphur atoms occupying the six faces of the cube ($RS^- = $ thiolate; $L = $ ligand).

Nitrogenase catalyses the reduction of a number of gases in addition to dinitrogen. Acetylene is reduced to ethylene, and nitrous oxide is reduced to dinitrogen. These substrates probably bind to different sites on the Mo–Fe protein. The reaction with dioxygen is also of great interest. As noted earlier, both Fe and Mo–Fe proteins are destroyed by dioxygen. This is thought to result from interaction with the metal clusters and the one-electron reductions of dioxygen to superoxide and then peroxide which oxidise the proteins irreversibly. In view of this, it is remarkable that the ability to fix dinitrogen is found in aerobic and facultative organisms. The most extreme case is the cyanobacteria, which evolve dioxygen photosynthetically and simultaneously fix dinitrogen (Stewart, 1980). The strategies adopted by microorganisms to protect nitrogenase from damage by dioxygen have been reviewed by Gallon (1981). The strict anaerobes such as *C. pasteurianum* clearly avoid dioxygen. The cyanobacteria locate their nitrogenase in specialised cells, heterocysts, which have a thick cell wall to exclude dioxygen. The aerobic $N_2$ fixer *Azotobacter* removes dioxygen through its high respiration rate. On exposure to high concentrations of $O_2$, *Azotobacter* ceases fixing dinitrogen, possibly through conformational change of nitrogenase into an inactive but $O_2$-insensitive form. Finally, certain cyanobacteria stop the synthesis of nitrogenase on exposure to high $O_2$ concentrations, but resume this on removal of excess dioxygen.

A further area of current interest is hydrogen production. Indeed, some cyanobacteria can generate $H_2$ from water and sunlight (Stewart, 1980). Dihydrogen production always takes place alongside $N_2$ fixation (Robson and Postgate, 1980). This was once thought to be competitive, but is now recognised to be involved in the overall reaction stoicheiometry, as there is never less than 1 mole of dihydrogen produced per mole of dinitrogen fixed. The simplest explanation is that the binding of dinitrogen to the molybdenum releases 1 mole of dihydrogen [Eqs. (22) and (23)], possibly implying side-on binding of the dinitrogen which displaces two hydrides as $H_2$ (Chatt, 1980). The other view

$$N_2 + 8H^+ + 8\,\epsilon \rightarrow 2NH_3 + H_2 \tag{22}$$

$$
\begin{array}{c}
\diagdown \qquad \diagup H \\
-Mo-H \\
\diagup \qquad \diagdown H
\end{array}
\xrightarrow{N_2}
\begin{array}{c}
\diagdown \qquad \diagup H \\
-Mo \\
\diagup \qquad \diagdown N_2
\end{array}
+ H_2 \tag{23}
$$

is that $N_2$ results from disproportionation of a partly reduced species. This proposal does not accommodate the fact that dihydrogen is evolved in the absence of dinitrogen. Some interesting experiments have been carried out in which nitrogenase functions under an atmosphere of 50:50 $D_2$:$H_2$, in which case HD is produced. This result is also consistent with a scheme involving molybdenum-bound hydrides.

$$
D_2 + Mo
\begin{array}{c}
\diagup H \\
- H \\
\diagdown H
\end{array}
\rightleftharpoons Mo
\begin{array}{c}
\diagup H \\
- D \\
\diagdown D
\end{array}
+ H_2
$$

$$\downarrow N_2 \qquad\qquad \downarrow N_2$$

$$
Mo
\begin{array}{c}
\diagup H \\
\diagdown N_2
\end{array}
+ H_2 \quad
Mo
\begin{array}{c}
\diagup N_2 \\
\diagdown D
\end{array}
+ HD
$$

## Conclusion

The material described in this chapter has illustrated on the one hand the advances that have been made in recent years in the understanding of certain electron transfer and other redox proteins, and on the other the urgent need for more work, particularly structural work, on the nickel proteins, which still present a confused and uncertain picture. The discussion of the mechanisms of some of the microbial gas reactions is probably premature in view of the lack of data, but these reactions do present a stimulating challenge to those of a mechanistic disposition. Each topic discussed represents an important area of research, but the work on the reactions of carbon monoxide seems to be particularly novel. Indeed, the results published so far on these reactions seem to contain enough unusual features to encourage others to take up this work.

## Acknowledgments

I am grateful to Professor O. Meyer for helpful comments, and to Dr. R. K. Poole for discussion.

## References

Akhtar, M. J., Lutz, C. A. and Bonner, F. T. (1979). Decomposition of sodium trioxodinitrate in the presence of added nitrite in aqueous solution. *Inorganic Chemistry* **18**, 2369–2374.

Antonio, M. R., Averill, B. A., Moura, I., Moura, J. J. G., Orme-Johnson, W. H., Teo, B. K. and Xavier, A. V. (1982a). Core dimensions in the 3Fe cluster of *Desulfovibrio gigas* ferredoxin II by extended X-ray absorption fine structure spectroscopy. *Journal of Biological Chemistry* **257**, 6646–6649.

Antonio, M. R., Teo, B. K., Orme-Johnson, W. H., Nelson, M. J., Groh, S. E., Lindahl, P. A., Kauzlarich, S. M. and Averill, B. A. (1982b). Iron EXAFS of the iron–molybdenum cofactor of nitrogenase. *Journal of the American Chemical Society* **104**, 4703–4705.

Averill, B. A. and Tiedje, J. M. (1982). The chemical mechanism of microbial denitrification. *FEBS Letters* **138**, 8–11.

Beinert, H., Emptage, M. H., Dreyer, J. L., Scott, R. A., Hahn, J. E., Hodgson, K. O. and Thomson, A. J. (1983). Iron–sulphur stoicheiometry and structure of iron–sulphur clusters in three-iron proteins: Evidence for 3Fe–4S clusters. *Proceedings of the National Academy of Sciences of the U.S.A.* **80**, 393–396.

Bell, S. H., Dickson, D. P. E., Johnson, C. E., Cammack, R., Hall, D. O. and Rao, K. K. (1982). Mössbauer spectroscopic evidence for the conversion of 4Fe4S clusters in *Bacillus stearothermophilus* ferredoxin into 3Fe3S clusters. *FEBS Letters* **142**, 143–146.

Buchler, J. W., Kokisch, W. and Smith, P. D. (1978). Cis, trans, and metal effects in transition metal porphyrins. *Structure and Bonding (Berlin)* **34**, 79–135.

Cammack, R., Patil, D., Aguirre, R. and Hatchikian, E. C. (1982). Redox properties of the ESR-detectable nickel in hydrogenase from *Desulfovibrio gigas*. *FEBS Letters* **142**, 289–292.

Chatt, J. (1980). Chemistry relevant to the biological fixation of nitrogen. *Annual Proceedings of the Phytochemical Society of Europe* **18**, 1–18.

Chatt, J., Dilworth, J. R. and Richards, R. L. (1978). Recent advances in the chemistry of nitrogen fixation. *Chemical Reviews* **78**, 589–625.

Christner, J. A., Münck, E., Janick, P. A. and Siegel, L. M. (1983). Mössbauer evidence for exchange-coupled siroheme and [4Fe–4S] prosthetic groups in *Escherichia coli* sulphite reductase. *Journal of Biological Chemistry* **258**, 11147–11156.

Collman, J. P., Brauman, J. I., Collins, T. J., Iverson, B. L., Long, G., Pettman, R. B., Sessler, J.

L. and Walters, M. A. (1983a). Synthesis and characterisation of the pocket porphyrins. *Journal of the American Chemical Society* **105**, 3038–3052.

Collman, J. P., Brauman, J. I., Iverson, B. L., Sessler, J. L., Morris, M. R. and Gibson, Q. H. (1983b). $O_2$ and CO binding to Fe(II) porphyrins. *Journal of the American Chemical Society* **105**, 3052–3064.

Cotton, F. A. and Wilkinson, G. (1980). "Advanced Inorganic Chemistry," 4th ed. Wiley (Interscience), New York.

Coucouvanis, D. (1981). Fe–M–S complexes derived from $MS_4^{2-}$ anions and their possible relevance as analogues for structural features in the Mo site of nitrogenase. *Accounts of Chemical Research* **14**, 201–209.

Cypionka, H. and Meyer, O. (1983). The cytochrome composition of carboxydotrophic bacteria. *Archives of Microbiology* **135**, 293–298.

Diekert, G. and Ritter, M. (1982). Nickel requirement of *Acetobacterium woodii*. *Journal of Bacteriology* **151**, 1043–1045.

Drake, H. L. (1982). Occurrence of nickel in carbon monoxide dehydrogenase from *Clostridium pasteurianum* and *Clostridium thermoaceticum*. *Journal of Bacteriology* **149**, 561–566.

Drake, H. L., Hu, S. L. and Wood, H. G. (1980). Purification of carbon monoxide dehydrogenase, a nickel enzyme from *Clostridium thermoaceticum*. *Journal of Biological Chemistry* **255**, 7174–7180.

Ellefson, W. L., Whitman, W. B. and Wolfe, R. S. (1982). Nickel-containing factor F430: Chromophore of the methylreductase of *Methanobacterium*. *Proceedings of the National Academy of Sciences of the U.S.A.* **79**, 3707–3710.

Ferenci, T., Strøm, T. and Quayle, R. (1975). Oxidation of carbon monoxide by *Pseudomonas methanica*. *Journal of General Microbiology* **91**, 79–91.

Gallon, J. R. (1981). The oxygen sensitivity of nitrogenase: A problem for biochemists and microorganisms. *Trends in Biochemical Sciences*, **6**, pp. 19–23.

Garber, A. E., Wehrli, S. and Hollocher, T. C. (1983). [15]N Tracer and N. M. R. studies on the pathway of denitrification. Evidence against trioxodinitrate but for nitroxyl as an intermediate. *Journal of Biological Chemistry* **258**, 3587–3591.

Ghosh, D., Furey, W., O'Donnell, S. and Stout, C. D. (1981). Structure of a 7Fe ferredoxin from *Azotobacter vinelandii*. *Journal of Biological Chemistry* **256**, 4185–4192.

Groves, J. T. and Nemo, T. E. (1983). Aliphatic hydroxylation catalysed by iron porphyrin complexes. *Journal of the American Chemical Society* **105**, 6243–6248.

Haines, R. I. and McAuley, A. (1981). Synthesis and reactions of nickel(III) complexes. *Coordination Chemistry Reviews* **39**, 77–119.

Hegeman, G. (1980). Oxidation of carbon monoxide by bacteria. *Trends in Biochemical Sciences*, **5**, pp. 256–259.

Hill, C. L., Renaud, J., Holm, R. H. and Mortenson, L. E. (1977). Synthetic analogues of the active sites of iron–sulphur proteins. 15. Comparative polarographic potentials of the $[Fe_4S_4(SR)_4]^{2-,3-}$ and *Clostridium pasteurianum* ferredoxin redox couples. *Journal of the American Chemical Society* **99**, 2549–2557.

Hill, H. A. O. (1981). Oxygen, oxidases and the essential trace metals. *Philosophical Transactions of the Royal Society of London, Series B* **294**, 119–128.

Hoffman, B., Roberts, J. E. and Orme-Johnson, W. H. (1982). [95]Mo and [1]H ENDOR spectroscopy of the nitrogenase MoFe protein. *Journal of the American Chemical Society* **104**, 860–862.

Hollocher, T. C., Garber, E., Cooper, A. J. L. and Reiman, R. E. (1980). [13]N, [15]N Isotope and kinetic evidence against hyponitrite as intermediate in denitrification. *Journal of Biological Chemistry* **255**, 5027–5030.

Holm, R. H. (1981). Metal clusters in biology: Quest for a synthetic representation of the catalytic site of nitrogenase. *Chemical Society Reviews* **10**, 455–490.

Hoyano, J. K., McMaster, A. D. and Graham, W. A. G. (1983). Activation of methane by iridium complexes. *Journal of the American Chemical Society* **105**, 7190–7191.

Hughes, M. N. (1981). "The Inorganic Chemistry of Biological Processes," 2nd ed. Wiley, Chichester.

Hughes, M. N. and Wimbledon, P. E. (1977). The chemistry of trioxodinitrates. Part 2. The effect of added nitrite on the stability of sodium trioxodinitrate in aqueous solutions. *Journal of the Chemical Society, Dalton Transactions*, pp. 1650–1653.

Johnson, J. L. (1980). The Molybdenum cofactor common to nitrate reductase, xanthine dehydrogenase and sulfite oxidase. *In* "Molybdenum and Molybdenum-containing Enzymes" (Ed. M. P. Coughlan), pp. 345–383. Pergamon Press, Oxford.

Johnson, M. K., Czernuszewicz, R. S., Spiro, T. G., Ramsay, R. R. and Singer, T. P. (1983). Resonance Raman studies of beef heart aconitase and a bacterial hydrogenase. *Journal of Biological Chemistry* **258**, 12771–12774.

Kent, T. A., Moura, I., Moura, J. J. G., Lipscomb, J. D., Huynh, B. H., LeGall, J., Xavier, A. V. and Münck, E. (1982). Conversion of [3Fe–3S] into [4Fe–4S] clusters in a *Desulfovibrio gigas* ferredoxin and isotopic labeling of iron–sulphur subsites. *FEBS Letters* **138**, 55–58.

Kochi, J. K. (1979). "Organometallic Mechanisms and Catalysis." Academic Press, London and New York.

Lancaster, J. R. (1982). New biological paramagnetic center: Octahedrally coordinated nickel(III) in the methanogenic bacteria. *Science* **216**, 1324–1325.

Malmström, Bo. G. (1982). Enzymology of oxygen. *Annual Review of Biochemistry* **51**, 21–59.

Meyer, O. and Schlegel, H. G. (1980). Carbon monoxide:methylene blue oxidoreductase from *Pseudomonas carboxydovorans*. *Journal of Bacteriology* **141**, 74–80.

Meyer, O. and Schlegel, H. G. (1983). Biology of aerobic carbon monoxide-oxidising bacteria. *Annual Review of Microbiology* **32**, 277–310.

Moore, G. R. and Williams, R. J. P. (1977). Structural basis for the variation in redox potential of cytochromes. *FEBS Letters* **79**, 229–232.

Moura, J. J. G., Moura, I., Huynh, B. H., Kruger, H. J., Teixeira, M., DuVarney, R. C., DerVartanian, D. V., Xavier, A. V., Peck, H. D. and LeGall, J. (1982). Unambiguous identification of the nickel E.P.R. signal in $^{61}$Ni-enriched *Desulfovibrio gigas* hydrogenase. *Biochemical and Biophysical Research Communications* **108**, 1388–1393.

Muetterties, E. L. and Krause, M. J. (1983). Catalysis by molecular metal clusters. *Angewandte Chemie, International Edition in English* **22**, 135–148.

Nakamura, A. (1979). "Principles and Applications of Homogeneous Catalysis". Wiley (Interscience), New York and Chichester.

Nicholls, P. and Chanady, G. A. (1981). Interaction of cytochrome $aa_3$ with oxygen and carbon monoxide. *Biochimica et Biophysica Acta* **634**, 256–265.

Palermo, R. E. and Holm, R. H. (1983). Reactivity properties of single $MoFe_3S_4$ cubane-type clusters. *Journal of the American Chemical Society* **105**, 4310–4318.

Pawlik, M., Hoq, M. F. and Shepherd, R. E. (1983). Catalysis of the water gas shift reaction by $Ru(TPPS)CO^{4-}$. *Journal of the Chemical Society, Chemical Communications*, pp. 1467–1468.

Peisach, J., Beinert, H., Emptage, M. H., Mims, W. B., Fee, J. A., Orme-Johnson, W. H., Rendina, A. R. and Orme-Johnson, N. R. (1983). Characterisation of 3-iron ferredoxins by means of the linear electric field effect in E.P.R. *Journal of Biological Chemistry* **258**, 13014–13016.

Pfaltz, A., Jaun, B., Fassier, A., Eschenmoser, A., Jaenchen, R., Gilles, H. H., Diekert, G. and Thauer, R. K. (1982). Zur kenntnis des faktors F430 aus methanogenen bakterien: Struktur des porphinoiden ligandsystems. *Helvetica Chimica Acta* **65**, 828–865.

Poole, R. K. (1983). Bacterial cytochrome oxidases. A structurally and functionally diverse group of electron-transfer proteins. *Biochimica et Biophysica Acta* **726**, 205–243.

Purcell, K. F. and Kotz, J. C. (1977). "Inorganic Chemistry". Saunders, Philadelphia, Pennsylvania.

Ragsdale, S. W., Ljungdahl, L. G. and DerVartanian, D. V. (1982). E.P.R. evidence for nickel–substrate interaction in carbon monoxide dehydrogenase from *Clostridium thermoaceticum*. *Biochemical and Biophysical Research Communications* **108**, 658–663.

Robson, R. L. and Postgate, J. R. (1980). Oxygen and hydrogen in biological nitrogen fixation. *Annual Review of Microbiology* **34**, 183–207.

Ruzicka, F. J. and Beinert, H. (1978). The soluble 'high potential' type iron–sulphur protein from mitochondria is aconitase. *Journal of Biological Chemistry* **253**, 2514–2517.

Scheidt, W. R. and Reed, C. A. (1981). Spin-state/stereochemical relationships in iron porphyrins: Implications for the hemoproteins. *Chemical Reviews* **81**, 543–555.

Shriver, D. F. (1983). Activation and reduction of carbon monoxide. *Chemistry in Britain*, pp. 482–487.

Spratt, H. G. and Hubbard, J. S. (1981). Carbon monoxide metabolism in roadside soils. *Applied and Environmental Microbiology* **41**, 1192–1201.

Stewart, W. D. P. (1980). Some aspects of structure and function in $N_2$-fixing cyanobacteria. *Annual Review of Microbiology* **34**, 497–536.

Teo, B. K., Antonio, M. R., Coucouvanis, D., Simhon, E. D. and Stremple, P. P. (1983). Mo, W, and Fe EXAFS of the $[Cl_2FeS_2MS_2FeCl_2]^{2-}$ dianions. A comparison with the Mo EXAFS of nitrogenase. *Journal of the American Chemical Society* **105**, 5767–5770.

Theil, E., Eicchorn, G. L. and Marzilli, L. G. (Eds.) (1983). "Iron-binding Proteins without Cofactors or Sulphur Clusters." American Elsevier, New York.

Traylor, T. G., White, D. K., Campbell, D. H. and Berzinis, A. P. (1981a). Electronic effects on the binding of dioxygen and carbon monoxide to hemes. *Journal of the American Chemical Society* **103**, 4932–4936.

Traylor, T. G., Mitchell, M. J., Tsuchiya, S., Campbell, D. H., Stynes, D. V. and Koga, N. (1981b). Cyclophane hemes.4. Steric effects on dioxygen and carbon monoxide binding to hemes and hemoproteins. *Journal of the American Chemical Society* **103**, 5234–5236.

Ueyama, N., Terakawa, T., Nakata, M. and Nakamura, A. (1983). Positive shift of redox potential of $[Fe_4S_4(Z-Cys-Gly-Ala-OMe)_4]^{2-}$ in dichloromethane. *Journal of the American Chemical Society* **105**, 7098–7102.

Uffen, R. L. (1981). Metabolism of carbon monoxide. *Enzyme and Microbial Technology* **3**, 197–206.

Ward, B., Wang, C. B. and Chang, C. K. (1981). Non-bonding steric effect on CO and $O_2$ binding to hemes. Kinetics of ligand binding in iron–copper cofacial-diporphyrins and strapped hemes. *Journal of the American Chemical Society* **103**, 5236–5238.

Watson, P. L. (1983). Methane exchange reactions of lanthanide and early transition metal methyl complexes. *Journal of the American Chemical Society* **105**, 6491–6493.

White, R. E. and Coon, M. J. (1980). Oxygen activation by cytochrome P-450. *Annual Review of Biochemistry* **49**, 315–356.

Xavier, A. V., Moura, J. J. G. and Moura, I. (1981). Novel structures in iron–sulphur proteins. *Structure and Bonding (Berlin)* **43**, 187–213.

Yoch, D. C. and Carithers, R. P. (1979). Bacterial iron–sulphur proteins. *Microbiological Reviews* **43**, 384–421.

Young, L. J., Choc, M. G. and Caughey, W. S. (1979). Role of oxygen and cytochrome *c* oxidase in the detoxification of CO by oxidation to $CO_2$. *In* "Biochemical and Clinical Aspects of Oxygen" (Ed. W. S. Caughey), pp. 355–361. Academic Press, New York.

# Part II
# Oxygen

# 2

# Microbial Metabolism of Oxygen: The Binding and Reduction of Oxygen by Bacterial Cytochrome Oxidases

ROBERT K. POOLE,* BALDEV S. BAINES,*,[1]
JULIA A. M. HUBBARD*,† AND HUW D. WILLIAMS*

*Departments of *Microbiology and †Chemistry, Queen Elizabeth College,
University of London, London, United Kingdom*

## Introduction

The binding and activation of molecular oxygen in biological processes are accomplished by three general classes of protein (for a review, see Malmström, 1982). The most familiar are the blood pigments, such as haemoglobin and myoglobin. Second, certain enzymes such as catalase and superoxide dismutase have oxygen as one of their products. Third, there are now more than 200 enzymes known that have oxygen as one of their substrates, and which may be subdivided into the oxygenases and oxidases. Although the oxygen-carrying blood pigments are in many ways distinct from the last class, they will be seen later to have certain features in common with the oxidases with regard to the binding of the ligand.

Oxygenases catalyse reactions in which both atoms (in the case of dioxygenases) or only one atom (monooxygenases) of molecular oxygen are inserted into a substrate. Methane monooxygenase is an example of the latter group and is described by Dalton and Leak (Chapter 11, this volume). In oxidase-catalysed reactions, however, the dioxygen molecule functions as an electron acceptor only and is thereby reduced to $O_2^-$ (superoxide), $H_2O_2$ (hydrogen peroxide) or $2H_2O$. Although cytochrome oxidases, the subject of this contribution, represent only one sub-group of the structurally very diverse oxidases, they are of paramount importance, being responsible for $> 90\%$ of biological $O_2$ consumption.

---

[1]Present address: Biotechnology Department, Microbiology Division, Glaxo Group Research Ltd., Greenford Road, Greenford, Middlesex UB6 0HE, United Kingdom.

MICROBIAL GAS METABOLISM:
MECHANISTIC, METABOLIC
AND BIOTECHNOLOGICAL ASPECTS

All the enzymes involved in biological oxygen utilization are conjugated proteins, in which the prosthetic groups are flavins, copper or iron. Cytochrome oxidases are all haem proteins in which the first step in oxygen activation is the binding of $O_2$ to a low-valence form of the metal prosthetic group ($Fe^{2+}$ or $Cu^{1+}$). One, but more commonly two, metallic centre(s) may be involved in subsequent electron transfer reactions to the ligand (Hughes, Chapter 1, this volume).

It is axiomatic that life as we know it is dependent on the biological "burning" of oxygen. The prominent role of oxygen as terminal electron acceptor in aerobic respiratory electron transfer arises from (1) its high reduction potential and (2) its relative unreactivity. It is also only sparingly soluble in aqueous media. Thus, at physiological temperatures it reacts only in the presence of a catalyst. This contribution is addressed to the properties of one such group of catalysts, the terminal cytochrome oxidases of aerobic respiratory chains, especially of bacteria, with special emphasis on the mechanisms of oxygen activation.

## Historical Background

It was no great revelation to the early bacteriologists that most bacteria required oxygen for growth, since it was believed that life could be supported only when a free supply of oxygen was available. Nevertheless, the history of microbial respiration has not been without its surprises. Eukaryotic and prokaryotic microorganisms were amongst the first cell types to be examined in the pioneering work of Warburg, Keilin and others, and ever since have allowed fascinating comparisons with the "classical" respiratory systems of mitochondria in higher eukaryotic cells.

Eukaryotic microbes have provided a rich hunting ground for those seeking subtle variations in respiratory chain compositions and, most notably, a bewildering array of systems exhibiting respiratory capacity in the presence of classical inhibitors such as cyanide. The contributions of algae, yeasts and other fungi, and Protozoa to such studies are tabulated by Lloyd and Edwards (1977) and described in detail by Lloyd (1974).

In contrast, bacterial respiratory chains are characterized by the diversity of respiratory carriers found, especially cytochromes, complex branched pathways of electron transport to oxygen as well as other electron acceptors and variable efficiencies of energy transduction associated with these pathways. Here, we stress those aspects of oxygen utilization by bacterial oxidases that differ most strikingly from the analogous reactions in eukaryotic cells.

Table 1 amplifies the lists given by Lloyd and Edwards (1977) and Wikström *et al.* (1981). It serves to highlight some of the more significant discoveries

**Table 1.**  *Selected advances in the understanding of bacterial cytochrome oxidases*

| Date | Author(s) | Findings |
|------|-----------|----------|
| 1925 | Keilin | "Cytochrome" present in *Bacillus subtilis* but not *Clostridium sporogenes*. |
| 1927 | Keilin | "Typical and somewhat modified absorption spectra of cytochrome" recorded in several bacteria.[a] |
| 1928 | Yaoi and Tamiya | Cytochrome $a_2$ (d) discovered in several bacteria, including *Escherichia coli*. |
| 1932 | Kubowitz and Haas ⎫ | 589-nm pigment of *Acetobacter pasteurianum* identified as |
| 1933 | Warburg *et al*.  ⎭ | "$O_2$-transporting ferment". |
| 1927–1933 | | Warburg opposed view that cytochrome plays a part in normal respiratory activity of the cell.[a] |
| 1934 | Yamagutchi | Classification of bacteria into four groups on basis of sensitivity to cyanide and CO. |
| 1934 | Negelein and Gerischer | 632-nm band of *Azotobacter chroococcum* attributed to "$O_2$-transporting ferment". |
| 1952 | Moss | Effect of $O_2$ concentration on formation of cytochrome *d* described. |
| 1953 | Chance *et al*. | Application of improved spectrophotometric techniques. |
| 1953 | Chance | "CO binding pigment" described, later to be called cytochrome *o*. |
| 1954 | Vernon and Kamen | CO-binding *Rhodospirillum* haem protein (RHP) thought to be an oxidase. |
| 1954 | Tissières | Demonstration of respiratory activity of small membrane fragments. |
| 1955 | Castor and Chance | Photochemical action spectrum of "CO-binding pigment". |
| 1956 | Barrett | Prosthetic group of cytochrome $a_2$ described as chlorin. |
| 1958 | Horio | Description of soluble cytochrome $cd_1$ in *Pseudomonas aeruginosa*. |
| 1959 | Castor and Chance | Extension of photochemical action spectroscopy and renaming of "CO-binding pigment" as "cytochrome *o*" ("*o*" for oxidase). |
| 1964 | White and Smith | Variable stoicheiometry of cytochromes in *Haemophilus parainfluenzae* membranes. |
| 1966 | Chance *et al*. | Oxidase role for RHP refuted. |
| 1966 | Webster and Hackett | Purification of cytochrome *o* from *Vitreoscilla*. |
| 1966 | Iwasaki | First report of "oxygenated" oxidase-type compound. |
| 1967 | Jones and Redfearn | Physical resolution of functionally distinct respiratory assemblies. |
| 1970 | Smith *et al*. | Rapid flow studies indicate kinetic competence of cytochromes *o* and *d* as oxidases. |
| 1975 | Jones *et al*. | Correlation between presence of high potential cytochrome *c* and energy transduction in terminal region of respiratory chain. |
| 1975–present | | Extensive kinetic studies on mechanism of action of cytochrome $cd_1$ by Greenwood and others.[b] |

(*continued*)

**Table 1.**   (*Continued*)

| Date | Author(s) | Findings |
|------|-----------|----------|
| 1979 | Watanabe *et al.* | Purification of a then-novel *bd* complex from *Photobacterium.* |
| 1979 | Sone *et al.* | Claim for purification of single-subunit *aa*$_3$ oxidase from thermophile. |
| 1979a | Poole *et al.* | Low temperature trapping of cytochrome *o* oxygen intermediate. |
| 1980 | Ludwig and Schatz | Purification of two-subunit *aa*$_3$ oxidase from *Paracoccus.* |
| 1981 | Powers *et al.* | X-ray studies of purified *aa*$_3$-type oxidase. |
| 1982 | Kent *et al.* | Mössbauer studies of $^{57}$Fe-enriched *Thermus* oxidase. |
| 1983a | Poole *et al.* | Reassignment of cytochrome "*d*$_{650}$" to stable oxygenated form. |

[a]Keilin (1970).
[b]See Poole (1983).


relating to these enzymes but, more importantly, illustrates the changes in direction that research in this area has taken. Prior to 1934, there were violent disagreements, chiefly between Keilin and Warburg, about the significance of cytochromes in cellular respiration. It was only in 1933, on making direct spectroscopic observations of the autooxidizable cytochromes $a_1$ and $d$ in *Acetobacter* and *Azotobacter* that Warburg and his collaborators waived their earlier objections and accepted the central role of cytochromes (Keilin, 1970). During this important period, cytochromes $a_1$ and $d$ were both correctly identified as being directly involved ("$O_2$ transporting ferment") in reactions with $O_2$, and the work of Moss provided the first systematic study of the conditions (now a greatly extended list) that favour appearance of cytochrome $d$.

Improved spectrophotometric techniques resulted in the 1950s and 1960s in the identification of additional CO-binding pigments. One of these (cytochrome $o$) is now recognized as the most widespread of bacterial oxidases, while other haemoproteins believed to have a similar role [for example *Rhodospirillum* haem protein (RHP)] were shown not to be functional oxidases. There are many lessons to be learnt from this early work.

With the advent of these techniques, research in this area has taken three main directions. First, there have been extensive studies of the spectral, thermodynamic and functional properties of the oxidases *in situ* in membrane preparations and intact cells. Second, some of the oxidases have been purified, allowing descriptions of their structure (subunit composition, haem and metal content) and functions (notably electron transfer reactions). Until recently, success was confined to those few cytochrome oxidases that can be readily solubilized, for example the *Vitreoscilla* cytochrome $o$ and *Pseudomonas aeruginosa* cytochrome $cd_1$. Considerable impetus was given to this approach by the finding that the $aa_3$-type oxidases from certain bacteria had relatively simple subunit compositions (see later). The availability of such enzymes is now being exploited using advanced biophysical techniques such as X-ray absorption and Mössbauer spectroscopy. For an example, see Sato *et al.* (1983). Third, and most recent, is the application of low-temperature techniques to identify the intermediate stages in oxygen binding, reduction and electron transfer to the oxidase.

### Mitochondrial Cytochrome Oxidase

The mitochondrial oxidase (E.C. 1.9.3.1) is undoubtedly the best understood and most intensively studied of cytochrome oxidases. As such, it has assumed the role of a benchmark in studies of bacterial systems, partly because of the increasing evidence for close analogies with those bacterial oxidases (the $aa_3$ type) having similar redox centres, and also because the diverse bacterial oxidases have broadly similar functions, so that a comparative study should yield information about the essential features of cytochrome oxidase function.

Figure 1 is a distillation of current concepts of the structure and topography of the mitochondrial oxidase [see Wikström *et al.* (1981), for an excellent survey, and Brunori and Wilson (1982), for a short review]. The enzyme is highly asymmetrical with regard to both its shape and its disposition in the phospholipid

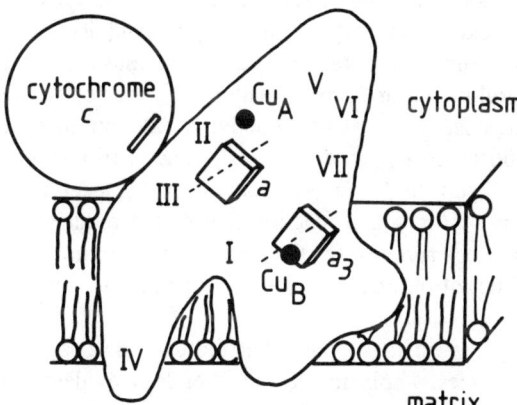

**Fig. 1.** Structure and topography of mitochondrial cytochrome $c$ oxidase. The shape of the protein mass is derived from electron microscopic analysis of a crystal form obtained by deoxycholate extraction. The topography of the subunits (I to VII) is based on chemical and immunological labelling. The square planar structures are the A haems, and the circles, the two copper atoms. $Cu_B$ and haem $a_3$ are in very close proximity, probably coupled by a bridging ligand. Electron paramagnetic resonance (EPR) studies with dysprosium–EDTA as probe suggest that haem $a$ is close to the cytoplasmic side of the membrane, whilst haem $a_3$ is less accessible. Optical and EPR studies of vesicles and multilayers show that both haems are oriented with their planes perpendicular to the membrane plane. Estimates of haem $c$–haem $a$ (25 Å) and haem $a$–haem $a_3$ (10 Å) have been made. For further details, see Wikström *et al.* (1981).

bilayer of the inner mitochondrial membrane. The protein protrudes extensively into the intermembrane space ("cytoplasmic" side), where it is in close proximity to its immediate electron donor, cytochrome $c$. Four metal centres are present, two haems ($a$ and $a_3$) and, associated with each, a copper atom, called (by some) $Cu_A$ and $Cu_B$ respectively. Haem $a_3$ and $Cu_B$ constitute the $O_2$-reactive couple, which appears closer to the matrix aspect than are haem $a$ and $Cu_A$.

The subunit composition of the enzyme is dauntingly complex: at least 7 and as many as 13 different subunits have been reported. A lively controversy surrounds the precise number [see Saraste (1983) and Kadenbach *et al.* (1983), and references therein]. It is not clear which of these subunits binds the haems and copper atoms, but there is attractive experimental (Winter *et al.*, 1980) and circumstantial evidence (see below) that they are associated with subunits I and II. The roles of the other subunits are obscure; the approximate locations of those proposed in the simplest model are shown in Fig. 1.

The electron transfer from cytochrome $c$ to $O_2$ catalysed by cytochrome $c$ oxidase is highly exergonic (96 KJ/2e$^-$) and the energy released is conserved primarily as an electrochemical gradient across the inner mitochondrial membrane. There is now strong support for the idea that the mitochondrial enzyme acts as a proton pump (not shown in Fig. 1), translocating two electric charge equivalents across the membrane per transferred electron. The mechanism, however, is ambiguous (for a review, see Wikström *et al.*, 1981).

The structure and topography of the enzyme and its remarkable function present severe experimental obstacles. The most serious are: (1) the tight association of this integral membrane protein with lipids, making purification difficult, and the lipid dependence of oxidase activity; (2) antiferromagnetic coupling between $Cu_B$ and haem $a_3$, rendering both centres EPR-undetectable; (3) the confusion that still surrounds interpretation of optical spectra of the enzyme; (4) the remarkable $O_2$ affinity and reaction rate of the reduced enzyme with $O_2$, frustrating kinetic experiments; (5) the belief that the minimal functional unit *in situ* may be a dimer of the structure in Fig. 1; and (6) the subunit complexity of the enzyme.

There is every reason to suppose that these properties, and the associated experimental obstacles, apply to the mitochondrial oxidases of all eukaryotic cells.

## Bacterial Oxidases of the $aa_3$ Type

*Introduction*

The past five years have seen an upsurge of interest in bacterial cytochrome oxidases of the $aa_3$ type, due mainly to the finding that certain of them have extremely simple subunit compositions by comparison with the mitochondrial enzymes. This topic has been reviewed by Ludwig (1980) and Poole (1983), who emphasized the structural and functional similarities of the mitochondrial and bacterial enzymes. The most recently published data serve to reinforce such analogies. For example, there is now substantial evidence that the PS3 oxidase is an electrogenic proton pump (Sone and Hinkle, 1982; Yanagita *et al.*, 1983). Similar observations have been made with the enzymes from *Paracoccus denitrificans* (Solioz *et al.*, 1982), *Thermus thermophilus* (Sone *et al.*, 1983a) and the cyanobacterium *Anacystis nidulans* (Peschek, 1983), whilst the *Nitrobacter* enzyme does not pump protons in a reconstituted system (Sone *et al.*, 1983a).

Recently, the two subunits of the *Paracoccus* enzyme have been isolated, partially sequenced and found to be strikingly homologous to subunits I and II of mitochondrial cytochrome *c* oxidase (Steffens *et al.*, 1983). This strongly suggests that the corresponding subunits of the latter also bear the redox-active metal centres, as indeed has been claimed by Winter *et al.* (1980). Since subunit II is also the binding site for cytochrome *c* in eukaryotic oxidases, it would seem that all electron transport functions can be performed with only two of the 7 to 13 subunits.

The lack of subunit III in the *Paracoccus* enzyme is probably responsible for the insensitivity of proton pumping to dicyclohexylcarbodiimide (Püttner *et al.*, 1983), a feature of the mitochondrial oxidase.

### The aa₃-Type Oxidase of the Thermophilic Bacterium PS3

*The $aa_3$-Type Oxidase of the Thermophilic Bacterium PS3*

Studies of PS3 (believed to be *Bacillus stearothermophilus*) were the first to suggest that the $aa_3$-type oxidases of bacteria may have dramatically simpler subunit compositions than the analogous enzyme purified from the mitochondria of higher and lower eukaryotes. Sone *et al.* (1979) reported the oxidase as having only one subunit of apparent $M_r$ 38,000. Later, however, (N. Sone, cited in Ludwig, 1980) an additional subunit of apparent $M_r$ 50,000 to 55,000 was revealed and most recently three subunits have been reported by Sone and Yanagita (1982). The discrepancies probably arose from aggregation during the supposed dissociation in sodium dodecyl sulphate (SDS). Our own work (Baines and Poole, Chapter 3, this volume) supports the existence of three subunits of $M_r$ (depending on the gel system used) (I) 50,000 to 58,000, (II) 28,000 and (III) 22,000 to 27,000, respectively. Subunit II appears somewhat smaller than in the preparation of Sone and Yanagita (1982).

Earlier work on the membrane-bound enzyme failed to detect any perturbation by CO of the α band of the oxidase (Fee *et al.*, 1978). Low-temperature spectroscopy (E. Yang and N. Sone, cited in Ludwig, 1980) also suggested that cytochrome $a_3$ (by definition, the CO-binding component of the oxidase) was missing and that the complex contained haems *a*, *o* and *c* in molar ratios 5:1:5. It was suggested that the stoicheiometry and published haem *a*-to-protein ratio could be explained by a novel configuration, namely the association of haems *a* and $c$ with different polypeptides, but both of $M_r$ 38,000, clustering an $o$-type cytochrome, which was not detected in the original preparation. These data and the model now require reassessment in view of the revised subunit composition (see above) and the following considerations:

1. It is widely recognized that the contribution of haem $a_3$ to the reduced cytochrome $aa_3$ band in the $\alpha$ region is small (the "classical" contention; Wikström *et al.*, 1981, pp. 74–83). Since ligand binding to cytochrome $a_3$ causes no large changes in the optical spectrum of cytochrome $a$, the CO-difference spectra of Poole (1981) suggest that the contribution of haem $a_3$ to the peak of the $aa_3$ complex may be even smaller in PS3, helping to explain the data of Fee *et al.* (1978).

2. Cytochrome $o$ is indeed synthesized by PS3 (Poole, 1981) and can function as a terminal oxidase (Poole *et al.*, 1982c) but its appearance is regulated by growth conditions, for example the onset of stationary phase and $O_2$ limitation, both of which favour cytochrome $o$ synthesis at the expense of $a$-type cytochromes in both defined (B. S. Baines and R. K. Poole, unpublished) and complex (Sone *et al.*, 1983b) growth media. Thus cytochromes $o$ and $a$ (or $a_3$) are not synthesized coordinately.

3. Cytochrome $a_3$ can be demonstrated by its binding of CO and other ligands in either membranes (Poole, 1981; Poole *et al.*, 1983c) or purified preparations (Sone *et al.*, 1979, 1983b; Sone and Yanagita, 1982; Baines and Poole, Chapter 3, this volume).

4. Potentiometric titrations of PS3 membranes reveal two components contributing to the $\alpha$ band of cytochromes $a$ (Poole *et al.*, 1984) with midpoint potentials of $+340$ to $+360$ mV (33 to 40% total) and $+190$ to $+200$ mV (60 to 67% total). These values are close to those reported for the mitochondrial enzyme.

5. Photochemical action spectroscopy shows that cytochromes $o$ and $a_3$ are functional oxidases (Poole *et al.*, 1982c).

6. Flash photolysis of the CO-liganded enzymes at sub-zero temperatures and in the presence of dissolved $O_2$ (for details of techniques, see later) reveals directly the binding of $O_2$ to cytochrome $a_3$ (Poole *et al.*, 1982c).

7. Copper has been reported in the purified enzyme (Sone *et al.*, 1979; Baines and Poole, Chapter 3, this volume); cytochromes of the $aa_3$ type are the only cytochrome oxidases known to possess this metal.

In summary, this thermophilic bacterium should no longer be regarded as a special case, but as an organism containing a "conventional" bacterial $aa_3$-type oxidase (and a less well characterized cytochrome $o$, see later) of unusual stability to high temperatures (Sone *et al.*, 1979). The enzyme is surprisingly labile at 4°C, however (Baines and Poole, Chapter 3, this volume). Like the $aa_3$-type oxidases from other thermophiles, the enzyme bears a tightly bound cytochrome $c$ (Poole, 1983).

*Studies of the Catalytic Mechanism*

The functional properties of the bacterial $aa_3$-type oxidases, and in particular the kinetics and catalytic mechanism, remain to be described in detail. Broad similarity with the mitochondrial enzyme has been reported, however (see Poole, 1983, for a review), with respect to $O_2$ affinity, the half-times of reoxidation of the reduced form by $O_2$ and the kinetic and spectral identification of two intermediates in the reaction with $O_2$. Such preliminary findings make it tempting to speculate that the stages of $O_2$ binding and reduction are similar in bacterial and mitochondrial cytochrome oxidases of the $aa_3$ type.

It is inappropriate to discuss in detail the chemistry of $O_2$ reduction here. The most germane features are:

1. One-electron transfer to $O_2$ gives superoxide, a highly reactive species, and the reaction has a very low $E_{m7}$. In addition, there is strong evidence that the active site of $O_2$ reduction by cytochrome oxidase is a binuclear centre (haem$a_3$/Cu$_B$). Thus, on all grounds, one-electron reduction appears unlikely to be the first step.
2. Assuming peroxide to be the first reduction product, further donation of one electron would give the HO· radical, again highly reactive and rendering this mechanism unlikely.

The generally accepted scheme for $O_2$ reduction by cytochrome oxidase is the result of many different approaches, but especially kinetic measurements at room temperature and very low temperatures; the latter will be emphasized here.

Introduction of the so-called triple-trapping technique by Chance *et al.* (1975a,b) has allowed temporal resolution of the individual steps in $O_2$ reduction ("freezing the biological burning"; Chance, 1975). The technique, which relies on ligand exchange in the frozen state after photolysis of the reduced oxidase–CO complex, has been described elsewhere and is shown in Fig. 2.

On photolysis of the reduced enzyme, in the presence of, typically, 400 $\mu M$ $O_2$, and at about $-100°C$, the 590-nm band of the $a_3$–CO species is bleached and a band appears at higher wavelengths, near 612 nm for the mitochondrial enzyme (Chance *et al.*, 1975a; Poole *et al.*, 1979b) and apparently at slightly shorter wavelengths (603 to 609 nm) in PS3 (Poole *et al.*, 1982c; R. K. Poole and B. Chance, unpublished). As the temperature is raised, spectral changes that are roughly the inverse of photolysis ensue, suggesting ligand binding to the reduced oxidase to give the structure $a_3^{2+}$–$O_2$ Cu$_B^{1+}$. This "oxygenated" species (compound A) can be readily observed after trapping at sub-zero temperatures.

When the experiment is repeated in the absence of $O_2$, photolysis is followed by a slow recombination of CO with the oxidase. In fact, the recombination appears anomalously slow by comparison with CO binding to other haemoproteins. It has been suggested (Orii, 1978) that this is due to the displacement at

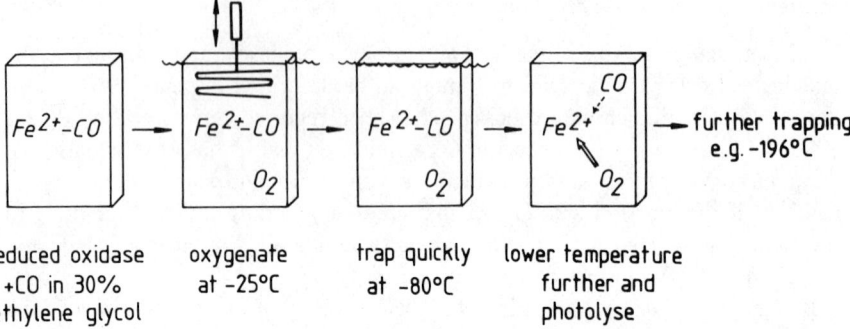

reduced oxidase    oxygenate    trap quickly    lower temperature
+CO in 30%       at -25°C      at -80°C        further and
ethylene glycol                              photolyse

**Fig. 2.** Steps in the low-temperature ligand-exchange technique for studying the reactions of cytochrome oxidases with $O_2$ at sub-zero temperatures. An anoxic sample of substrate-reduced enzyme in co-solvent is bubbled with CO and equilibrated for 5 min in an ethanol–dry ice bath at $-23$ to $-25$°C. Oxygen is added by vigorous stirring or addition of an appropriate volume of air-saturated buffer containing co-solvent. At such temperatures, the half-time for dissociation of CO from the oxidase (assuming $E_a = 10$ kcal) is about 100 s, so that rapid $O_2$ addition and prompt trapping at lower temperatures (generally in an ethanol–dry ice bath) are required to give an acceptably small degree of dissociation of the CO compound. On photolysis of the CO compound at the experimental temperature, the rate of reaction of the reduced enzyme with $O_2$ greatly exceeds that of CO recombination, although exceptions do occur at much lower temperatures (see text). Modifications to the scheme shown include further trapping of the reaction stages in liquid $N_2$ or addition of ferricyanide or hexachloroiridate just before $O_2$ to oxidise components of appropriate potential, thus forming a mixed-valence state. For further details, see Chance (1978).

very low temperatures of CO from ferrous iron to cuprous copper, a proposal confirmed by the detection, using Fourier transform infrared spectroscopy, of a copper–carbonyl species (Fiamingo *et al.*, 1982). However, it is important to note that similarly slow ligand-binding kinetics have been observed for cytochrome *o* (Poole *et al.*, 1979a), which is almost certainly devoid of copper.

The subsequent reactions of compound A have not yet been described in detail for a bacterial cytochrome $aa_3$, but it is probable that they will resemble the steps described in detail by Wikström *et al.* (1981). The next-formed intermediate in the reaction of the fully reduced enzyme with $O_2$ is the peroxidic species in which two electrons are donated to the bound $O_2$, probably one from each of $a_3$ and $Cu_B$, although quadrivalent iron and retention of $Cu_B$ in the cuprous state cannot be ruled out. The terminology of this species is confused. In the reaction with $O_2$ of half-reduced enzyme (that is, only haem $a_3$ and $Cu_B$ reduced) a different intermediate is formed, the so-called compound C. Wikström *et al.* (1981) discuss the possibility that this is identical to "intermediate II" (Clore *et al.*, 1980) and that the so-called compound B is actually heterogeneous.

*Summary*

In conclusion, those bacterial oxidases of the $aa_3$ type that have been studied in detail show close analogies with the mitochondrial system. It follows that the experimental obstacles presented by the latter are not circumvented. However, the bacterial systems offer certain now-familiar advantages, including the possibility of isolating mutants defective in oxidase structure and/or function, thus opening up the techniques of molecular genetics. Phenotypic variants resulting, for example, from growth in Fe- or Cu-deficient media, or the presence of respiratory inhibitors and at various $O_2$ tensions should also be available. Copper-limited growth is an especially attractive possibility since this could provide direct information on the contribution of $Cu_A$ and $Cu_B$ to optical and EPR spectra, their roles in ligand binding and $O_2$ reduction, and whether the copper sites can be filled with other metals. Such work is in progress in this laboratory.

## Cytochrome *o*

### Purification and Structure

Of the four or five major classes of cytochrome oxidase found in bacteria, cytochrome *o* is the most widespread. Its main distinguishing feature is the CO difference or photodissociation spectrum (Fig. 3) which reveals protohaem to be the CO-binding prosthetic group.

Despite its common occurrence, cytochrome *o* proved difficult to isolate and purify until the last few years, when a number of reports have appeared (Table 2). No completely unifying picture appears, although copper appears to be absent, and two haems (two *b* types or a *b* and *c* type) seem common.

The first cytochrome *o* to be isolated, and still the best studied, was that from the myxobacterium *Vitreoscilla*. This could be regarded as a special case, however, since (1) the enzyme is water-soluble and (2) the remarkable stability of its oxygenated intermediates resembles the properties of an $O_2$ carrier more than an oxidase. Very recently, DeMaio et al. (1983) described an additional cytochrome "*o'*" in whole cells of *Vitreoscilla,* having kinetic characteristics for $O_2$ binding and a low dissociation constant very similar to that of the membrane-bound cytochromes $o_{432}$ and $o_{436}$ of *Escherichia coli* (see later). It is suggested that both membrane-bound and soluble cytochromes *o* in *Vitreoscilla* may be involved in electron transport to $O_2$; the former enzyme has not been isolated.

The PS3 cytochrome *o,* shown to be a functional, terminal oxidase (Poole et al., 1982c), appears to contain only one subunit type ($M_r \approx 47,500$), but the polydispersity of the Triton-solubilized enzyme has precluded estimation of its native size (Baines and Poole, Chapter 3, this volume).

It is of great interest that the purified *o*-type oxidases lack copper, a conclusion tentatively predicted from the absence of optical signals (Poole et al., 1983b) analogous to those seen around 830 nm in the mitochondrial enzyme (for references, see Wikström et al., 1981). However, two haems seem to be a common

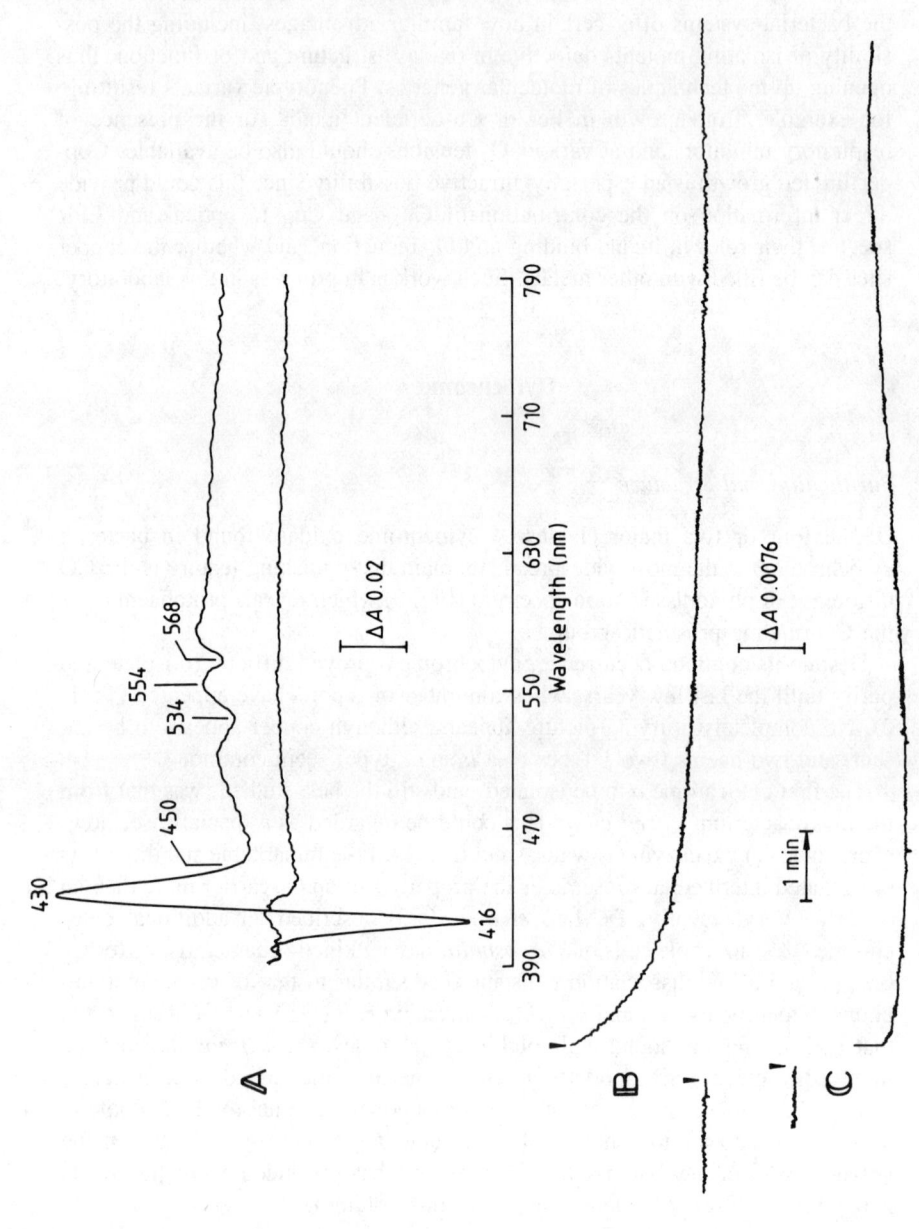

**Table 2.** *Composition of purified cytochromes* o

| Organism | Haem types | Subunits (major) | Copper present | Reference |
|---|---|---|---|---|
| *Azotobacter vinelandii* | *b, c* | 1 type | — | Yang and Jurtshuk (1978a,b); Yang *et al.* (1979); Jurtshuk *et al.* (1981) |
| *Escherichia coli* | *b, b* | 4 | — | Kranz and Gennis (1983); Matsushita *et al.* (1983) |
| *Methylophilus methylotropus*[a] | *b, c* | 2 (+ minor components?) | — | Carver and Jones (1983) |
| *Pseudomonas aeruginosa*[a] | *b, c* | 4? | Variable | Matsushita *et al.* (1982); Yang (1982) |
| *Rhodopseudomonas palustris*[a] | *b, c* | 4 | — | King and Drews (1976) |
| *R. capsulata*[b] | atypical *c* or *b* | 1 | — | Hüdig and Drews (1982a,b) |
| Thermophile PS3 | *b, b* | 1 type | None | Baines and Poole (Chapter 3, this volume) |
| *Vitreoscilla*[c] | *b, b* | 2 identical | None | Tyree and Webster (1978) |

[a]Oxidase role not demonstrated by photochemical action spectroscopy.
[b]CO-insensitive.
[c]Water-soluble.

feature of the purified cytochrome *o*-containing complexes, which in concert might provide the binuclear centre for $O_2$ reduction.

A detailed description of the catalytic mechanism of cytochrome *o* is problematic in the absence, for any organism except *Vitreoscilla,* of detailed information about the numbers and types of haems and their midpoint potentials. Some

**Fig. 3.** Spectral and kinetic properties of cytochrome oxidase *o* in *E. coli* at sub-zero temperatures. In (A), the spectrum of a succinate-reduced, CO-liganded suspension of intact cells in 30% (v/v) ethylene glycol was scanned and stored in the digital memory of a dual-wavelength spectrophotometer at −124°C. The reference wavelength was 500 nm; this spectrum was subtracted from all recorded subsequently. The next scan gave the CO-reduced minus CO-reduced baseline (unlabelled). Photolysis was achieved with light from a 150-W projector bulb focussed onto an afferent limb of a bifurcated light guide, which transmitted light to the sample and gave the photodissociation spectrum shown. The scan rate was 2.86 nm $s^{-1}$, spectral band width 8 nm. In (B), photolysis preceded measurements in the dual-wavelength mode (430 minus 450 nm) of the kinetics of CO recombination at −71°C. The photolysing light was switched on and off at the first and second arrows, respectively. (C) A similar experiment in which $O_2$ (~400 μ*M*) was stirred into the sample at −23°C before freeze trapping. Note the rapid oxidation of *b*-type cytochromes after photolysis in (C), followed by slow, partial re-reduction. (Unpublished experiment of H. D. Williams and R. K. Poole.)

of the reported values (Poole, 1983) are surprisingly low. In our previous potentiometric study of PS3 (Poole *et al.*, 1983c), the CO-binding cytochrome $b_{557}$
($E_{m7}$ = +104 mV) was tentatively identified as cytochrome $o$, a suggestion
supported by the subsequent purification of the oxidase (Baines and Poole,
Chapter 3, this volume).

*Identification of "Oxygenated" and Other Intermediate Species*

Low-temperature, ligand-exchange and trapping techniques applied to cytochrome $o$ in *E. coli* provided the first evidence that an oxygenated compound,
equivalent to compound A in mitochondria, was an early, and probably the first,
intermediate in the reaction of a bacterial oxidase with $O_2$. Independent evidence
from studies of other bacteria (Table 3) supports the existence of such intermediates. It is interesting that many, for example *E. coli* cytochrome $o$, but not all of
these have spectral properties similar to the CO compound, in this respect resembling compound A of the mitochondrial oxidase, bacterial $aa_3$-type oxidases and
the oxygenated form of haemoglobin (see Table V of Poole, 1983). One explanation of why such an intermediate is not always found is that the species described
are not the primary ones and may not be homogeneous, representing a mixture of
intermediate forms. In most cases where low-temperature trapping has been
used, spectral equivalence of the CO and oxy compounds has been observed,
although cytochrome $d$ remains a notable exception (see later). Also inexplicable
at present is the behaviour in low-temperature experiments of cytochrome $o$ in
PS3 membranes; photolysis at $-103°C$ in the presence of $O_2$ (Poole *et al.*,
1982c) fails to reveal the expected Soret band of the ferrous haem at about 436
nm, perhaps due to unexpectedly fast formation of the oxygenated compound.
More detailed kinetic experiments at lower temperatures are required.

Whatever the spectral properties of the presumptive oxygenated intermediates
of cytochrome $o$, the kinetics of their formation are quite distinct from those for
cytochrome $a_3$ (Table 4). Strikingly similar results have been reported for
cytochrome $o$ of *E. coli* and *Vitreoscilla* cytochrome $o'$. The limited evidence
available suggests a different strategy for $O_2$ trapping in $o$-type oxidases. Indeed,
the structure of cytochrome $a_3$ is consistent with the observed low affinity at sub-
zero temperatures ($K_d \approx 0.3$–$0.48$ m$M$), since the substituent formyl group on the
porphyrin provides strong electron-withdrawing power when conjugated with the
ring (Wikström *et al.*, 1981). The much higher affinity observed at physiological
temperatures may be attributed to rapid electron donation to compound A.

Table 4 also illustrates another interesting feature of the *E. coli* system, namely the appearance, during $O_2$-limited growth, of a cytochrome $o$ with enhanced
kinetic competence for reaction with $O_2$ (Poole and Chance, 1981). However,
the adaptation to such conditions, *par excellence,* appears to be the synthesis of

**Table 3.** *Compounds described as stabilized intermediates in the reactions of bacterial cytochrome oxidases with oxygen*[a]

| Bacterium | Oxidase type | Temperature of observation (°C) | Intermediates observed | Absorption maxima (nm) | References |
|---|---|---|---|---|---|
| PS3 | $aa_3$ | −103 | Compound A | [b,c]430, 588 | Poole et al. (1982c) |
| Acetobacter suboxydans | "Haemoprotein 558" | Room temperature | Not named | [b]423, 538, 572 / 417, 540, 582 | Iwasaki (1966) |
| Acetobacter pasteurianum | o | −84 | Not named | [b,c]417, 534, 568 | H. D. Williams and R. K. Poole (unpublished) |
| Vitreoscilla | o | 0 | Compound D / "Oxygenated compound" | 418, 548, 576 / 414, 543, 576 | Tyree and Webster (1979) |
| | o' | −90 to −125 / −80 | I' / I' | [b,c]428, 530–534, 564 / [c]422, 534, 564 | DeMaio et al. (1983) |
| E. coli | o | −90 to −110 / −80 | Compound A-like / Not named | [b,c]415, 531, 567 / [c]428, 531, 567 | Poole et al. (1979a); Poole and Chance (1981) |
| E. coli (A. vinelandii has similar properties) | d | Room temperature or −90 to −130 | "$d_{650}$" | [c]650–653 | Poole et al. (1983a,b) |
| Thiobacillus ferrooxidans | $a_1$ | −106 | Compound A | [b,c]426, 592 | De Fonseka et al. (1980) |
| | | | Compound B | — — | K. De Fonseka, G. A. Reid, W. J. Ingledew, and B. Chance (unpublished) |
| Acetobacter pasteurianum | $a_1$ | −150 | Compound A | ? , 592 | H. D. Williams and R. K. Poole (unpublished) |
| Pseudomonas aeruginosa | $cd_1$ | Room temperature | A / B | Indistinct in α / ~ 640 } | Greenwood et al. (1978) |
| | | | "Oxygenated" | 646 | Shimada and Orii (1976) |

[a]Note: DeMaio et al. (1983) have chosen to label the maxima in their photodissociation spectra, while Poole et al. (1979a) describe the minima to emphasize the similarity with the CO-liganded form.
[b]Resembling CO form.
[c]With CO reduced form as reference spectrum.

**Table 4.** *Comparison of the kinetics of the reactions with $O_2$ of cytochromes o and $aa_3$ at sub-zero temperatures*

| Oxidase | Temperature (°C) | Forward velocity constant, $K_{+1}$ ($M^{-1}$ $s^{-1}$) | Dissociation constant, $K_d$ ($\mu M$) | Reference |
|---|---|---|---|---|
| *E. coli* cytochrome $o_{432}$ | −101 | 0.91 | 15 | Poole *et al.* (1979a) |
| *E. coli* cytochrome $o_{436}$ | −100 | 10.9 | n.e.[a] | Poole and Chance (1981) |
| *E. coli* cytochrome $o_{436}$ | −90 | 21 | 7 | Poole and Chance (1981) |
| *Vitreoscilla* cytochrome o' | −99 | 9.1 | 11 | DeMaio *et al.* (1983) |
| Mitochondrial cytochrome $aa_3$ | −91 | 685 | 480 | Chance *et al.* (1975a) |

[a]n.e., not estimated.

cytochrome *d*. For a review of the diversity and adaptability of the *E. coli* respiratory system, see Ingledew and Poole (1984).

## Cytochrome *d*

### Function and Structure

Cytochrome *d* offers an example of a third haem type in bacterial oxidases, chlorin. Despite the discovery of cytochrome *d* (then named "$a_2$") more than 50 years ago (Table 1), there has, until recently, been very little information that might contribute to an understanding of its catalytic mechanism; instead, workers in this field have produced an extensive catalogue of those conditions that favour cytochrome *d* synthesis in *E. coli* and other bacteria (Poole, 1983). Even so, we are not much closer to an understanding of the regulation of its synthesis, although it is significant that it is produced at the expense of cytochrome $aa_3$ during copper-limited growth.

The most pertinent properties that may contribute to a description of its function are as follows. First, its role as an oxidase has been clearly established by photochemical action spectroscopy (Castor and Chance, 1959) and rapid kinetic measurements support this conclusion (Smith *et al.*, 1970; Haddock *et al.*, 1976). The band of the reduced form titrates as a single component with a midpoint potential of +260 to +280 mV (Pudek and Bragg, 1976; Reid and Ingledew, 1979). In *E. coli*, there is evidence that the oxidase's affinity for $O_2$ is higher than that of cytochrome $o$ (Rice and Hempfling, 1978) but the reverse appears to be the case in *Azotobacter vinelandii* (Hoffman *et al.,* 1979).

Cytochrome $d$-containing complexes have been purified by Reid and Ingledew (1980) and Miller and Gennis (1983). The latter preparation consists of two subunit types ($M_r$ = 57K and 43K) yet, interestingly, contains cytochrome $d$ and $b$ and also exhibits an optical band attributed at present to cytochrome $a_1$.

### Spectral Properties and the Oxygenated Intermediate

Our own work in this area stemmed from a desire to study the catalytic mechanism of cytochrome $d$ at sub-zero temperatures and, by identifying intermediates in the reaction, delineate the pathways of electron transfer to $O_2$. In practice, the results initially led to our questioning the correctness of conventional assignments of the optical bands.

It had long been recognized that cytochrome $d$ had an $\alpha$ band in the reduced form that was at unusually long wavelengths, in the red region of the spectrum. Attention was drawn to this property by White and Sinclair (1970): "Microbiology's greatest gift to biochemists interested in cytochromes is cytochrome oxidase $a_2$ (*sic*), whose spectral maximum at 631 nm is uncomplicated by other pigments". Figure 4 confirms the wisdom of this remark but also illustrates how understated it was. The band of the reduced form at about 630 nm is well separated from that of $a$-, $b$-, and $c$-type cytochromes. Furthermore, CO (and certain other ligands; Hubbard *et al.,* 1983, also Chapter 16, this volume) bind to the reduced form and shift its band even further to the red. The most remarkable feature is the band at about 650 nm that is seen in aerated preparations and which Keilin (1970) described as "the narrow band of its ferric, i.e. oxidized, form". The work that follows questions this assignment and has exploited the availability of two lasers. The HeNe laser (line at 632.8 nm) is almost perfect for selective photolysis of the CO compound and has been used to demonstrate the oxidation of $b$-type cytochromes, distinct from cytochrome $o$ (whose CO complex is not photolysed by this laser), following the reaction of cytochrome $d$ with $O_2$ (Poole and Chance, 1981; Poole *et al.,* 1983b). The krypton ion laser (line at 647.1 nm) is equally convenient for probing the nature of the 650-nm form by vibrational spectroscopy (Fig. 4).

Our first experiments with this system were extremely puzzling. We chose to work with cells from $O_2$-limited cultures of *E. coli,* since these are quite green with high cytochrome $d$ levels. Anoxic suspensions of substrate-reduced cells or membranes were reacted with CO and taken to low temperatures in the presence of 30% (v/v) ethylene glycol. Photolysis was attempted at $\approx -130°C$ with a xenon arc lamp, with surprising results. In the absence of $O_2$ in the frozen samples, no signals attributable to cytochrome $d$, i.e. between 610 and 860 nm, were seen in the photodissociation spectrum (Poole *et al.,* 1983b). Having elimi-

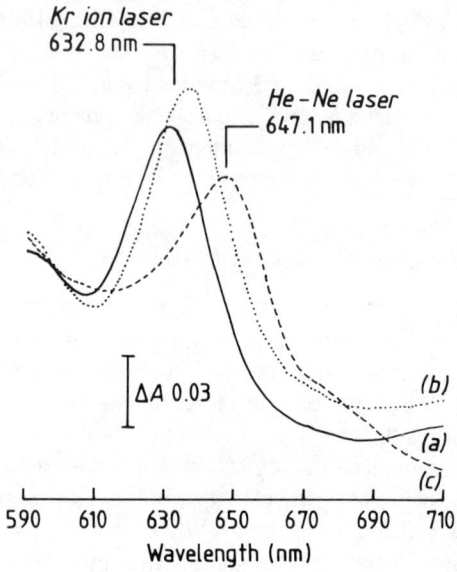

**Fig. 4.** Absorption bands of cytochrome *d* in *Escherichia coli* and their relation to lines from the krypton ion and helium–neon lasers. Washed cells from an $O_2$-limited culture were suspended in buffer to a protein concentration of 18 mg ml$^{-1}$. Difference spectra (with milk in the reference cuvette) were scanned at 0.5 nm s$^{-1}$ for samples (a) reduced with 12 m$M$ succinate for 5 min, (b) reduced then bubbled with CO for 2 min and (c) shaken vigorously with a few grains of potassium ferricyanide. Spectra have been aligned at the isosbestic point, 600 nm.

nated the possibility that the CO had been lost from the oxidase during the trapping procedure, photolysis at much lower temperatures (4 K) was performed to test the hypothesis that, like certain other haemoproteins, CO recombines readily even below liquid nitrogen temperatures. The first photodissociation spectrum of cytochrome *d* (Poole *et al.*, 1982a) thus obtained confirmed this suggestion and also demonstrated the weak Soret band predicted by Chance (1953). In the presence of oxygen, however, flash photolysis at $-130°C$ of substrate-reduced, CO-liganded membranes generated a sharp, symmetrical band at 650–652 nm (where the reference spectrum is the CO-liganded reduced form) (Poole *et al.*, 1983a,b). A similar difference spectrum is produced at room temperature by aerating membranes, but not by treatment with ferricyanide. EPR spectroscopy of the "cytochrome $d_{650}$" trapped at this temperature (Poole *et al.*, 1983a) showed that its spectrum was indistinguishable from that of the CO-liganded form and did not reveal resonances in the $g = 6$ region attributed by Dervartanian *et al.* (1973) to high-spin ferric haem *d*. Based on these results, and considering the curious potentiometric and ligand-binding properties of cytochrome $d_{650}$, Poole *et al.* (1983a) suggested that the fully oxidized form of cytochrome *d* has no characteristic absorbance in the red region of the spectrum and is thus equivalent to the cytochrome $d_x$ or $d*$ forms proposed by others, and analogous to methaemoglobin, whereas cytochrome $d_{650}$ is an early intermediate in the reaction of reduced oxidase with oxygen and analogous to oxyhaemoglobin. This hypothesis is the converse of previous proposals, which invoke the existence of a hypothetical, invisible intermediate (Pudek and Bragg, 1974) and an oxidized configuration with unusually distinct optical properties. Based on the necessity of oxygen for the formation of cytochrome $d_{650}$, Pudek and Bragg (1976) originally proposed that it "might be analogous to the 'oxygenated' cytochrome oxidase of mitochondria," which at that time was also poorly defined.

This interpretation of the optical spectrum has important consequences for all previous studies in which cytochrome $d$ concentration or kinetic behaviour was monitored with 650 nm as the reference wavelength.

Evidence that $d_{650}$ is of the form $d^{2+} \cdot O_2$ (or $d^{3+} \cdot O_2$) has come from vibrational spectroscopy of aerated cell suspensions, conditions under which the optically distinct form is stable for hours. In resonance Raman scattering, which can provide direct information on electronic configuration, a molecule is irradiated with light corresponding to an electronic transition, for example the peak of a visible absorption band, and the intensity of certain Raman bands can be greatly enhanced. We were fortunate again in having an appropriate laser line (from the krypton ion laser) for excitation of the 650-nm absorbing form. The resonance Raman spectrum (Poole et al., 1982b) showed signals at about 1100 cm$^{-1}$ (wave numbers) very similar to those observed for oxymyoglobin and oxyhaemoglobin.

This oxidase then appears to have several characteristics of an $O_2$-carrying protein, including (1) rapid ligand binding below liquid nitrogen temperatures and (2) formation of a stable oxygenated form. One may speculate that it is an effective $O_2$ scavenger for which speed of oxygen trapping and storage is of greater immediate significance than $O_2$ reduction. Indeed, there is some evidence (for example, Meyer and Jones, 1973a) that cytochrome $d$ terminates non-phosphorylating respiratory pathways in bacteria so that there is no thermodynamic gain from such pathways, although they may also serve to dispose of excess reducing equivalents. The ligand-binding behavior of cytochrome $d$ also justifies the idea that it serves a role in respiratory protection for the nitrogenase in *Azotobacter*.

## Cytochrome $a_1$

*Definition and Possible Function*

The synthesis of cytochrome $d$ (previous section) is accompanied in *E. coli* by the appearance of a band between 585 and 595 nm, assumed to represent cytochrome $a_1$. Its function is obscure and almost every aspect of its study has been the subject of controversy. The only evidence for its being a terminal oxidase is a photochemical action spectrum (Edwards *et al.*, 1981), which conflicts with an earlier experiment of this type (Castor and Chance, 1959) and with unpublished results (R. I. Scott and D. Lloyd, personal communication) in which no band attributable to cytochrome $a_1$ could be shown. Cytochrome $a_1$ does appear to bind CO, however (Poole *et al.*, 1981). In the one detailed kinetic study made, cytochrome $a_1$ was reoxidized sluggishly ($t_{1/2} \approx 25$ ms) and was considered kinetically incompetent to act as a terminal oxidase (Haddock *et al.*, 1976). Its midpoint potential is around $+147$ to $+160$ mV (e.g., Reid and Ingledew, 1979); a second higher potential described by these authors was probably due to cytochrome *d*.

In a few bacteria, its role as an oxidase has been demonstrated, however, notably in *Acetobacter pasteurianum*, on which the early work of Warburg, Keilin and Chance (see Table 1 and Poole, 1983) was performed. In retrospect, *Acetobacter* proved to be a fortunate choice; not only is this organism one of the few in which an oxidase role for cytochrome $a_1$ has been established, but the strain used by Chance appeared to lack all other CO-binding pigments. Our attempts to locate such a strain have been unsuccessful and studies of the $O_2$ and CO reactions in our laboratory are being pursued with a strain in which cytochromes *o* and *d* are also present (Fig. 5).

Since this field was last reviewed (Poole, 1983), new observations have been published. An *a*-type cytochrome, described as cytochrome $a_1$, has been shown by photochemical action spectra to be a terminal oxidase in *Pseudomonas carboxydovorans* (cited in Meyer and Schlegel, 1983) and to occur in other carboxydobacteria. As Ingledew (1977) has pointed out, however, the definition of cytochrome $a_1$ is problematic, relying solely on the position of the $\alpha$ band of the reduced enzyme at room temperature. The *a*-type cytochromes in certain carboxytrophic bacteria, tabulated by Cypionka and Meyer (1983), are very close to 600 nm and could arguably be described as of the $aa_3$ type.

*Paracoccus denitrificans* appears to be an interesting case, since it is one of the few species in which both $aa_3$ and $a_1$ have been reported to coexist (van Verseveld *et al.*, 1983). However, an inevitable consequence is that a role for the presumptive cytochrome $a_1$ is difficult to define since (1) a band at about 590 nm can be attributed to numerous other haemoproteins (examples given in Poole, 1983) and (2) the CO compounds of cytochromes $a_1$ and $aa_3$ are almost indistinguishable (peaks at 590 nm). Thus the photodissociation spectrum of nitrate-limited cells (van Verseveld *et al.*, 1983), which purports to show cytochrome $a_1$, could equally well be attributed to cytochrome $aa_3$. The CO sensitivity of these cells was too low for recording the photochemical action spectrum; no other evidence for the role of "cytochrome $a_1$" as an oxidase in *P. denitrificans* has yet been published.

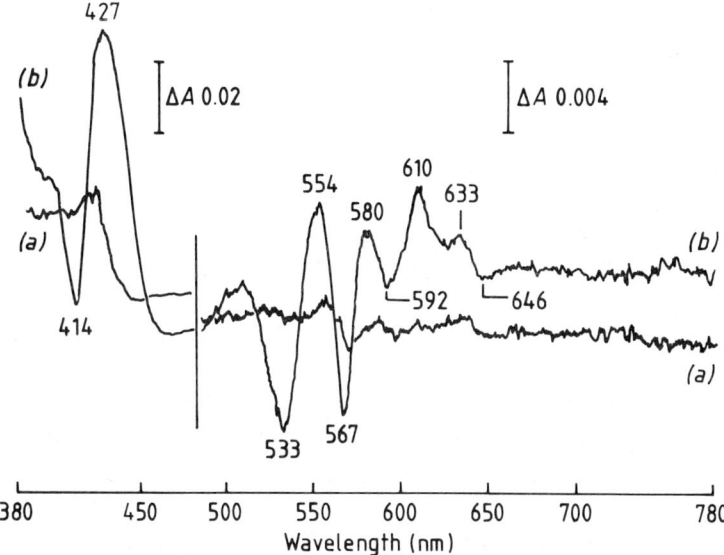

**Fig. 5.** Photodissociation spectrum of *Acetobacter pasteurianus* at $-150°C$. The spectrum of a dithionite-reduced, CO-supplemented suspension of membranes in 30% (v/v) ethylene glycol (to which no $O_2$ had been added) was scanned and stored in the memory of a dual-wavelength scanning spectrophotometer. The reference wavelength was 500 nm. The sample was rescanned to give the baseline (a). After photolysis as described in the legend to Fig. 3, the sample was scanned again to give the photolysed minus CO-reduced difference spectrum (b). Troughs represent the positions of absorbance of the CO complexes of cytochrome oxidases *o* (414, 533, 567 nm), $a_1$ (592 nm) and *d* (646 nm). The scanning speed was 2.86 nm s$^{-1}$ and spectral band width 8 nm. (Unpublished experiment of H. D. Williams and R. K. Poole.)

An *a*-type cytochrome has recently been purified as an $a_1c_1$ complex from *Nitrobacter agilis* (Tanaka *et al.*, 1983). Unlike *E. coli*, the $a_1$ component in *N. agilis* does not bind CO and probably functions as a nitrite oxidase rather than an $O_2$ reductase.

It should be clear that there are problems of nomenclature of such cytochromes and the present working definition, that is any component with an α maximum in the reduced state between about 585 and 596 nm, requires reassessment. It seems reasonable that any presumptive cytochrome $a_1$ should at least be shown to contain haem *a*, yet such reports are rarely made. An interesting feature of the cytochrome $a_1$ in *Thiobacillus ferrooxidans*, which has been studied in detail by Ingledew and Cobley (1980) and De Fonseka *et al.* (1980), is the apparent lack of involvement of copper (although detectable in membranes) in the mechanism of $O_2$ reduction. It is premature to suggest that this is a universal property of cytochrome $a_1$, however, and we must await further data that reveal other characteristics of the *a*-type cytochromes absorbing at 585 to 595 nm.

*The Soluble* $a_1$-*Like Haemoprotein from* E. coli

The apparent diversity of "cytochromes $a_1$" is further emphasised by our recent demonstration that, under diverse growth conditions, a cytochrome $a_1$-like haemoprotein can be easily and extensively solubilized from *E. coli*.

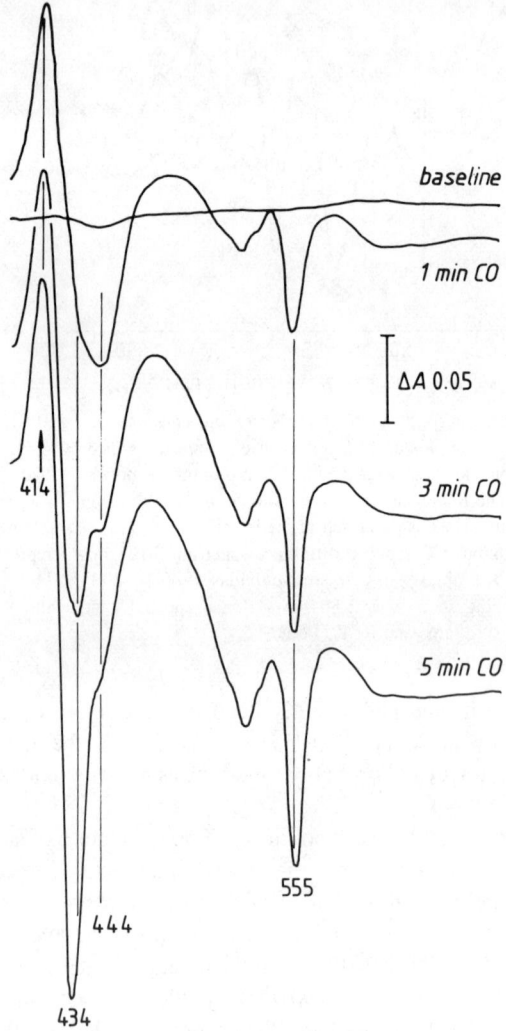

**Fig. 6.** Cytochromes in a soluble fraction from *Escherichia coli*, grown anaerobically with glycerol and fumarate. Washed cells were disrupted by sonication and centrifuged (see Poole *et al.*, 1983a) to give a fraction non-sedimentable at approximately $8.8 \times 10^{10}$ rad$^2$ s$^{-1}$. The baseline is a reduced (with dithionite) minus reduced difference spectrum; the effects of increasing the time of bubbling with CO are shown. The path length was 10 mm and the scanning speed of the split beam instrument 4 nm s$^{-1}$. (Unpublished experiment of H. D. Williams and R. K. Poole.)

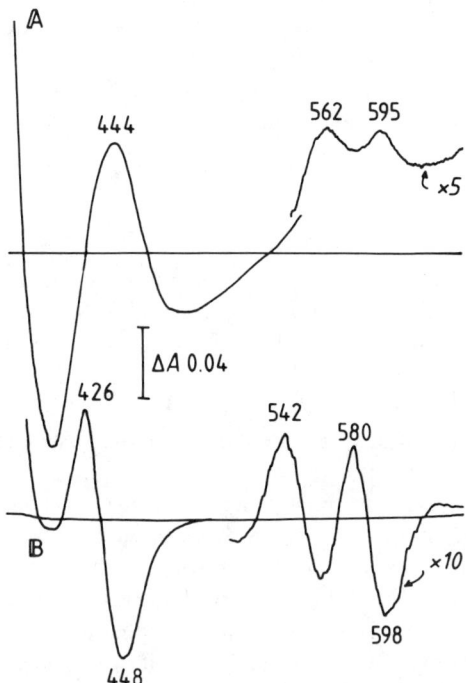

**Fig. 7.** Difference spectra of a partially purified haemoprotein complex from anaerobically grown *E. coli*. A high-speed supernatant fraction (see Fig. 6) was brought to 45% saturation with $(NH_4)_2SO_4$, and the pellet recovered by centrifugation and dialysed. Chromatography on DEAE–Sephadex A50 yielded the preparation shown. (A) Reduced (with dithionite) minus oxidized (with persulphate) difference spectrum and an oxidized minus oxidized baseline. (B) CO difference spectrum (reduced plus CO minus reduced) and a reduced minus reduced baseline. Spectra were obtained in a dual-wavelength scanning spectrophotometer with 475 nm as the reference wavelength. The path length was 10 mm, the scanning rate 5.71 nm s⁻¹ and the spectral band width 4 nm. Protein concentration was 1.45 mg ml⁻¹. (Unpublished experiment of B. S. Baines and R. K. Poole.)

Ultrasonication of cells grown either anaerobically with nitrate or fumarate, or under $O_2$-limited conditions, yields up to 90% of the "cytochrome $a_1$" in a soluble form. Although highly unusual, this is not unprecedented for a cytochrome oxidase, as witnessed by the *Vitreoscilla* cytochrome *o* and *Pseudomonas* cytochrome $cd_1$ (Poole, 1983). The solubilized haemoprotein binds CO readily and also reacts with nitrite or perhaps some other ligand derived from it, such as NO. Also present in the supernatant are a *c*-type cytochrome and a small amount of CO-binding cytochrome *b* (Fig. 6).

We have now partially purified from such supernatants a complex containing a haemoprotein that resembles cytochrome $a_1$ in having bands in the reduced minus oxidized spectrum (Fig. 7) at 444 and 595 nm (Baines *et al.*, 1984). Haem assays show haem of the *b*-, but not *a*-, type. The preparation has catalase and cytochrome *c* peroxidase activities. In view of these properties and the spectral similarities of the $a_1$-like component to well-known high-spin haem *b* proteins, we propose the nomenclature "haemoprotein b-590" until its properties have been better characterized (Poole *et al.*, 1984).

## Conclusions

The ability to bind $O_2$ and reduce it is not restricted to cytochrome *c* oxidase (E.C. 1.9.3.1). Although certain prokaryotes employ such enzymes as terminal oxidases, enzymes having quite distinct structures and catalytic mechanisms frequently supplement or replace the $aa_3$-type oxidases. In only a few cases is a rationalization of this phenomenon possible (Poole, 1983), but this should not detract from the value of the available oxidases in understanding mechanisms of $O_2$ utilization.

It is perhaps dangerous to draw general conclusions about the properties of each of the alternative cytochrome oxidases present in bacteria, based on the rather scattered information available. Drawing on the examples offered by the three organisms studied in our laboratory, however, cytochrome *o* (with protohaem as prosthetic group) apparently lacks copper and "traps" $O_2$ relatively slowly but irreversibly. In *E. coli* and PS3, it seems likely that subsequent electron transfer to the bound ligand is effected by other *b*-type haems, although cytochromes *c* appear to be associated with *o*-containing complexes in other organisms. Cytochrome *d* has remarkable spectral and functional properties, especially its ability to form an unusually stable "oxy" form. Again, the electron transport pathways are currently obscure. "Cytochrome $a_1$" is noteworthy for its apparent diversity and the fact that in *E. coli* and several other bacteria no function has been reported.

The diversity of bacterial cytochrome oxidases, illustrated by the examples selected here and those described elsewhere (Poole, 1983), is intriguing. Even if the speculator restricts his attention to those cytochromes proven to be functional terminal oxidases, it is not clear why four or five structurally quite distinct haemoproteins should be extant if they have in common oxygen reduction. The phenotypic adaptability of bacteria, exemplified by the changes in cytochrome content with growth conditions (for example, nutrient, including oxygen, limitation; growth rate; temperature) suggests the better suitability of certain oxidases under certain conditions. Rationalization of these observations requires more complete information about the structure, function and biosynthesis of the enzymes.

An alternative or additional explanation for the present diversity might lie in the evolutionary origin of these enzymes. Oxygen appeared in the earth's originally anoxic atmosphere rather more than 2 billion ($2 \times 10^9$) years ago. Siliceous rocks of this age on the north shore of Lake Superior, part of the Gunflint Iron Formation, are rich in unequivocally contemporaneous microbial fossils, resembling cyanobacteria and budding bacteria (Broda and Peschek, 1979; Cloud, 1983). The cyanobacteria contain $aa_3$-type oxidases, similar in many respects to the mitochondrial enzyme (Peschek et al., 1982), suggesting that this is the oldest class of such enzymes. Between then and the time that the first eukaryotic cells appeared, 0.6 billion years later, the free atmospheric oxygen was still less than 3% of the present level, and it is likely that all bacterial oxidases appeared during this time.

Rather than looking to global oxygen concentrations as the moulding force for evolution of oxidases, it may be more profitable to investigate microenvironments and the advantages accruing from certain functional attributes of extant oxidases. For example, the very rapid rates of electron transfer to oxygen seen in A. vinelandii, and for which cytochrome $d$ is largely responsible, may have evolved in response to the requirements for maintaining anoxia for $N_2$ fixation. There is some evidence (Meyer and Jones, 1973b) that certain cytochromes $a_1$ have an unusually high $O_2$ affinity and capacity for $O_2$ uptake and would be favoured in $O_2$-poor environments. The finding that certain bacterial oxidases, including $o$, $d$, and $cd_1$ types, can form relatively stable oxygenated compounds perhaps demonstrates their ability to act as oxygen-storage or oxygen-transport proteins. Why, from this plethora of oxidases, cytochromes of the $aa_3$ type assumed the position of the sole cytochrome oxidase in most eukaryotes and multicellular organisms must remain a subject for conjecture.

## Acknowledgments

Work in this laboratory was supported generously by the Science and Engineering Research Council, the Royal Society, the Nuffield Foundation and the University of London Central Research Fund. We wish to thank the many who have contributed in diverse ways to this work, especially Dr. M. N. Hughes.

## References

Baines, B. S., Williams, H. D., Hubbard, J. A. M. and Poole, R. K. (1984). Partial purification and characterization of a soluble haemoprotein, having spectral properties similar to cytochrome $a_1$, from anaerobically grown Escherichia coli. FEBS Letters 171, 309–314.

Barrett, J. (1956). The prosthetic group of cytochrome $a_2$. Biochemical Journal 64, 626–639.

Broda, E. and Peschek, G. A. (1979). Did respiration or photosynthesis come first? Journal of Theoretical Biology 81, 201–212.

Brunori, M. and Wilson, M. T. (1982). Cytochrome oxidase. Trends in Biochemical Sciences 7, 295–299.

Carver, M. A. and Jones, C. W. (1983). The terminal respiratory chain of the methylotrophic bacterium *Methylophilus methylotrophus*. *FEBS Letters* **155**, 187–191.

Castor, L. N. and Chance, B. (1955). Photochemical action spectra of carbon monoxide-inhibited respiration. *Journal of Biological Chemistry* **217**, 453–465.

Castor, L. N. and Chance, B. (1959). Photochemical determinations of the oxidases of bacteria. *Journal of Biological Chemistry* **234**, 1587–1592.

Chance, B. (1953). The carbon monoxide compounds of the cytochrome oxidases. I. Difference spectra. *Journal of Biological Chemistry* **202**, 383–396.

Chance, B. (1975). "Freezing the Biological Burning. The President's Lectures." University of Pennsylvania, Philadelphia.

Chance, B. (1978). Cytochrome kinetics at low temperatures: Trapping and ligand exchange. *In* "Methods in Enzymology" Vol. 54 (Eds. S. Fleischer and L. Packer), pp. 102–111. Academic Press, New York.

Chance, B., Smith, L. and Castor, L. (1953). New methods for the study of the carbon monoxide compounds of respiratory enzymes. *Biochimica et Biophysica Acta* **12**, 289–298.

Chance, B., Horio, T., Kamen, M. D. and Taniguchi, S. (1966). Kinetic studies on the oxidase systems of photosynthetic bacteria. *Biochimica et Biophysica Acta* **112**, 1–7.

Chance, B., Saronio, C. and Leigh, J. S. (1975a). Functional intermediates in the reaction of membrane-bound cytochrome oxidase with oxygen. *Journal of Biological Chemistry* **250**, 9226–9237.

Chance, B., Graham, N. and Legallais, V. (1975b). Low temperature trapping method for cytochrome oxidase oxygen intermediates. *Analytical Biochemistry* **61**, 552–579.

Clore, G. M., Andréasson, L.-E., Karlsson, B., Aasa, R. and Malmström, B. G. (1980). Characterization of the low-temperature intermediates of the reaction of fully-reduced soluble cytochrome oxidase with oxygen by electron paramagnetic resonance and optical spectroscopy. *Biochemical Journal* **185**, 139–154.

Cloud, P. (1983). The biosphere. *Scientific American* **249**, September, pp. 132–144.

Cypionka, H. and Meyer, O. (1983). The cytochrome composition of carboxydotrophic bacteria. *Archives of Microbiology* **135**, 293–298.

De Fonseka, K., Reid, G. A. and Ingledew, W. J. (1980). Detection and kinetics of functional intermediates in the reaction of *Thiobacillus ferro-oxidans* cytochrome $a_1$ with oxygen at low temperatures. *Federation Proceedings, Federation of American Societies for Experimental Biology* **39**, 6.

DeMaio, R. A., Webster, D. A. and Chance, B. (1983). Spectral evidence for the existence of a second cytochrome *o* in whole cells of *Vitreoscilla*. *Journal of Biological Chemistry* **258**, 13768–13771.

Dervartanian, D. V., Iburg, L. K. and Morgan, T. V. (1973). EPR studies on phosphorylating particles from *Azotobacter vinelandii*. *Biochimica et Biophysica Acta* **305**, 173–178.

Edwards, C., Beer, S., Siviram, A. and Chance, B. (1981). Photochemical action spectra of bacterial *a*- and *o*-type oxidases using a dye laser. *FEBS Letters* **128**, 205–207.

Fee, J. A., Findling, K. L., Lees, A. and Yoshida, T. (1978). Respiratory proteins of some extremely thermophilic bacteria. *In* "Frontiers of Biological Energetics" Vol. 1 (Eds. P. L. Dutton, J. S. Leigh and A. Scarpa), pp. 118–126. Academic Press, New York.

Fiamingo, F. G., Altschuld, R. A., Moh, P. P. and Alben, J. O. (1982). Dynamic interactions of CO with $a_3Fe$ and $Cu_B$ in cytochrome *c* oxidase in beef heart mitochondria studied by Fourier transform infrared spectroscopy at low temperatures. *Journal of Biological Chemistry* **257**, 1639–1650.

Greenwood, C., Barber, D., Parr, S. R., Antonini, E., Brunori, M. and Colosimo, A. (1978). The reaction of *Pseudomonas aeruginosa* cytochrome $c_{551}$ oxidase with oxygen. *Biochemical Journal* **173**, 11–17.

Haddock, B. A., Downie, J. A. and Garland, P. B. (1976). Kinetic characterization of the membrane-bound cytochromes of *Escherichia coli* grown under a variety of conditions using a stopped-flow dual-wavelength spectrophotometer. *Biochemical Journal* **154**, 285–294.

Hoffman, P. S., Morgan, T. V. and Dervartanian, D. V. (1979). Respiratory-chain characteristics of mutants of *Azotobacter vinelandii* negative to tetramethyl-*p*-phenylenediamine oxidase. *European Journal of Biochemistry* **100**, 19–27.

Horio, T. (1958). Terminal oxidation system in bacteria. I. Purification of cytochromes from *Pseudomonas aeruginosa*. *Journal of Biochemistry* **45**, 195–205.

Hubbard, J. A. M., Hughes, M. N. and Poole, R. K. (1983). Nitrate, but not silver, ions induce spectral changes in *Escherichia coli* cytochrome *d*. *FEBS Letters* **164**, 241–243.

Hüdig, H. and Drews, G. (1982a) Isolation of a *b*-type cytochrome oxidase from membranes of the phototrophic bacterium *Rhodopseudomonas capsulata*. *Zeitschrift für Naturforschung, C: Biosciences* **37C**, 193–198.

Hüdig, H. and Drews, G. (1982b) Characterization of a *b*-type cytochrome oxidase of *Rhodopseudomonas capsulata*. *FEBS Letters* **146**, 389–392.

Ingledew, W. J. (1977). Cytochrome $a_1$ as an oxidase? *In* "Functions of Alternative Terminal Oxidases" (Eds. H. Degn, D. Lloyd and G. C. Hill), pp. 79–87. Pergamon Press, Oxford.

Ingledew, W. J. and Cobley, J. G. (1980). A potentiometric and kinetic study on the respiratory chain of ferrous-iron-grown *Thiobacillus ferrooxidans*. *Biochimica et Biophysica Acta* **590**, 141–158.

Ingledew, W. J. and Poole, R. K. (1984). The respiratory chains of *Escherichia coli*. *Microbiological Reviews* **48**, 222–271.

Iwasaki, H. (1966). Lactate oxidation system in *Acetobacter suboxydans*, with special reference to carbon monoxide-binding pigment. *Plant and Cell Physiology* **7**, 199–216.

Jones, C. W. and Redfearn, E. (1967). Preparation of red and green electron transport particles from *Azotobacter vinelandii*. *Biochimica et Biophysica Acta* **143**, 354–362.

Jones, C. W., Brice, J. M., Downs, A. J. and Drozd, J. W. (1975). Bacterial respiration-linked proton translocation and its relationship to respiratory-chain composition. *European Journal of Biochemistry* **52**, 265–271.

Jurtshuk, P., Mueller, T. J. and Wong, T. Y. (1981). Isolation and purification of the cytochrome oxidase of *Azotobacter vinelandii*. *Biochimica et Biophysica Acta* **637**, 374–382.

Kadenbach, B., Ungibauer, M., Jarausch, J., Büge, U. and Kuhn-Nentwig, L. (1983). The complexity of respiratory complexes. *Trends in Biochemical Sciences* **8**, 398–400.

Keilin, D. (1925). On cytochrome, a respiratory pigment, common to animals, yeast and higher plants. *Proceedings of the Royal Society of London, Series B* **98**, 312–339.

Keilin, D. (1927). Le cytochrome, pigment respiratoire intracellulaire commun aux micro-organismes, aux plantes et aux animaux. *Comptes Rendus des Seances de la Société de Biologie et de ses Filiales* **96**, S.P. 39–S.P. 68.

Keilin, D. (1970). "The History of Cell Respiration and Cytochrome." Cambridge University Press, London and New York.

Kent, T. A. Münck, E., Dunham, W. R., Filter, W. F., Findling, K. L., Yoshida, T. and Fee, J. A. (1982). Mössbauer study of a bacterial cytochrome oxidase: Cytochrome $c_1aa_3$ from *Thermus thermophilus*. *Journal of Biological Chemistry* **257**, 12489–12492.

King, M. T. and Drews, G. (1976). Isolation and partial characterization of the cytochrome oxidase from *Rhodopseudomonas palustris*. *European Journal of Biochemistry* **68**, 5–12.

Kranz, R. G. and Gennis, R. B. (1983). Immunological characterization of the cytochrome *o* terminal oxidase from *Escherichia coli*. *Journal of Biological Chemistry* **258**, 10614–10621.

Kubowitz, F. and Haas, E. (1932). Ausbau der photochemischen Methoden zur Untersuchung des sauerstoffübertrangenden Ferments. (Anwendung auf Essigbakterien und Hefezellen). *Biochemische Zeitschrift* **255**, 247–277.

Lloyd, D. (1974). "The Mitochondria of Microorganisms". Academic Press, London.

Lloyd, D. and Edwards, S. W. (1977). Electron transport pathways alternative to the main phosphorylating respiratory chain. *In* "Functions of Alternative Terminal Oxidases" (Eds. H. Degn, D. Lloyd and G. C. Hill), pp. 1–10. Pergamon Press, Oxford.

Ludwig, B. (1980). Heme $aa_3$-type cytochrome $c$ oxidases from bacteria. *Biochimica et Biophysica Acta* **594**, 177–189.

Ludwig, B. and Schatz, G. (1980). A two-subunit cytochrome $c$ oxidase (cytochrome $aa_3$) from *Paracoccus denitrificans*. *Proceedings of the National Academy of Sciences of the U.S.A.* **77**, 196–200.

Malmström, B. G. (1982). Enzymology of oxygen. *Annual Reviews of Biochemistry* **51**, 21–59.

Matsushita, K., Shinagawa, E., Adachi, O. and Ameyama, M. (1982). $o$-Type cytochrome oxidase in the membrane of aerobically-grown *Pseudomonas aeruginosa*. *FEBS Letters* **139**, 255–258.

Matsushita, K., Patel, L., Gennis, R. B. and Kaback, H. R. (1983). Reconstitution of active transport in proteoliposomes containing cytochrome $o$ oxidase and *lac* carrier protein purified from *Escherichia coli*. *Proceedings of the National Academy of Sciences of the U.S.A.* **80**, 4889–4893.

Meyer, D. J. and Jones, C. W. (1973a). Oxidative phosphorylation in bacteria which contain different cytochrome oxidases. *European Journal of Biochemistry* **36**, 144–151.

Meyer, D. J. and Jones, C. W. (1973b). Reactivity with oxygen of bacterial cytochrome oxidases $a_1$, $aa_3$ and $o$. *FEBS Letters* **33**, 101–105.

Meyer, O. and Schlegel, H. G. (1983). Biology of aerobic carbon monoxide-oxidizing bacteria. *Annual Reviews of Microbiology* **37**, 277–310.

Miller, M. J. and Gennis, R. B. (1983). The purification and characterization of the cytochrome $d$ terminal oxidase complex of the *Escherichia coli* aerobic respiratory chain. *Journal of Biological Chemistry* **258**, 9159–9165.

Moss, F. (1952). The influence of oxygen tension on respiration and cytochrome $a_2$ formation of *Escherichia coli*. *Australian Journal of Experimental Biology and Medical Science* **30**, 531–540.

Negelein, E. and Gerischer, W. (1934). Direkter spektroskopischer Nachweis des sauerstoffübertragenden Ferments in Azotobakter. *Biochemische Zeitschrift* **268**, 1–7.

Orii, Y. (1978). Photochemical reactions of cytochrome oxidase at low temperatures. *In* "Advances in Biophysics" (Ed. M. Kotani), pp. 285–308. University Park Press, Baltimore, Maryland.

Peschek, G. A. (1983). Proton pump coupled to cytochrome $c$ oxidase in the cyanobacterium *Anacystis nidulans*. *Journal of Bacteriology* **153**, 539–542.

Peschek, G. A., Schmetterer, G., Lauritsch, G., Nitschmann, W. H., Kienzl, P. F. and Muchl, R. (1982). Do cyanobacteria contain "mammalian-type" cytochrome oxidase? *Archives of Microbiology* **131**, 262–265.

Poole, R. K. (1981). Ligand-binding cytochromes $a_3$, $c$ and $o$ in membranes from the thermophilic bacterium PS3. *FEBS Letters* **133**, 255–259.

Poole, R. K. (1983). Bacterial cytochrome oxidases. A structurally and functionally diverse group of electron-transfer proteins. *Biochimica et Biophysica Acta* **726**, 205–243.

Poole, R. K. and Chance, B. (1981). The reaction of cytochrome $o$ in *Escherichia coli* K12 with oxygen. Evidence for a spectrally and kinetically distinct cytochrome $o$ in cells from oxygen-limited cultures. *Journal of General Microbiology* **126**, 277–287.

Poole, R. K., Waring, A. J. and Chance, B. (1979a). The reaction of cytochrome $o$ in *Escherichia coli* with oxygen. Low-temperature kinetic and spectral studies. *Biochemical Journal* **184**, 379–389.

Poole, R. K., Lloyd, D. and Chance, B. (1979b). The reaction of cytochrome oxidase with oxygen in the fission yeast *Schizosaccharomyces pombe* 972h⁻. Studies at subzero temperatures and measurement of apparent oxygen affinity. *Biochemical Journal* **184**, 555–563.

Poole, R. K., Scott, R. I. and Chance, B. (1981). The light-reversible binding of carbon monoxide to cytochrome $a_1$ in *Escherichia coli* K12. *Journal of General Microbiology* **125**, 431–438.

Poole, R. K., Sivaram, A., Salmon, I. and Chance, B. (1982a). Photolysis at very low temperatures

2. BACTERIAL CYTOCHROME OXIDASES

of CO-liganded cytochrome oxidase (cytochrome *d*) in oxygen-limited *Escherichia coli*. *FEBS Letters* **141**, 237–241.

Poole, R. K., Baines, B. S., Hubbard, J. A. M., Hughes, M. N. and Campbell, N. J. (1982b). Resonance Raman spectroscopy of an oxygenated intermediate species of cytochrome oxidase *d* from *Escherichia coli*. *FEBS Letters* **150**, 147–150.

Poole, R. K., Scott, R. I., Baines, B. S., Salmon, I. and Lloyd, D. (1982c). Identification of cytochromes *o* and *a₃* as functional terminal oxidases in the thermophilic bacterium PS3. *FEBS Letters* **150**, 281–285.

Poole, R. K., Kumar, C., Salmon, I. and Chance, B. (1983a). The 650 nm chromophore in *Escherichia coli* is an 'oxy' or oxygenated compound, not the oxidized form of cytochrome oxidase *d*: An hypothesis. *Journal of General Microbiology* **129**, 1335–1344.

Poole, R. K., Salmon, I. and Chance, B. (1983b). The reaction with oxygen of cytochrome oxidase (cytochrome *d*) in *Escherichia coli* K12: Optical studies of intermediate species and cytochrome *b* oxidation at sub-zero temperatures. *Journal of General Microbiology* **129**, 1345–1355.

Poole, R. K., van Wielink, J. E., Baines, B. S., Reijnders, W. N. M., Salmon, I. and Oltmann, L. F. (1983c). The membrane-bound cytochromes of an aerobically grown, extremely thermophilic bacterium, PS3. Characterization by spectral deconvolution coupled with potentiometric analysis. *Journal of General Microbiology* **129**, 2163–2173.

Poole, R. K., Baines, B. S., Curtis, S. J., Williams, H. D. and Wood, P. M. (1984). Haemoprotein *b*-590 (*Escherichia coli*): Redesignation of a bacterial "cytochrome a₁." *Journal of General Microbiology* **130**, 3055–3058.

Powers, L., Chance, B., Ching, Y. and Angiolilli, P. (1981). Structural features and the reaction mechanism of cytochrome oxidase. Iron and copper X-ray absorption fine structure. *Biophysical Journal* **34**, 465–498.

Pudek, M. R. and Bragg, P. D. (1974). Inhibition by cyanide of the respiratory chain oxidases of *Escherichia coli*. *Archives of Biochemistry and Biophysics* **164**, 682–693.

Pudek, M. R. and Bragg, P. D. (1976). Trapping of an intermediate in the oxidation–reduction cycle of cytochrome *d* in *Escherichia coli*. *FEBS Letters* **62**, 330–333.

Püttner, I., Solioz, M., Carafoli, E. and Ludwig, B. (1983). Dicyclohexylcarbodiimide does not inhibit proton pumping by cytochrome *c* oxidase of *Paracoccus denitrificans*. *European Journal of Biochemistry* **134**, 33–37.

Reid, G. A. and Ingledew, W. J. (1979). Characterization and phenotypic control of the cytochrome content of *Escherichia coli*. *Biochemical Journal* **182**, 465–472.

Reid, G. A. and Ingledew, W. J. (1980). The purification of a respiratory oxidase complex from *Escherichia coli*. *FEBS Letters* **109**, 1–4.

Rice, C. W. and Hempfling, W. P. (1978). Oxygen-limited continuous culture and respiratory energy conservation in *Escherichia coli*. *Journal of Bacteriology* **134**, 115–124.

Saraste, M. (1983). How complex is a respiratory complex? *Trends in Biochemical Sciences* **8**, 139–142.

Sato, M., Tanaka, N., Kakiuchi, K., Fukumori, Y., Yamanaka, T., Kasai, N. and Kakudo, M. (1983). Small-angle X-ray scattering study on cytochrome *aa₃*-type terminal oxidase derived from *Nitrobacter agilis*. *Biochemistry International* **7**, 345–352.

Shimada, H. and Orii, Y. (1976). Oxidation–reduction behaviour of the heme *c* and heme *d* moieties of *Pseudomonas aeruginosa* nitrite reductase and the formation of an oxygenated intermediate at heme *d*. *Journal of Biochemistry* **80**, 135–140.

Smith, L., White, D. C., Sinclair, P. and Chance, B. (1970). Rapid reactions of cytochromes of *Hemophilus parainfluenzae* on addition of substrates or oxygen. *Journal of Biological Chemistry* **245**, 5096–5100.

Solioz, M., Carafoli, E. and Ludwig, B. (1982). The cytochrome *c* oxidase of *Paracoccus denitrificans* pumps protons in a reconstituted system. *Journal of Biological Chemistry* **257**, 1579–1582.

Sone, N. and Hinkle, P. C. (1982). Proton transport by cytochrome *c* oxidase from the thermophilic bacterium PS3 reconstituted in liposomes. *Journal of Biological Chemistry* **257**, 12600–12604.

Sone, N. and Yanagita, Y. (1982). A cytochrome $aa_3$-type terminal oxidase of a thermophilic bacterium. Purification, properties and proton pumping. *Biochimica et Biophysica Acta* **682**, 216–226.

Sone, N., Ohyama, T. and Kagawa, Y. (1979). Thermostable single-band cytochrome oxidase. *FEBS Letters* **106**, 39–42.

Sone, N., Yanagita, Y., Hon-nami, K., Fukumori, Y. and Yamanaka, T. (1983a). Proton-pump activity of *Nitrobacter agilis* and *Thermus thermophilus* cytochrome *c* oxidases. *FEBS Letters* **155**, 150–154.

Sone, N., Kagawa, Y. and Orii, Y. (1983b). Carbon monoxide-binding cytochromes in the respiratory chain of the thermophilic bacterium PS3 grown with sufficient or limited aeration. *Journal of Biochemistry* **93**, 1329–1336.

Steffens, G. C. M., Buse, G., Oppliger, W. and Ludwig, B. (1983). Sequence homology of bacterial and mitochondrial cytochrome *c* oxidases. *Biochemical and Biophysical Research Communications* **116**, 335–340.

Tanaka, Y., Fukumori, Y. and Yamanaka, T. (1983). Purification of cytochrome $a_1c_1$ from *Nitrobacter agilis* and characterisation of nitrite oxidation system of the bacterium. *Archives of Microbiology* **135**, 265–271.

Tissières, A. (1954). Role of high-molecular-weight components in the respiratory activity of cell-free extracts of *Aerobacter aerogenes*. *Nature (London)* **174**, 183–184.

Tyree, B. and Webster, D. A. (1978). The binding of cyanide and carbon monoxide to cytochrome *o* purified from *Vitreoscilla*. Evidence for subunit interaction in the reduced protein. *Journal of Biological Chemistry* **253**, 6988–6991.

Tyree, B. and Webster, D. A. (1979). Intermediates in the reaction of reduced cytochrome *o* (*Vitreoscilla*) with oxygen. *Journal of Biological Chemistry* **254**, 176–179.

van Verseveld, H. W., Braster, M., Boogerd, F. C., Chance B. and Stouthamer, A. H. (1983). Energetic aspects of growth of *Paracoccus denitrificans:* Oxygen-limitation and shift from anaerobic nitrate-limitation to aerobic succinate-limitation. Evidence for a new alternative oxidase, cytochrome $a_1$. *Archives of Microbiology* **135**, 229–236.

Vernon, L. D. and Kamen, M. D. (1954). Haematin compounds in photosynthetic bacteria. *Journal of Biological Chemistry* **211**, 643–662.

Warburg, O., Negelein, E. and Haas, E. (1933). Spektroskopischer Nachweis des sauerstoffübertragenden Ferments neben Cytochrom. *Biochemische Zeitschrift* **266**, 1–8.

Watanabe, H., Kamita, Y., Nakamura, T., Takimoto, A. and Yamanaka, T. (1979). The terminal oxidase of *Photobacterium phosphoreum*. A novel cytochrome. *Biochimica et Biophysica Acta* **547**, 70–78.

Webster, D. and Hackett, D. P. (1966). The purification and properties of cytochrome *o* from *Vitreoscilla*. *Journal of Biological Chemistry* **241**, 3308–3315.

White, D. C. and Sinclair, P. R. (1970). Branched electron-transport systems in bacteria. *Advances in Microbial Physiology* **5**, 173–211.

White, D. C. and Smith, L. (1964). Localization of the enzymes that catalyze hydrogen and electron transport in *Hemophilus parainfluenzae* and the nature of the respiratory chain system. *Journal of Biological Chemistry* **239**, 3956–3963.

Wikström, M., Krab, K. and Saraste, M. (1981). "Cytochrome Oxidase—A Synthesis". Academic Press, London.

Winter, D. B., Bruyninckx, W. J., Foulke, F. G., Grinich, N. P. and Mason, H. S. (1980). Location of heme a on subunits I and II and copper on subunit II of cytochrome *c* oxidase. *Journal of Biological Chemistry* **255**, 11408–11414.

Yamagutchi, S. (1934). Über die Beeinflussung der Sauerstoffatmung von verschiedenen Bakterien durch Blausäure und Kohlenoxyd. *Acta Phytochimica* **10**, 171–198.

Yanagita, Y., Sone, N. and Kagawa, Y. (1983). Proton pumping and oxidase activity of thermophilic cytochrome oxidase remain after its extensive proteolysis. *Biochemical and Biophysical Research Communications* **113**, 575–580.

Yang, T. (1982). Tetramethyl-*p*-phenylenediamine oxidase of *Pseudomonas aeruginosa*. *European Journal of Biochemistry* **121**, 335–341.

Yang, T. Y. and Jurtshuk, P. (1978a). Studies on the red oxidase (cytochrome *o*) of *Azotobacter vinelandii*. *Biochemical and Biophysical Research Communications* **81**, 1032–1039.

Yang, T. Y. and Jurtshuk, P. (1978b). Purification and characterization of cytochrome *o* from *Azotobacter vinelandii*. *Biochimica et Biophysica Acta* **502**, 543–548.

Yang, T., O'Keefe, D. and Chance, B. (1979). The oxidation–reduction potentials of cytochrome *c* + *c₄* and cytochrome *o* purified from *Azotobacter vinelandii*. *Biochemical Journal* **181**, 763–766.

Yaoi, H. and Tamiya, H. (1928). On the respiratory pigment, cytochrome, in bacteria. *Proceedings of the Imperial Academy (Tokyo)* **4**, 436–439.

# Discussion

*R. Cammack:* Is it clear whether these oxidases reduce oxygen directly to water, or is $H_2O_2$ produced as an intermediate? Since some of them lack copper, I am interested, in connection with alternative oxidases such as those in plants, to know which types of components are necessary for four-electron reduction.

*R. K. Poole:* This is an important point. We have attempted (D. O'Hara and R. K. Poole, unpublished work) to determine whether peroxide is a free intermediate in the reduction of oxygen by cytochrome *o* or cytochrome *d* in *E. coli*. Membranes from aerobically grown (cytochrome $o_{432}$ and cytochrome *d*) or oxygen-limited cells (cytochrome $o_{436}$ and high levels of cytochrome *d*) also have catalase activity, but which can be totally inhibited by 0.1 to 0.2 m*M* KCN. Such cyanide concentrations do not inhibit oxygen uptake by either membrane type, indicating that the formation of peroxide by the oxidase, followed by catalase action, is not the mechanism involved [for rationale, see J. A. Downie and P. B. Garland, *Biochemical Journal* **134**, 1051–1061 (1973)]. Also, measurements of (1) the relative rates of $O_2$ uptake and NADH oxidation or (2) the stoicheiometric relationship between succinate or NADH oxidation and $O_2$ uptake indicate four-electron transfer to $O_2$. However, *Vitreoscilla* cytochrome *o* does reduce $O_2$ to peroxide stoicheiometrically [for details, see R. K. Poole, *Biochimica et Biophysica Acta* **726**, 205–243 (1983)]. In the case of *aa₃*-type oxidases, the product is probably water. The stoicheiometry of ferrocytochrome *c* oxidized to oxygen reduced is 4:1 for the purified PS3 cytochrome *aa₃* and exogenous catalase is without effect on the kinetics [N. Sone and Y. Yanagita, *Biochimica et Biophysica Acta* **682**, 216–226 (1982)]. With respect to the role of copper, the *Thiobacillus* cytochrome $a_1$ warrants more attention, since a role for copper has not been established in low-temperature experiments [W. J. Ingledew and J. G. Cobley, *Biochimica et Biophysica Acta* **590**, 141–158 (1980); W. J. Ingledew, personal communication]. In the absence of copper, pairs of haems could provide the necessary binuclear centres for $O_2$ reduction.

*O. Meyer:* How do you explain the solubility of some terminal oxidases? Are they fully functional?

*Poole:* First, those oxidases that appear to be soluble have not been identified as functional terminal oxidases by photochemical action spectroscopy, except for the *Vitreoscilla* cytochrome *o*. Even this example is now complicated by the finding of an additional, membrane-bound *o* [R. A. DeMaio, D. A. Webster and B. Chance, *Journal of Biological Chemistry* **258**, 13768–13771 (1983)]. Secondly, there seems to be no *a priori* reason why a soluble protein should not reduce $O_2$ to water; the *Pseudomonas* $cd_1$, which is very loosely associated with the membrane, has of course been studied in great detail. If by "fully functional" you mean "proton-translocating", then that is a different matter.

*P. J. Henderson:* Do individual oxidase enzymes translocate $H^+$ across the bacterial membrane?

*Poole:* The $aa_3$-type oxidases (despite some early reports to the contrary) do seem to be proton pumps. In the case of cytochrome $o$, there are claims that the cytochrome $b_{562}-o$ segment of the chain acts as a coupling site, based on the formation of a membrane potential ($\Delta\psi$) (interior negative) by the complex reconstituted with liposomes [K. Kita, M. Kasahara and Y. Anraku, *Journal of Biological Chemistry* **257**, 7933–7935 (1982)] and the generation of $\Delta\mu_{H^+}$ (interior negative and alkaline), which can drive lactose uptake [K. Matsushita, L. Patel, R. B. Gennis and H. R. Kaback, *Proceedings of the National Academy of Sciences of the U.S.A.* **80**, 4889–4893 (1983)]. The additional presence of cytochrome $d$ has no effect on respiration-driven proton extrusion or molar growth yield [C. W. Rice and W. P. Hempfling, *Journal of Bacteriology* **134**, 115–124 (1978)].

*C. W. Jones:* If I may comment on your aside that cytochrome $o$ reconstituted into proteoliposomes is capable of proton translocation. I would be rather wary of this, as there is no evidence from whole cell studies that cytochrome $o$ is capable of proton pumping.

# 3

# Cytochrome Oxidases from the Thermophilic Bacterium PS3: Solubilization with Various Detergents, Purification and Preliminary Characterization

BALDEV S. BAINES[1] AND ROBERT K. POOLE

*Department of Microbiology, Queen Elizabeth College, University of London, London, United Kingdom*

## Introduction

In common with other tightly bound membrane proteins, the purification of bacterial cytochrome oxidases in high yield is fraught with problems, relating especially to (1) achieving soluble species that retain biological activity and their native structure (Helenius and Simons, 1975; Tanford and Reynolds, 1976) and (2) avoiding artefactual aggregation of the solubilized membrane components, whose association has no biological relevance. Here, we outline the solubilization by Triton X-100 and characterization of two cytochrome oxidases from PS3 and then draw attention to the advantages of alternative detergents both for this organism and more generally.

The thermophile PS3 has at least three CO-binding cytochromes (Poole, 1981), of which cytochromes $o$ and $aa_3$ are functional terminal oxidases (Poole *et al.*, 1982). Previous reports of the purification of the latter (Sone *et al.*, 1979; Ludwig, 1980; Sone and Yanagita, 1982), albeit differing in detail, demonstrate the relative simplicity of the enzyme's subunit composition. Cytochrome $o$ has not been purified from a thermophile before and the structural properties of this class of oxidase are poorly understood (Poole, 1983; Poole *et al.*, Chapter 2, this volume).

[1]Present address: Biotechnology Department, Microbiology Division, Glaxo Group Research Ltd., Greenford Road, Greenford, Middlesex UB6 0HE, United Kingdom.

MICROBIAL GAS METABOLISM:
MECHANISTIC, METABOLIC
AND BIOTECHNOLOGICAL ASPECTS

## Purification of the PS3 Oxidases Extracted with Triton
## X-100

The organism was grown aerobically at 65°C in a 10-litre fermenter on a chemically defined medium with 90 m$M$ Na glutamate as major carbon and nitrogen source. Membranes, isolated after digestion of harvested cells with lysozyme and EDTA (Poole, 1981), were initially extracted as described ("steps 1–3") by Sone and Yanagita (1982). Subsequent steps are shown in Fig. 1. It is noteworthy, with regard to our subsequent discussion of detergents, that those fractions at step 4(b), eluted with gradient 3, were composed entirely of various aggregated cytochromes (not shown).

## Properties of the Isolated Oxidases

### Properties of Cytochrome caa₃

*Properties of Cytochrome* caa$_3$

The properties (Table 1) are similar to those described by Sone and Yanagita (1982), except for the somewhat lower $M_r$ of subunit II in our preparation. Like the $aa_3$-type oxidases from other thermophilic bacteria (Poole, 1983; Yoshida *et al.*, 1984), this enzyme contains a firmly associated cytochrome $c$, the $E_{m7}$ value probably being +229 mV (Poole *et al.*, 1983). At 77 K, but not at room temperature, the oxidase preparation shows a small spectral peak at 561 nm which may be due in part to a degradation product (B. S. Baines and R. K. Poole, unpublished). The spectral data, haem and copper contents suggest the composition 1 haem $a$, 1 haem $a_3$, 2 Cu and 1 haem $c$. The copper:haem $a$ ratio falls on storage, coincident with loss of the 604-nm peak (results not shown). Fuller studies of the catalytic activities, especially at sub-zero temperatures, are in progress.

### Properties of Cytochrome o

*Properties of Cytochrome* o

Reduced minus oxidized and CO difference spectra (Table 1) reveal only $b$-type cytochrome(s). The $E_{m7}$ of $b_{557}$ is +104 mV (Poole *et al.*, 1983), tentatively identifying this component as the ligand-binding cytochrome $o$. The one subunit type found has an $M_r$ somewhat larger than for other purified cytochromes $o$ (Poole *et al.*, Chapter 2, this volume). The band is unlikely to be the result of incomplete dissociation, however, since two independent methods for dissociation, namely the acid/phenol system and sodium dodecyl sulphate/urea at room temperature, gave identical results. The polydispersity of the purified protein has precluded measurement of its native size. The iron:haem $b$ ratio was 1:1 and copper was absent.

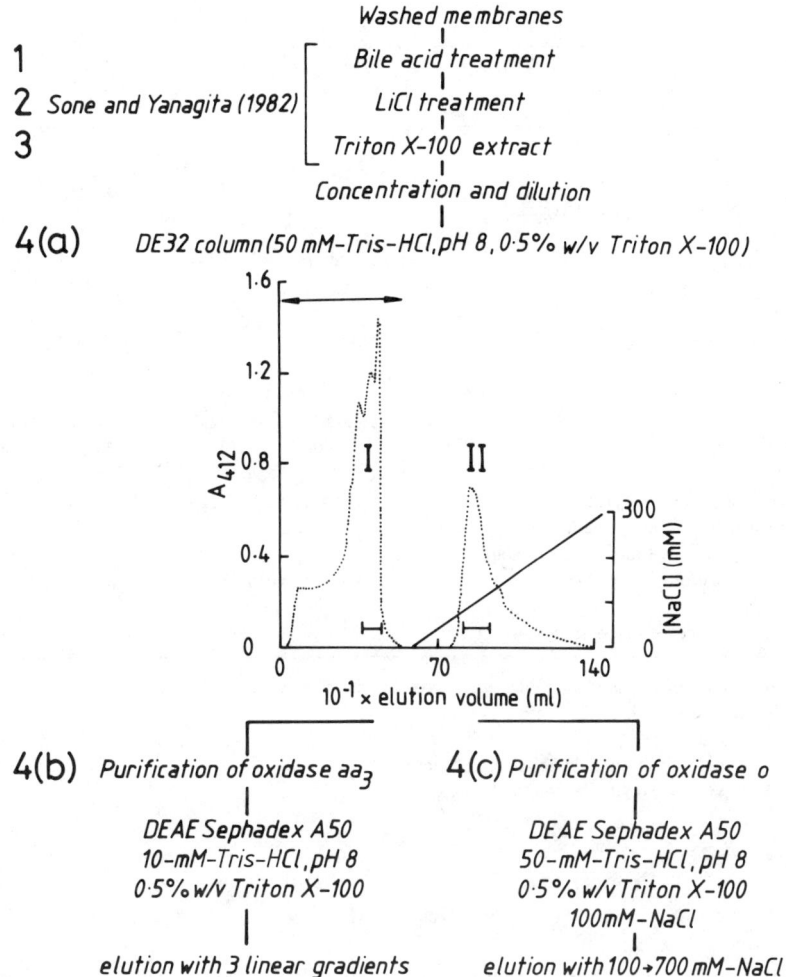

**Fig. 1.** Flow chart for purification of cytochromes $caa_3$ and $o$ from PS3 membranes. The material loaded onto each column in step 4 was first concentrated by ultrafiltration and diluted with distilled water. In 4(a), loading was followed by washing with buffer (horizontal arrow), until $A_{412} \geqslant 0.001$, and a 1-litre NaCl gradient as shown. Pooled fractions I and II (horizontal bar lines) were used for subsequent steps. In 4(b) the first gradient was 10 to 100 m$M$ Tris; the second, 0 to 100 m$M$ NaCl (all in 100 m$M$ Tris); and the third, 100 m$M$ to 1 $M$ NaCl (all in 100 m$M$ Tris). All gradients contained 0.5% (w/v) Triton and were pH 8. Material eluted with the second gradient was selected for the cytochrome $aa_3$ preparation. In 4(c) the NaCl gradient contained 50 m$M$ Tris-HCl (pH 8) and 0.5% (w/v) Triton. Material from the single major band was selected for the cytochrome $o$ preparation.

**Table 1.**   *Some properties of cytochrome oxidases purified from PS3*

| Property | $caa_3$ complex | | Cytochrome $o$-containing complex | |
|---|---|---|---|---|
| Spectral properties (nm) | | | | |
| Reduced minus oxidized ($\alpha$ peaks, 77 K) | 547, [a](561), 604 | | 554, 557 | |
| CO-reduced minus reduced (293 K) | n.d.[b] | | | |
| Peaks | 427, (432), 540, 593 | | 412 | |
| Troughs | 444, (560), 612 | | 430, 560 | |
| Haem groups [nmol (mg protein)$^{-1}$] | | | | |
| $b$ | —[c] | | 20.2 | |
| $c$ | 8.05 | | — | |
| $a$ | 14.8 | | — | |
| Subunits | (i)[d] | (ii)[e] | | |
| I | 54,950 | 58,200 | (i) | (ii) |
| II | 27,850 | 28,200 | 47,300 | 48,700 |
| III | 22,400 | 26,650 | | |
| Cu content [nmol (mg protein)$^{-1}$] | 13.6 | | 1.7 | |
| Fe content [nmol (mg protein)$^{-1}$] | n.d.[b] | | 21.7 | |
| Catalytic activities [min$^{-1}$ (mg protein)$^{-1}$] | | | | |
| Ferrocytochrome $c \rightarrow O_2$ | 217 nmol cytochrome $c$ oxidized | | 62 nmol cytochrome $c$ oxidized | |
| Hydroquinone $\rightarrow$ cytochrome $c \rightarrow O_2$ | 162 nmol $O_2$ reduced | | 29 nmol $O_2$ reduced | |

[a]Small amount of CO-binding $b$-like haemoprotein which increases on storage.
[b]n.d., not determined.
[c]—, none found.
[d](i) SDS/urea polyacrylamide gel electrophoresis.
[e](ii) Acid polyacrylamide gel electrophoresis.

## Properties and General Applications of Other Detergents

Such observations of artefactual aggregation and polydispersity in Triton X-100-solubilized preparations led us to consider alternative detergents. The properties of such surfactants have been reviewed in detail (Helenius and Simons, 1975; Tanford and Reynolds, 1976; Helenius *et al.*, 1979).

### Non-Ionic Detergents

These detergents, particularly the polyoxyethylene ether derivatives (for example, the Triton X and Brij series), have been widely used, since they are effective in solubilization yet give retention of catalytic activity. Unfortunately, they do not readily break protein–protein interactions, leading to the danger of artefactual aggregation of the solubilized proteins. In addition, commercial samples are frequently impure and are difficult to purify. They may also contain varying amounts of additives, there are batch-to-batch variations and, after prolonged storage, the composition of liquid non-ionic detergents may be heterogeneous through the sample. Their ultraviolet opacity precludes $A_{280}$ measurements of column eluates. A special disadvantage in purification of membrane-bound proteins is that the detergents are extremely difficult to separate from the isolated proteins. This is due in part to their low critical micelle concentration (CMC) and in part to the high affinities that they have for membrane proteins (Baron and Thompson, 1975). For practical purposes, CMC can be defined as the highest monomeric detergent concentration obtainable (Helenius et al., 1979). The CMC value for Triton X-100 is 0.24 m$M$ and the average micellar relative molecular weight is 90,000. A high value of CMC is desirable whenever dialysis across membranes or rapid removal or displacement of detergent is required. In gel filtration experiments, micelle size is important since the separation of different membrane proteins according to size is easier with a detergent of smaller micelle size (Helenius et al., 1979). In spite of all these limitations, Triton X-100 is still used with considerable success for bacterial oxidases (see Sone and Yanagita, 1982).

The β-D-alkyl glucosides (Baron and Thompson, 1975) have superior properties (Stubbs et al., 1976; Helenius et al., 1981), including a high CMC (25 m$M$ for β-D-octylglucoside), a lower value for micellar relative molecular weight (50,000), UV transparency and chemical homogeneity. Only De Vrij et al. (1983) appear to have exploited these advantages with a bacterial cytochrome oxidase, although the lauryl derivative has been used to prepare fully active mitochondrial cytochrome $c$ oxidase, but lacking in subunit III (Thompson and Ferguson-Miller, 1983).

*Ionic Detergents*

Alkyl ionic detergents such as the anionic sodium dodecyl sulphate, sodium dodecyl-$N$-sarcosine (used for *E. coli* membranes by Kita and Anraku, 1981) and the cationic cetyltrimethylammonium bromide are nearly always denaturants at the concentrations and temperatures required for complete solubilization (Helenius et al., 1979).

Zwitterionic detergents have been gaining favour for solubilization of membrane-bound components (Gonenne and Ernst, 1978; Hjelmeland, 1980; Bailyes et al., 1982). The most widely used are based on sulphobetaine, the synthetic amphiphile 3-(alkyldimethylammonio)-1-propanesulphonate, designated SB$_n$, where $n$ is the number of carbon atoms of the alkyl chain. These detergents retain their zwitterionic character over the entire pH range, due to the presence of both a strongly basic quaternary ammonium ion and an acidic sulphonate ion of equal strength. Thus, they are ideally suited for purification procedures involving ion-exchange chromatography, since they are unlikely to bind irreversibly to either anion- or cation-exchange resins. Like ionic surfactants, zwitterionic detergents are characterized by relatively high CMC values and low micellar relative molecular weights, but which vary with the length of the alkyl side chain and ionic strength. In contrast to earlier claims that zwitterionic detergents are generally non-denaturing in their interaction with membrane-bound proteins, Hjelmeland *et al.* (1979) have demonstrated that $SB_{14}$, in concentrations sufficient to solubilize endoplasmic reticulum, completely denatures cytochrome P-450. The solubilization and purification of *E. coli* cytochrome *d*, dispersed in $SB_{12}$, has recently been reported (Miller and Gennis, 1983). The apparent stability of the complex may result from the use of a detergent concentration inadequate to effect full solubilization (Hjelmeland *et al.*, 1979). This emphasises the importance of controlling and quoting detergent:protein ratios.

Bile salts, for example sodium cholate and sodium deoxycholate (Helenius *et al.*, 1979; Helenius and Simons, 1975), have been extensively used in the purification of mammalian cytochrome *c* oxidase (Kuboyama *et al.*, 1972). Their main features are the low CMC and micellar relative molecular weight values (but highly dependent on pH, ionic strength and other additives). Unfortunately, these surfactants have proved ineffective in solubilizing the tightly membrane-bound bacterial oxidases and so their use has been limited to pre-extraction of unwanted proteins (Fig. 1) (Sone and Yanagita, 1982).

In summary, the choice of detergents in this area has been largely empirical and, in view of the properties of Triton X-100, an unfortunate one.

*Effectiveness of Various Detergents in Solubilizing Membrane Proteins from PS3*

Bailyes *et al.* (1982) have proposed three increasingly stringent criteria for optimal solubilization and applied them to 5'nucleotidase. These are (1) non-sedimentation of the solubilized enzyme after high-speed centrifugation, (2) penetration of the solubilized enzyme(s) into polyacrylamide gels and (3) determination of molecular size of the detergent–enzyme complex. The first criterion has been employed in the preliminary studies reported here.

Figure 2 illustrates the effectiveness of both main classes of detergents considered above in solubilizing PS3 cytochromes. Triton X-100 (Fig. 2a) is the least effective of the four detergents in solubilizing cytochrome $aa_3$, with only 5% of the oxidase released at 1% (w/v) detergent. Up to 0.4% (w/v) Triton, the *c*-type cytochromes solubilized are predominantly CO-binding (Poole, 1981). Conversely, CO-reactive *b*-type cytochrome (presumably cytochrome *o*) is solubilized only above 0.2% (w/v) detergent.

β-D-Dodecyl maltoside (Fig. 2b) is substantially more effective in solubilizing cytochrome $aa_3$; furthermore, the bulk of the cytochrome *o* is released at, or below, 0.8% (w/v) detergent (not shown).

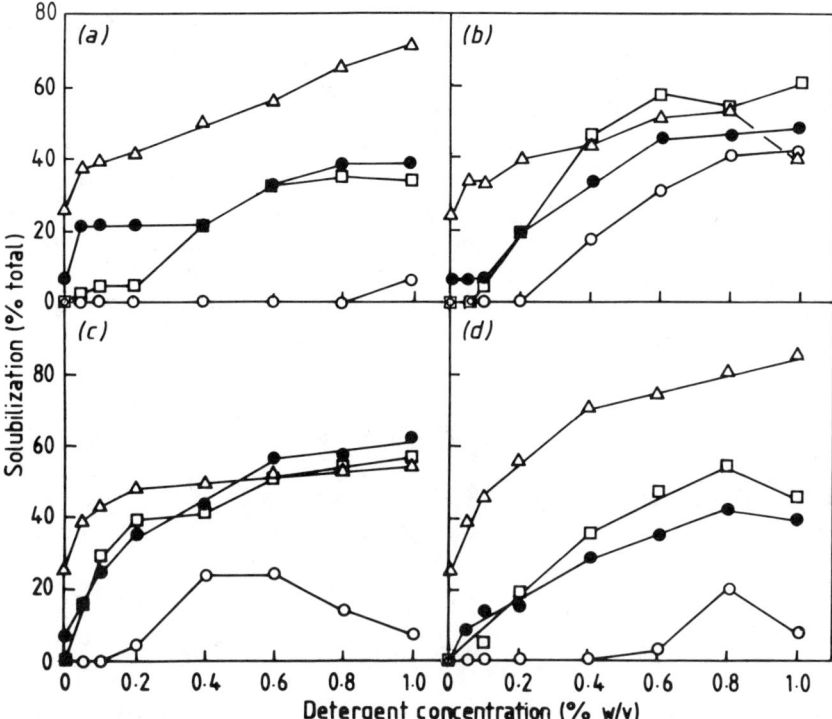

**Fig. 2.** Solubilization of cytochrome oxidase and other cytochromes from PS3 membranes. Washed membranes (Poole, 1981) were suspended in 50 m$M$ Tris-HCl buffer (pH 8) at 0.9 to 1 mg protein ml$^{-1}$. To 5-ml aliquots were added small volumes of concentrated detergent solutions in the same buffer to give the final concentrations shown. After 30 min incubation at room temperature, suspensions were centrifuged at 100,000 $g$ for 1 hr and the amounts of cytochromes released determined from reduced minus oxidized difference spectra, using the wavelength pairs given by Smith (1978). Protein was estimated by the Lowry method. The detergents were (a) Triton X-100, (b) β-D-dodecylglucoside, (c) Zwittergent 3-12 and (d) sodium dodecyl-$N$-sarcosinate. Shown are protein (△), cytochrome oxidase $aa_3$ (○), cytochrome $b$ (including cytochrome $o$; □) and $c$-type cytochromes (●).

Zwitterionic and ionic surfactants (Fig. 2c and 2d, respectively) are generally less effective than β-D-dodecylmaltoside and their use seems to be characterized by a decrease in the amount of cytochromes $aa_3$ solubilized. Denaturation (see earlier section) may contribute to this phenomenon.

On the basis of these preliminary experiments, we conclude that β-D-dodecyl-maltoside is the detergent of choice, but its expense may preclude widespread use. Bacterial oxidases have generally been purified by protein fractionation following extensive solubilization of membrane proteins. This approach is potentially troublesome with bacterial as opposed to mitochondrial systems, however, because of the diversity of membrane proteins present in the former and the consequent low specific concentration of the oxidases (Ludwig, 1980; Poole, 1983). The finding (Fig. 2) that certain detergents exert some specificity of solubilization suggests that the method of sequential fractionation (Kuboyama *et al.*, 1972) may be usefully extended to bacterial enzymes. For example, pre-extraction of PS3 membranes with low concentrations of Zwittergent 3-12 (Fig. 2c) removes 40% of the *b*- and *c*-type cytochromes, half the protein, but less than 5% of cytochrome $aa_3$. Re-extraction of the pelleted material (e.g. with β-D-dodecylmaltoside) should yield the bulk of the oxidase, leading to substantial improvements in yields. Such an approach is under investigation and will be reported elsewhere.

## Acknowledgments

This work was supported by the Royal Society, SERC, and the University of London Central Research Fund. We are grateful to Julia Hubbard for the iron and copper analyses.

## References

Bailyes, E. M., Newby, A. C., Siddle, K. and Luzio, J. P. (1982). Solubilization and purification of rat liver 5′-nucleotidase by use of a zwitterionic detergent and a monoclonal-antibody immunoadsorbent. *Biochemical Journal* **203**, 245–251.

Baron, C. and Thompson, T. E. (1975). Solubilization of bacterial membrane proteins using alkyl glucosides and dioctanoyl phosphatidylcholine. *Biochimica et Biophysica Acta* **382**, 276–285.

De Vrij, W., Azzi, A. and Konings, W. N. (1983). Structural and functional properties of cytochrome *c* oxidase from *Bacillus subtilis* W23. *European Journal of Biochemistry* **131**, 97–103.

Gonenne, A. and Ernst, R. (1978). Solubilization of membrane proteins by sulfobetaines, novel zwitterionic surfactants. *Analytical Biochemistry* **87**, 28–38.

Helenius, A. and Simons, K. (1975). Solubilization of membranes by detergents. *Biochimica et Biophysica Acta* **415**, 29–79.

Helenius, A., McCaslin, D. R., Fries, E. and Tanford, C. (1979). Properties of detergents. *In* "Methods in Enzymology Vol. 56 (Eds. S. Fleischer and L. Packer), pp. 734–749. Academic Press, New York

Helenius, A., Sarvas, M. and Simons, K. (1981). Asymmetric and symmetric membrane reconstitution by detergent elimination. Studies with Semliki-Forest-virus spike glycoprotein and penicillinase from the membrane of *Bacillus lichenformis*. *European Journal of Biochemistry* **116**, 27–35.

Hjelmeland, L. M. (1980). A nondenaturing zwitterionic detergent for membrane biochemistry: Design and synthesis. *Proceedings of the National Academy of Sciences of the U.S.A.* **77**, 6368–6370.

Hjelmeland, L. M., Nebert, D. W. and Chrambach, A. (1979). Electrofocusing of integral membrane proteins in mixtures of zwitterionic and nonionic detergents. *Analytical Biochemistry* **95**, 201–208.

Kita, K. and Anraku, Y. (1981). Composition and sequence of $b$ cytochromes in the respiratory chain of aerobically grown *Escherichia coli* K-12 in the early exponential phase. *Biochemistry International* **2**, 105–112.

Kuboyama, M., Yong, F. C. and King, T. E. (1972). Studies on cytochrome oxidase. VIII.Preparation and some properties of cardiac cytochrome oxidase. *Journal of Biological Chemistry* **247**, 6375–6383.

Ludwig, B. (1980). Heme $aa_3$-type cytochrome $c$ oxidases from bacteria. *Biochimica et Biophysica Acta* **594**, 177–189.

Miller, M. J. and Gennis, R. B. (1983). The purification and characterization of the cytochrome $d$ terminal oxidase complex of the *Escherichia coli* aerobic respiratory chain. *Journal of Biological Chemistry* **258**, 9159–9163.

Poole, R. K. (1981). Ligand-binding cytochromes $a_3$, $c$ and $o$ in membranes from the thermophilic bacterium PS3. *FEBS Letters* **133**, 255–259.

Poole, R. K. (1983). Bacterial cytochrome oxidases. A structurally and functionally diverse group of electron-transfer proteins. *Biochimica et Biophysica Acta* **726**, 205–243.

Poole, R. K., Scott, R. I., Baines, B. S., Salmon, I. and Lloyd, D. (1982). Identification of cytochromes $o$ and $aa_3$ as functional terminal oxidases in the thermophilic bacterium PS3. *FEBS Letters* **150**, 281–285.

Poole, R. K., Van Wielink, J. E., Baines, B. S., Reijnders, W. N. M., Salmon, I. and Oltmann, L. F. (1983). The membrane-bound cytochromes of an aerobically grown, extremely thermophilic bacterium, PS3: Characterization by spectral deconvolution coupled with potentiometric analysis. *Journal of General Microbiology* **129**, 2163–2173.

Smith, L. (1978). Bacterial cytochromes and their spectral characterization. *In* "Methods in Enzymology" Vol. 53 (Eds. S. Fleischer and L. Packer), pp. 202–212. Academic Press, New York.

Sone, N. and Yanagita, Y. (1982). A cytochrome $aa_3$-type terminal oxidase of a thermophilic bacterium. Purification, properties and proton pumping. *Biochimica et Biophysica Acta* **682**, 216–226.

Sone, N., Ohyama, T. and Kagawa, Y. (1979). Thermostable single-band cytochrome oxidase. *FEBS Letters* **196**, 39–42.

Stubbs, G. W., Smith, H. G. and Litman, B. J. (1976). Alkyl glucosides as effective solubilizing agents for bovine rhodopsin. A comparison with several commonly used detergents. *Biochimica et Biophysica Acta* **425**, 45–56.

Tanford, C. and Reynolds, J. A. (1976). Characterization of membrane proteins in detergent solutions. *Biochimica et Biophysica Acta* **457**, 133–170.

Thompson, D. A. and Ferguson-Miller, S. (1983). Lipid and subunit III depleted cytochrome $c$ oxidase purified by horse cytochrome $c$ affinity chromatography in lauryl maltoside. *Biochemistry* **22**, 3178–3187.

Yoshida, T., Lorence, R. M., Choc, M. G., Tarr, G. E., Findling, K. L. and Fee, J. A. (1984). Respiratory proteins from the extremely thermophilic aerobic bacterium, *Thermus thermophilus*. Purification procedures for cytochromes $c_{552}$, $c_{555, 549}$ and $c_1aa_3$ and chemical evidence for a single subunit cytochrome $aa_3$. *Journal of Biological Chemistry* **259**, 112–123.

# Part III
# Hydrogen

# 4

# Hydrogenases: Structure and Applications in Hydrogen Production

RICHARD CAMMACK, DAVID O. HALL AND K. KRISHNA RAO

*Department of Plant Sciences, King's College London, University of London, London, United Kingdom*

## Occurrence and Properties of Hydrogenases

Recently there has been an upsurge in interest in artificial and photobiological systems for hydrogen gas production (Braun, 1983). The biological approach has centred around the enzyme hydrogenase (Krasna, 1979). There were initial problems due to its instability, which have since been partly alleviated by the isolation of more stable hydrogenases. These enzymes are now also being incorporated into hybrid systems with photochemical catalysts. The potential applications of hydrogenases have led to a renewed interest in the catalytic mechanism of the enzyme.

Hydrogenases catalyze the simplest reaction in biology: the reversible combination of two protons with two electrons.

$$2H^+ + 2e^- = H_2$$

Unlike nitrogenase, an enzyme which shows remarkable similarity in molecular properties in all organisms from which it has been isolated, hydrogenases are very diverse in their relative molecular mass, cofactor composition and spectroscopic properties (Adams *et al.*, 1981; Mayhew and O'Connor, 1982). These differences probably reflect the different functions of the enzyme in different microorganisms. Therefore the term "hydrogenase" refers not to a single enzyme but to a class of enzymes. Hydrogenases are found in many species of bacteria, both aerobic and anaerobic, and in eukaryotic algae.

Hydrogenases can be subdivided into those that produce hydrogen and those that consume it. We consider it best to make the distinction on the basis of the physiological function of the enzyme *in the cell*. Sometimes hydrogenases have been classified as "hydrogen-producing hydrogenases" if they catalyze the evo-

MICROBIAL GAS METABOLISM:
MECHANISTIC, METABOLIC
AND BIOTECHNOLOGICAL ASPECTS

lution of hydrogen with reduced methyl viologen ($E_m$, pH 7.0 = $-440$ mV) as electron donor, "hydrogen-uptake hydrogenases" if they consume hydrogen with dyes such as methylene blue ($E_m$, pH 7.0 = $-11$ mV) as acceptor and "bidirectional" hydrogenases if they catalyze both reactions. However, these operational definitions may reflect only the acceptor specificity of the enzymes towards the mediator dyes, which is not directly related to their physiological function.

## Studies on the Mechanism of Hydrogenase

Despite its simplicity and the fact that it has been known for 50 years, little was known about the catalytic centre of the enzyme and its mechanism of action until 5 years ago. The enzyme which was most thoroughly characterized was the $H_2$-

**Table 1.** *Hydrogenases in which nickel has been implicated*

| Source of enzyme[a] | Ni required for synthesis | Ni detected in purified protein | Ni ESR signals | Reference[b] |
|---|---|---|---|---|
| *Alcaligenes eutrophus* | | | | |
| MBH | Yes | Yes | Yes | 1 |
| SolH | Yes | Yes | No | 2 |
| *Methanobacterium thermoautotrophicum*, Marburg strain | Yes | Yes | Yes | 3 |
| *Desulfovibrio desulfuricans*, strain 27774 | Yes | | Yes | 4 |
| *Desulfovibrio gigas* | Yes | Yes | Yes | 5 |
| *Desulfovibrio desulfuricans* Norway strain | | | | |
| MBH | | Yes | Yes | 6 |
| SolH | | Yes | No | 7 |
| *Desulfovibrio vulgaris* | | Yes | No | 8 |
| *Azotobacter chroococcum* | Yes | | | 9 |
| *Rhodopseudomonas capsulata* | Yes | Yes | | 10 |
| *Escherichia coli* | Yes | Yes | Yes | 11 |
| *Nocardia opaca* | | Yes | No | 12 |
| *Vibrio succinogenes* | | Yes | | 13 |
| *Azospirillum brasilense* | Yes | | | 14 |
| *Azotobacter lipoferum* | Yes | | | 14 |
| *Derxia gummosa* | Yes | | | 14 |

[a]MBH, membrane-bound hydrogenase; SolH, soluble hydrogenase.
[b]References: 1, Friedrich *et al.* (1981); 2, Schneider *et al.* (1984a); 3, Graf and Thauer (1981); 4, Krüger *et al.* (1982); 5, Cammack *et al.* (1982b); 6, Lalla-Maharajh *et al.* (1982); 7, Rieder *et al.* (1984); 8, Grande *et al.* (1983); 9, Partridge and Yates (1982); 10, Takakuwa and Wall (1981); 11, D. Hallahan, R. Cammack and D. O. Hall, unpublished; 12, Schneider *et al.* (1984b); 13, Unden *et al.* (1982); 14, Pedrosa and Yates (1983).

producing hydrogenase from *Clostridium pasteurianum* (Chen, 1978). This is a highly active enzyme, but is extremely sensitive to oxygen, which causes irreversible inactivation. The only cofactors known to be present in *C. pasteurianum* hydrogenase are iron–sulphur clusters (Chen, 1978). The more stable hydrogenases which have been studied are uptake hydrogenases. These are also iron–sulphur proteins, but in addition almost all of them have been found to contain the unusual trace element nickel. As shown in Table 1, this has been shown either indirectly, by the requirement for nickel for hydrogenase synthesis, or directly, by the detection of significant amounts of nickel in the purified enzyme and the detection of a characteristic electron spin resonance (ESR) signal in the purified hydrogenase (Fig. 1). Apart from iron, acid-labile sulphur and nickel, two hydrogenases have also been found to contain selenium (Table 2), although its function is unknown.

Table 2 lists a number of hydrogenases that have been isolated, together with the analytical data for $M_r$ and their content of trace metals and other cofactors. These values should not be taken as absolute because, as well as the errors inherent in trace metal analysis of proteins, the values rely on having an accurate estimate of the $M_r$ and having a pure sample of intact protein. This latter criterion is particularly difficult to ensure, since a protein which appears homogeneous on polyacrylamide gel electrophoresis may contain a proportion of "crippled" molecules, in which the metal centres have been damaged. This difficulty must also be borne in mind in the interpretation of spectroscopic measurements.

## A Model for Hydrogenase Structure

In order to try to classify the diverse hydrogenases so far isolated, we consider a general model (Fig. 2) for their structure. This comprises three sites or domains with recognizably different functions:

1. The hydrogen-activating (*H*) site. We work on the assumption that since all these enzymes use the substrate hydrogen, they contain one type, or at most a few types, of such sites. Because the highly unusual metal nickel is found in most if not all of these enzymes, it is a strong candidate for a component of the *H* site.
2. The acceptor (*A*) site. It is probably here that the different hydrogenases show their greatest diversity, because of the different electron acceptors, such as ferredoxins, flavodoxins, deazaflavin, cytochromes, NAD(P)$^+$, that they use.
3. The regulatory (*R*) site. We postulate the existence of a separate site to account for the reversible inactivation of the enzymes under oxidizing conditions. On this interpretation, the *R* site occupies an intermediate position between the *A* and *H* sites.

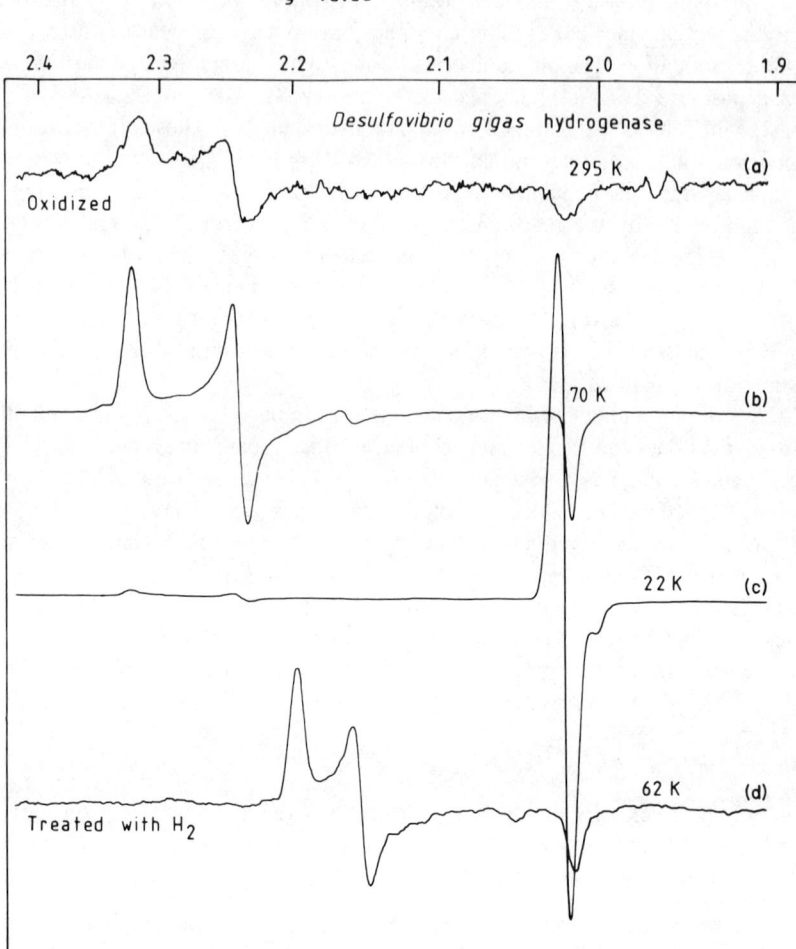

**Fig. 1.** ESR spectra of oxidized hydrogenase from *D. gigas.* (a) Nickel signal, recorded at room temperature. (b) Nickel signal at 70 K. (c) Spectrum recorded at 22 K and a lower gain; its most prominent feature is the $g = 2.02$ signal due to the [3Fe–xS] cluster, which is only detectable at low temperatures. (d) Modified spectrum of nickel which appears after treatment with hydrogen. (R. Cammack, E. C. Hatchikian and D. S. Patil, unpublished results.)

**Table 2.** Compositions of hydrogenase preparations

| Source of enzyme | Subunit composition (kD) | ESR signals[a] Oxidized 1 | 2 | 3 | Reduced 4 | 5 | 6 | Cofactor analysis[b] Ni | Fe | S | Se | Flavin | Reference[c] |
|---|---|---|---|---|---|---|---|---|---|---|---|---|---|
| *Chromatium vinosum* | 60 | + | + | + | – | – | – | 1.8 | 4 | 4 | 0 | 0 | 1 |
| *Desulfovibrio vulgaris* | | | | | | | | | | | | | |
| Strain Hildenborough | 50 | – | – | + | + | – | – | 0.7 | 12 | 12 | | | 2 |
| Strain Miyazaki | 59 + 28 | | | | | | | | 8 | 8 | | | 3 |
| *Desulfovibrio gigas* | 62 + 26 | + | – | + | – | – | – | 1.2 | 12 | 12 | 0 | 0 | 4 |
| *Desulfovibrio desulfuricans* Norway | | | | | | | | | | | | | |
| MBH | 60 + 27 | + | + | – | + | – | – | 0.5 | 6 | 6 | 0 | 0 | 5 |
| SolH | 56 + 29 | – | – | – | + | – | – | 0.5 | 6 | 6 | 0.5 | 0 | 6 |
| *Desulfovibrio desulfuricans* 27774 | (2, total 78) | + | – | + | – | – | – | 1 | 11 | 11 | | | 7 |
| *Escherichia coli* | 56 + 56 | + | – | + | + | – | – | | 12 | 12 | | | 8 |
| *Megasphaera elsdenii* | 50 | | – | + | + | – | – | 0.4 | 12 | 12 | | | 2 |
| *Proteus mirabilis* | 2×63 2×33 | | – | + | – | – | – | | 24 | 24 | | | 9 |
| *Alcaligenes eutrophus* | | | | | | | | | | | | | |
| MBH | 67 + 31 | + | + | + | + | – | – | 0.7 | 7–9 | 7–9 | 0 | 0 | 10 |
| SolH | 68 + 60 + 29 + 29 | – | + | + | + | + | + | 2 | 14 | 14 | 0 | 1 | 11 |
| *Pseudomonas pseudoflava* | 65 + 30 | – | + | + | + | – | – | | 6 | 6 | 0 | 0 | 12 |
| *Nocardia opaca* | 64 + 56 + 31 + 27 | – | + | + | + | + | + | 4 | 14 | 14 | 0 | 1 | 13 |
| *Methanococcus vannielii* | 42 + 35 + 27 + 27M | | | | | | | | | | 4 | | 14 |
| *Methanobacterium thermoautotrophicum* ΔH | 40 + 31 + 26 | + | – | – | – | – | + | 3 | 38 | 27 | | 2 | 15 |
| *Clostridium pasteurianum* | 60 | – | – | + | + | – | – | 0 | 12 | 12 | 0 | 0 | 16 |

[a] ESR signals: 1, Ni(III); 2, "interaction"; 3, [3Fe–xS]$^{3+}$; 4, [4Fe–4S]$^{1+}$; 5, [2Fe–2S]$^{1+}$; 6, flavin radical.

[b] Gaps indicate no values given.

[c] References: 1, Albracht *et al.* (1983); 2, Grande *et al.* (1983); 3, Yagi (1976); 4, Hatchikian *et al.* (1978); 5, Lalla-Maharajh *et al.* (1982); 6, Rieder *et al.* (1984); 7, Krüger *et al.* (1982); 8, Adams and Hall (1979); 9, Schoenmaker *et al.* (1979); 10, Schink and Schlegel (1979); 11, Schneider *et al.* (1979); 12, Weiss *et al.* (1980); 13, Schneider *et al.* (1984a); 14, Yamazaki (1982); 15, Kojima *et al.* (1983); 16, Erbes *et al.* (1975).

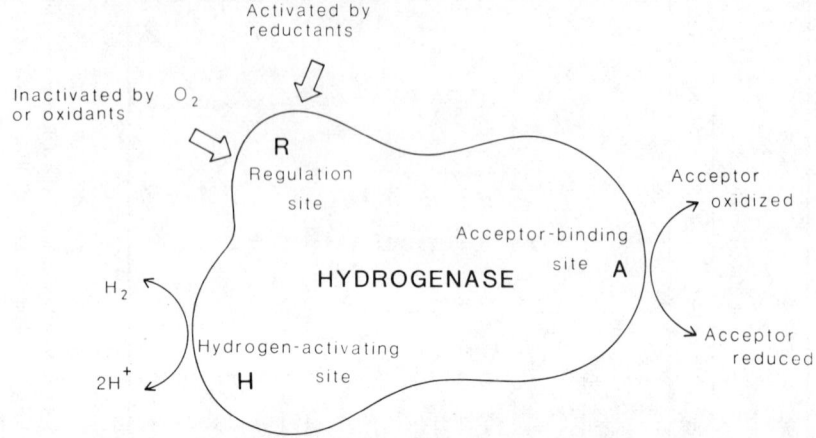

**Fig. 2.** General model of the hydrogenase molecule.

## Redox Centres

To study the iron–sulphur clusters and nickel centres in these proteins, some type of spectroscopic technique is necessary. For this purpose, optical absorption spectroscopy is rather uninformative. Greater resolution is provided by ESR spectroscopy. The disadvantages of ESR are that it only detects the centres in those oxidation states in which they are paramagnetic, and requires low temperatures which preclude kinetic measurements. Nevertheless, ESR has provided a number of clues as to the nature of the active site(s). The paramagnetic centres can be identified from their spectra, by means of a spectroscopic parameter known as the $g$ value, by their temperature dependence and by the effects of interactions with other electron and nuclear spins. The initial hopes, that one could detect the essential features that are common to all hydrogenases, were confounded. As can be seen from Table 2, there are no ESR signals that are observed in *all* of the hydrogenases isolated. However, several different types of centres in the hydrogenases can be distinguished, which occur in many of them.

*[3Fe–xS] cluster.* The most prominent ESR signal from many uptake hydrogenases is a relatively narrow spectrum at $g = 2.02$, observed in the oxidized form, for example Fig. 1 at very low temperatures, below 30 K. These signals, or the more complex "interaction" signals, have been found in most of the uptake hydrogenases isolated, though recently hydrogenases have been found which lack them, such as *Methanobacterium autotrophicum* (Albracht *et al.*, 1982) and *Desulfovibrio desulfuricans* (Rieder *et al.*, 1984). We have summarized the properties of these signals in a previous paper (Cammack *et al.*, 1982a).

At the time the paper was written (in 1980), it was not known if the signal originated from a [4Fe–4S] cluster, or one of the recently discovered [3Fe–xS] clusters. It has recently been assigned as a [3Fe–xS] cluster, on the basis of Mössbauer (Krüger *et al.*, 1982) and resonance Raman spectroscopic measurements (Johnson *et al.*, 1983). Since the clusters have redox potentials too positive to be involved in hydrogen production, we considered four possible explanations for the existence of these centres in the isolated enzymes:

1. A regulatory function, related to the activation/inactivation of hydrogenases under reducing and oxidizing conditions. An analogy may be drawn to the citric acid cycle enzyme aconitase, in which it has been suggested that the interconversion of a [3Fe–xS] cluster into a [4Fe–4S] cluster might control the activity (Kent *et al.*, 1982). So far, however, we have not been able to demonstrate such an interconversion related to hydrogen activity.
2. An oxygen-binding site. Although the removal of oxygen is a requirement for activation of hydrogenase, present evidence appears to favour the view that regulation of hydrogenase activity is a redox process, rather than a direct interaction with oxygen. For example, studies of the hydrogen–deuterium exchange of *Desulfovibrio gigas* hydrogenase (Berlier *et al.*, 1982) showed that removal of oxygen prevented the "lag" phase in the activity, but did not affect the rate of activation under hydrogen.
3. A secondary electron acceptor.
4. An oxidized form of a [4Fe–4S] cluster.

On present evidence, explanations 3 and 4 remain as plausible explanations for the "$g = 2.02$" centre. However there is at present some dispute as to whether *any* [3Fe–xS] clusters have a physiological role, or have been derived from [4Fe–4S] clusters by oxidative damage (Beinert and Thompson, 1983).

*Nickel centre.* The second type of ESR signal observed in hydrogenases is a rhombic spectrum at $g > 2$. Unlike the spectra of the iron–sulphur clusters, the signal is detected at high as well as low temperatures (Fig. 1). The assignment of this signal as due to nickel has been unequivocally established by the incorporation, by growth on a suitably enriched medium, of the stable isotope [61]Ni (Albracht *et al.*, 1982; Moura *et al.*, 1982). This isotope has nuclear spin $I = 3/2$, and causes a splitting of the ESR spectrum into four $[=(2I + 1)]$ lines (Fig. 3). (This experiment, incidentally, makes it less likely that the centre giving rise to the signal is a coupled Ni–Ni pair, which would give a more complex multiplet pattern.) The oxidation state of the nickel has been assigned as due to low-spin Ni(III), which is unexpectedly high for an enzyme operating at such a low redox potential. Although there is evidence that all of the uptake hydrogenases

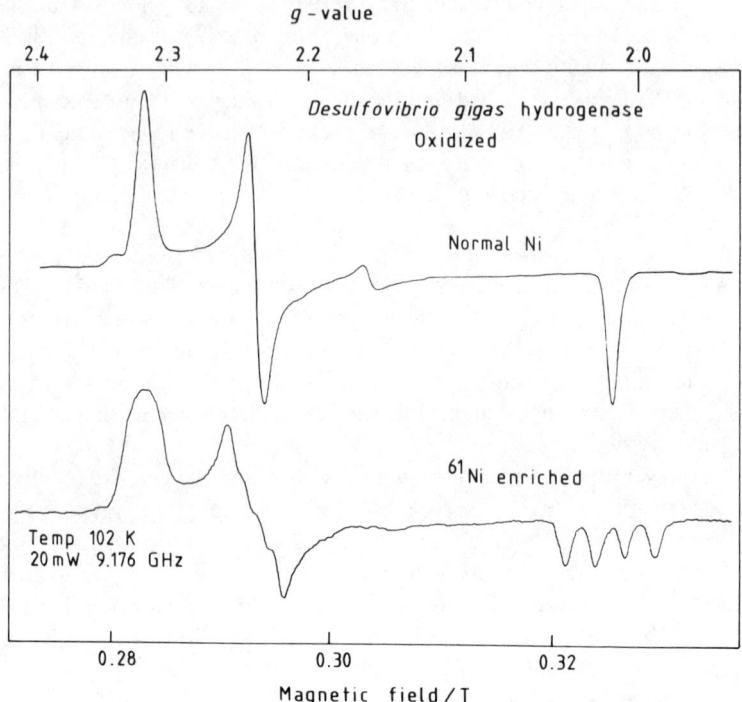

**Fig. 3.** ESR spectra of oxidized hydrogenase from *D. gigas.* (Upper spectrum) Native enzyme; (lower spectrum) enzyme isolated from cells grown on [61]Ni isotope. (R. Cammack, E. C. Hatchikian and D. S. Patil, unpublished results.)

contain nickel, not all the hydrogenases as isolated give rise to this ESR signal (Table 1).

*"Interaction signal"*. Some hydrogenases, including the membrane-bound hydrogenases of *Alcaligenes eutrophus* (Schneider *et al.,* 1983) and *Chromatium vinosum* (Albracht *et al.,* 1983), give more complex spectra around $g = 2$ which, like the spectra of the [3Fe–xS] clusters, are detectable only at very low temperatures. These signals, termed "interaction signals" in Table 2, have been assigned by Albracht *et al.* (1983) to a relatively weak spin–spin coupling between a [3Fe–xS] cluster and a nickel centre.

*[4Fe–4S] clusters.* In the reduced state, many hydrogenases give low-temperature spectra with average $g$ values less than 2.0, typical of [4Fe–4S] clusters. Once again, however, this is not observed in all hydrogenases, even when there is sufficient iron and sulphur present to form two or three [4Fe–4S] clusters, as in

the case of the *D. gigas* enzyme (Hatchikian *et al.*, 1978). The reason why some of the [4Fe–4S] clusters are ESR-silent is not clear at present. Possibly their ground state has an unusual high spin state, like the P clusters of nitrogenase (Johnson *et al.*, 1981), or the clusters are coupled by strong exchange interactions.

*[2Fe–2S] clusters.* A few hydrogenases, which use $NAD^+$ as electron acceptor, contain a [2Fe–2S] cluster which gives a signal with average $g$ values less than 2.0. This latter signal can be distinguished from those of the [4Fe–4S] clusters by its slower electron spin relaxation rate, which means that it can be detected up to somewhat higher temperatures, around 77 K (Schneider *et al.*, 1979).

*Flavin radicals.* The NAD-dependent hydrogenases, together with the de-azaflavin-reducing hydrogenases of methanogenic bacteria, contain flavin, which can be detected in the semiquinone state as a free-radical signal at $g = 2.00$ (Schneider *et al.*, 1979; Kojima *et al.*, 1983).

*Redox Potentials*

In the hydrogen-activating site, one would expect redox carriers with midpoint redox potential ($E_m$) values close to the $H^+/H_2$ couple, $-420$ mV at pH 7.0. Moreover, for a redox centre that reduces hydrogen, a pH dependence for $E_m$ of $-60$ mV per pH unit would be predicted. Measurements of the activity of hydrogenases of *C. pasteurianum* (Chen, 1978), *Megasphaera elsdenii* (Van Dijk and Veeger, 1981), *C. vinosum* (Fernandez *et al.*, 1982) and *D. gigas* (Fernandez *et al.*, 1984) as a function of applied redox potential indicate the existence of a centre, which determines activity, with $E_m$, pH 7.0, between $-360$ and $-420$ mV. It is clearly important to see if this can be identified with one of the ESR-detectable centres. The $E_m$ values for the different ESR-detectable centres can be determined by poising the protein at known redox potential values in the presence of suitable mediators at 25°C, then freezing samples for ESR measurements (Dutton, 1978).

The $g = 2.02$ signals attributed to [3Fe–xS] clusters show $E_m$ values too high to be involved in the hydrogen-activating site. The "interaction" signals show complex changes during reduction. First they change to a simpler $g = 2.02$ signal; during this stage Albracht *et al.* (1983) have shown that in *C. vinosum* hydrogenase a signal due to nickel also appears. This change occurs at relatively positive redox potentials, around $+100$ mV (Schneider *et al.*, 1983; R. Cammack, J. L. Serra and M. J. Llama, unpublished). At lower potentials the $g = 2.02$ signals are reduced.

The $E_m$ values of the signals attributed to nickel in the oxidized resting enzyme have been measured, in *D. gigas* hydrogenase, to be about $-150$ mV at pH 7.0 (Cammack *et al.*, 1982b). Significantly, they show the expected pH dependence for a hydrogen carrier. However the $E_m$ is too high for the active centre of an enzyme that can produce hydrogen. This difficulty may have been resolved by the detection of a new type of nickel signal, which appears in the enzyme after treatment with hydrogen (Teixeira *et al.*, 1983). This signal may represent an active form of nickel with a lower redox potential. We can postulate that the change in the state of the nickel takes place during the activation process.

The signals due to [4Fe–4S] clusters have been found to have varying $E_m$ values, between $-90$ and $-440$ mV (Schneider *et al.*, 1979, 1983). For reasons explained below, it seems likely that these ESR-detectable clusters represent secondary electron carriers associated with the acceptor site of hydrogenase. Similarly, the [2Fe–2S] clusters are probably associated with the flavin in the NAD-linked hydrogenases (Schneider *et al.*, 1979).

In summary, there is no "typical" hydrogenase from which conclusions about the essential components can be derived. The most likely candidate for the hydrogen-activating site at present is a modified form of the nickel centre. Since the nickel is sometimes found tightly coupled to a [3Fe–xS] cluster which might originate from oxidative damage to a [4Fe–4S] cluster, one attractive possibility is that the native enzyme contains a coupled pair, Ni–[4Fe–4S], which would be able to donate two electrons in the hydrogen-evolving reaction. Such a centre might not be detectable by ESR because of spin-coupling, like the Cu–haem pair in cytochrome oxidase (Griffith, 1971).

## Nature of the Acceptor Site

It has already been mentioned that one factor responsible for the considerable diversity of hydrogenases in their subunit and centre composition is the requirement to react with different electron acceptors. Some hydrogenases have only one subunit (Table 2), whereas the flavin-containing hydrogenases which react with $NAD^+$ or deazaflavin are particularly complex, with three or four different types of subunits (Schneider *et al.*, 1979; Kojima *et al.*, 1983). This complexity can be explained by the presence of a distinct flavin-binding domain, which is responsible for reaction with the acceptor.

A striking example of the separate nature of the acceptor-binding domain is provided by the NAD-dependent hydrogenase of *Nocardia opaca* (Schneider *et al.*, 1984a). This enzyme contains an unusually large amount of nickel, four atoms per molecule. It is also unusual in that two of them are relatively easily dissociated. The enzyme then falls into two parts, each containing two subunits, termed the "large dimer" (64 + 31 kD) and "small dimer" (56 + 27 kD). The two nickel atoms which apparently hold the dimers together can be replaced by high concentrations of other salts. Two nickel atoms remain in the small dimer.

The separated dimers exhibit partial activities. The small dimer has hydrogenase activity with methyl viologen as acceptor, while the larger dimer has diaphorase activity, reducing dyes at the expense of NADH. ESR spectroscopy showed a [3Fe–xS] cluster in the small dimer, while the flavin and ESR-detectable [4Fe–4S] and [2Fe–2S] clusters are present in the large dimer (Schneider et al., 1984b). In conclusion, therefore, the small dimer contains the $H$ site (and also presumably the $R$ site), while the large dimer contains the $A$ site. Moreover, the ESR-detectable [4Fe–4S] clusters ($E_m$ −420 mV) are found to be associated with the $NAD^+$-binding site, and not, as might be expected, the hydrogen-activating site.

The observations on $N.$ $opaca$ hydrogenase also demonstrate a problem in the assignment of functions to hydrogenases that have been isolated. In many cases the physiological acceptor is unknown, and hydrogenases are often isolated on the basis of the ability to produce hydrogen gas with reduced methyl viologen as electron donor. This assay only requires the hydrogen-activating part of the enzyme. If the acceptor side of the enzyme is loosely bound, it may be lost during purification. There is a possibility, therefore, that some of the hydrogenases isolated may therefore only represent modified forms, or only the hydrogen-activating domains of more complex hydrogenases.

## The Regulatory Site

During reactions in the reduced state, hydrogenases are particularly susceptible to damage by oxygen. If exposed to oxygen, all hydrogenases rapidly become inactive towards hydrogen. However, on transfer to reducing conditions again, activity is partially restored. Therefore two processes are involved in the loss of activity:

Reversible inactivation, by which the hydrogenase changes into a form which will no longer react with hydrogen, but which is relatively insensitive to oxygen damage. This is presumably a protection mechanism. Reactivation requires reducing conditions, and appears to take place on a site different from the hydrogen-activating site.

Irreversible damage, caused by reaction with oxygen. Superoxide ions have been implicated in this process (Henry et al., 1980; Schneider and Schlegel, 1981).

How much activity can be restored after exposure to oxygen, and how much is irreversibly destroyed, depends on the conditions, but particularly on the species of hydrogenase. The hydrogenases from the aerobic hydrogen-oxidizing bacteria are relatively resistant to oxygen damage. For example, the NAD-linked soluble hydrogenase of $A.$ $eutrophus$ is very stable in air, and oxidizing agents even have a preservative effect (Schneider and Schlegel, 1976). On the other hand, the hydrogen-producing hydrogenases of obligate anaerobes such as *C. pasteuria-num* are most sensitive. Van der Westen *et al.* (1980) noted that the hydrogenase of *Desulfovibrio vulgaris* (Hildenborough strain) suffered considerable irreversible inactivation if it was extracted from anaerobically grown cells and isolated in air, but much less if the cells, prior to extraction, were aerated or treated with sulphate. This observation supports the view already mentioned, that the reversible activation/inactivation involves a redox reaction but not necessarily direct reaction with oxygen.

Reactivation by reduction with hydrogen is usually slow. Hence it is observed that assays for hydrogen consumption show a lag phase of several minutes before activity is detected (Adams *et al.*, 1981). Activation by other reducing agents is often much faster.

A well-studied case is the soluble hydrogenase of *A. eutrophus*. The oxidized form of this enzyme will not immediately catalyze the hydrogen-dependent reduction of $NAD^+$; the reaction only begins after a lag phase. However, if catalytic amounts of NADH are added, activation takes place in seconds (Schneider and Schlegel, 1976). Presumably the NADH causes activation of the hydrogenase by reduction via the acceptor site, which leads to the "unmasking" of the hydrogen-activating site.

In Fig. 2, we refer to the region of the hydrogenase molecule, which controls the activity of the enzyme towards hydrogen, as the regulatory or *R* site. The process affects the *H* site, but has been shown to be initiated by reducing equivalents from the *A* site. Just what components are involved in the *R* site, and how it works, are at present unknown.

## Applications in Hydrogen Production

*Biophotolysis of Water*

*Hydrogen fuel.* Hydrogen has become a contender as an additional (or alternative) fuel in future energy systems. An energy economy based on hydrogen has many advantages: the raw material for its production, water, is readily available; it is a non-polluting and renewable fuel since the product of combustion of hydrogen is water; and hydrogen is used in the industrial production of methanol and ammonia, and for the hydrogenation of biomass for upgrading its energy quality. Hydrogen can be stored and transported in a number of ways, for example by mixing with natural gas and transmission through pipelines, by liquefaction in cylinders or as metal hydrides. Thus hydrogen is an optional energy carrier with advantages in storability, portability and efficiency (see Gregory and Pangborn, 1976).

*Photolysis of water.* The splitting of water to hydrogen and oxygen is an endothermic reaction and requires an energy input of 237 kJ mol$^{-1}$ of water decomposed. Hydrogen is produced from water in a number of ways, many of them consuming conventional sources of energy such as coal and electricity. Can we use solar energy to split water and generate hydrogen? Water is transparent to visible radiation and so direct decomposition of water by visible light is not practical. However, direct water photolysis can be achieved in the presence of photosensitizers (pigments) which can absorb visible light energy and transfer this energy for water oxidation. Such pigments are found in the photosystem II of $O_2$-evolving photosynthetic organisms, that is, blue-green algae (cyanobacteria), algae and plants. These organisms split water, in the light, to oxygen, electrons and protons; the electrons and protons are utilized to generate reducing power.

In the last few years, the possibilities have been explored of systems for water photolysis, based partly or completely on biological catalysts. Such systems are at present far from economic viability, but they may point the way to future alternative energy sources.

*Mimicking photosynthesis.* The capacity of plants and algae to split water during photosynthesis can be exploited in artificial biophotolytic systems which produce hydrogen and oxygen from water in the presence of added catalysts (Benemann *et al.*, 1973, Rao *et al.*, 1976). In the early systems, photosynthetic membranes (isolated chloroplasts) were used for the reduction of an electron mediator with water as electron source and subsequently the reduced mediator was coupled with a hydrogenase to liberate hydrogen (Rao and Hall, 1979, 1983; Hall *et al.*, 1981; Weaver *et al.*, 1980) as illustrated in Fig. 4. The rates and duration of $H_2$ evolution depended upon the nature of the chloroplasts and the electron mediator, and also on the stability of the hydrogenase. Usually $H_2$

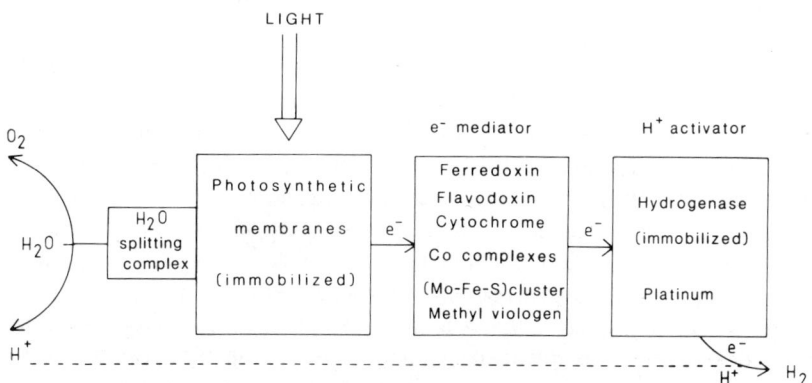

**Fig. 4.** Components of a possible hydrogen photoreactor.

evolution lasted for about 2–4 hr at rates varying from 10 to 30 $\mu$mol mg$^{-1}$ chlorophyll hr$^{-1}$. Obviously, such homogeneous systems are not efficient for the continuous photoproduction of hydrogen.

## Stability of Hydrogen Production Systems

The poor stability of the hydrogen production is due to various factors, including light inhibition of photosynthetic electron transport, autoxidation of the electron mediator, oxygen sensitivity of the chloroplasts and the hydrogenase and reversibility of the hydrogen evolution reaction. Some improvements in the efficiency of hydrogen production were achieved by modifications of the components in the reaction mixture and by the addition of stabilizing agents.

*Chloroplasts.* The least stable component of the photosynthetic hydrogen production system is the chloroplast. Photosynthetic electron transport activity of isolated chloroplast membranes decays in the dark as well as in the light. Dark inhibition could be partially prevented by the addition of Ficol 400 or reagents such as bovine serum albumin, which adsorb the fatty acids liberated by the action of galactolipases on membrane lipids during ageing (Morris *et al.,* 1982). Light-induced inhibition of chloroplast function is more pronounced in photosystem II, where water splitting occurs, than in photosystem I, where a strong reductant is the end product. When isolated chloroplasts are continuously exposed to light there appears to be a loss of a number of polypeptides which are integral components of the light-harvesting reaction centre, and electron transfer complexes of the photosynthetic apparatus, and which are essential for the activities of the thylakoid membrane. Santos and Hall (1982) studied chloroplast ageing in the dark and in the light and observed a correlation between the light-induced decay of photosystem II activity and a decrease in the content of membrane polypeptides with apparent $M_r$ of 36, 48, and 50 kD. These polypeptides, presumably, are associated with the photosystem II reaction centre.

*Immobilization of photosynthetic membranes.* Numerous studies have been carried out in recent years to stabilize the photosynthetic activity of isolated chloroplasts and of whole cells of cyanobacteria and algae by endogenous or exogenous immobilization of the photosynthetic apparatus (see Berezin and Varfolomeev, 1979; Rao and Hall, 1983). Some of these are listed in Table 3. As already mentioned, addition of certain reagents which can alter the environment of the chloroplasts, for example serum albumin, glycerol or polyethylene glycol, can also improve the stability of chloroplasts. Thus Zitzev (see Berezin and Varfolomeev, 1979) was able to extend the electron transport activity of isolated pea chloroplasts (normally 2 hr) to 360 hr by incubation in a medium containing 40% (v/v) glycerol and 0.1 m$M$ bovine serum albumin.

In our laboratory we have used agar (2%, w/v) and calcium alginate (2%, w/v) as entrapment gels for the endogenous immobilization of thylakoid membranes (Gisby et al., 1982). Rates of hydrogen evolution catalyzed by agar-immobilized chloroplasts were comparable to those of free chloroplasts; the kinetics of evolution of oxygen (measured as ferricyanide reduction) by the agar-immobilized chloroplasts paralleled the kinetics of evolution of $H_2$. Although the rates of hydrogen evolution catalyzed by alginate-entrapped chloroplasts were lower than those with free chloroplasts, the $H_2$ evolution catalyzed by the former lasted for a longer period. These gel-immobilized chloroplasts may prove useful in the construction of heterogeneous photosystems (with solid catalysts) for hydrogen production.

*Oxygen inhibition.* Reactivity of the catalysts with oxygen is another major drawback for continuous hydrogen production by water photolysis. Oxygen is the first product released in photosynthesis. In homogeneous, closed systems the oxygen released by photosystem II dissolves in the aqueous medium, reacts with the reduced electron carrier (for example ferredoxin or methyl viologen) generated by photosystem I and produces superoxide or hydrogen peroxide (Foyer and Hall, 1980). Usually the reduced electron carrier reacts more rapidly with the oxygen present than with the hydrogenase and thus the rate of hydrogen evolution is decreased markedly. Although a number of photoreducible, low-potential, organic and inorganic electron relays which can couple with hydrogenase have been synthesized recently (Cuendet and Gratzel, 1982), most of them are autoxidizable in their reduced state and therefore are not suitable catalysts for the simultaneous production of oxygen and hydrogen. Thus, simultaneous hydrogen and oxygen evolution is seldom observed from illuminated single-stage systems containing isolated chloroplasts (or algal cells), hydrogenase and an electron mediator. The hydrogen which is evolved is produced at the expense of oxidation of water to hydrogen peroxide. In many of the reported studies good rates of $H_2$ production from water were obtained only in the presence of oxygen and peroxide scavengers (usually a mixture of glucose, glucose oxidase, catalase and ethanol).

One should not conclude from this that water is not the electron source for hydrogen photoproduction. That water is the electron source in such biocatalytic systems has been demonstrated (1) by the observed lack of hydrogen evolution in the presence of the photosynthesis inhibitor 3-(3,4-dichlorophenyl)-1,1-dimethylurea (DCMU) (Rao et al., 1978b) and (2) by the simultaneous production of hydrogen and oxygen, in stoicheiometric amounts, catalyzed by isolated chloroplasts or algal cells, from a system wherein the two gases are continuously swept out as soon as they are released (Greenbaum, 1980; Greenbaum et al., 1983).

The second difficulty with oxygen-evolving hydrogen production systems is

**Table 3.** *Immobilization of photosynthetic systems*

| Organelle | Immobilization agent and conditions | Stability and photosynthetic activity after immobilization compared to free | Reference |
|---|---|---|---|
| Chloroplasts | 0.05% glutaraldehyde at 4°C | Higher stability in storage. No improvement in light stability. $H_2$ evolution with ferredoxin and hydrogenase. | Rao et al. (1976) |
| Chloroplasts (thylakoids) | Encapsulation in protamine + toluene diisocyanate | PSI activity retained. Loss of PSII activity. | Kitajima and Butler (1976) |
| Chloroplasts (thylakoids) | 0.37% glutaraldehyde + serum albumin at −30°C | 70% $O_2$ evolution activity. 27% ATP synthesis. Stability in light for 9 days. | Cocquempot et al. (1979) |
| Chloroplasts (thylakoids) | 3% (w/v) polyvinyl alcohol + serum albumin | 100% PSI and 80% PSII activity retained. Only 20% loss of activity in 5 wk storage | Ochiai et al. (1978) |
| Chloroplasts (thylakoids) | Agar in hollow-fibre reactor | NADP reduction for 2 hr; 31% activity after 1 wk at 5°C. | Karube et al. (1980) |
| Chloroplasts (thylakoids) | 2% (w/v) calcium alginate as films | 100% $O_2$ exchange, up to 67% $H_2$ evolution. Light stability increased to 7 hr compared to 4 hr for non-immobilized | Gisby and Hall (1980) |
| Chloroplasts (thylakoids) | 0.063% glutaraldehyde + 5% (w/v) gelatin Polyurethane + 5% (w/v) serum albumin | 45% of activity in ferricyanide reduction. 50% ferricyanide reduction activity; stability up to 400 hr. | Cocquempot et al. (1981) |
| Chloroplasts (thylakoids) | 2% (w/v) agar + 0.5% serum albumin | PSII and PSI stabilised without loss of activity. Better $H_2$ evolution rates in PSI-mediated $H_2$ evolution with other catalysts also immobilized | Rao et al. (1982b) |

| Organism | Immobilization method | Observations | Reference |
|---|---|---|---|
| *Mastigocladus laminosus* cells | 5% (w/v) calcium alginate deposited on $SnO_2$ electrode | 70% PSII and 50% PSI activity. Photocurrent generated on continuous illumination for 20 days. | Ochiai *et al.* (1980) |
| *Mastigocladus laminosus* cells | 2% (w/v) agar or 2% (w/v) calcium alginate | Continuous PSI-mediated $H_2$ production with methyl viologen and hydrogenase for more than 10 days. | Rao *et al.* (1982a) |
| *M. laminosus, Nostoc muscorum, Chlorogloea fritschii* | Polyurethane foam pieces (cyanobacteria cultured in the presence of foam) | Both PSII and PSI activity retained. Higher stability of PSI in light. PSI mediated $H_2$ production with exogenous hydrogenase in continuous light lasted more than 10 days. | Muallem and Hall (1982) |
| *Anabaena cylindrica* | Cells suspended in glass beads and cultured outdoors aerobically | Continuous $H_2$ production for more than 20 days. | Smith and Lambert (1981) |
| *Scenedesmus obliquus* | a) Polyurethane foam pieces and kept in nutrient flux b) Calcium alginate | $O_2$ evolution maintained for 1–3 months. $H_2$ evolution after anaerobic adaptation. | Brouers *et al.* (1983) |
| *Porphyridium purpureum* | a) Polyurethane foam pieces and kept in nutrient flux b) Calcium alginate | $O_2$ evolution maintained for 1–3 months. $H_2$ evolution after anaerobic adaptation. | Brouers *et al.* (1983) |
| *Anabaena* 27893 | Calcium alginate beads in a fluidised-bed column | No loss in $O_2$ evolution rate even after 48 days. Sustained photosynthesis and $N_2$ fixation over a 130-hr period. | Musgrave *et al.* (1983) |
| *Rhodospirillum rubrum* | Agar slabs | Continuous photo $H_2$ production with malate for 150 hr in a reactor. | Bennett and Weetall (1976) |
| *Rhodopseudomonas capsulata* chromatophores | 2% (w/v) barium alginate | 70% of ATP synthesis activity. | Paul and Vignais (1980) |
| *Chromatium* Miami PBS 1071 | Agar gel matrix | Continuous $H_2$ production for several weeks from sulphide compared to free cells, which lasted only for days. | Ikemoto and Mitsui (1983) |

that oxygen and its reduced forms, superoxide and hydrogen peroxide, if allowed to accumulate, can cause oxidation of the thylakoid membrane lipids and inhibition of photosynthetic electron transport, as well as inactivation of hydrogenase as already described.

*Immobilization of hydrogenases.* In early biophotolytic hydrogen evolution studies *C. pasteurianum* hydrogenase was used as the catalyst. Although this enzyme has a very high turnover rate for hydrogen evolution, it is readily and irreversibly deactivated by oxygen. On the other hand, the soluble hydrogenase from *A. eutrophus,* which is exceptionally stable in oxygen, showed low rates of hydrogen production when used with chloroplasts and methyl viologen. Also this enzyme, which is stable in oxygen in the oxidized form, is vulnerable to oxygen attack once reduced (Schneider and Schlegel, 1981). After a detailed study of hydrogen evolution in the presence of oxygen scavengers and chloroplast-stabilizing agents, we have identified the combinations of electron relays and hydrogenases for maximum yield of hydrogen (see Rao and Hall, 1983).

The activity of many enzymes can be protected against various factors such as oxygen, pH and proteolysis by immobilization in solid matrices (see Klibanov, 1983). Immobilization studies have been performed on many hydrogenases mainly to improve their oxygen stability. In some cases the enzyme was coimmobilized with chloroplasts and a suitable electron carrier and then used in hydrogen production reactions. Immobilization of hydrogenases generally resulted in a marked decrease in specific activity of the enzyme. However, once immobilized, the activity of the matrix-bound enzyme was much more stable to oxygen than that of the free, soluble enzyme. Some of the procedures used in immobilization of hydrogenases are listed in Table 4. As is the case with other types of immobilized enzymes, immobilized hydrogenases possess the advantage that the enzyme can be used as a component of the fluidized-bed type of photoreactor.

*Hydrogen Production Using Photosystem I of Plants and Algae*

We have seen that the two important factors which limit the rate and duration of biophotolytic hydrogen production from water are (1) the relative instability of the oxygen-evolving (water-splitting) complex at photosystem II and (2) the oxygen sensitivity of the catalysts used. The effect of these two factors on hydrogen production can be eliminated if water is not used as an electron donor. This approach then utilizes only the photosystem I of isolated chloroplast membranes or cyanobacterial cells for light-harvesting and electron transport. The system is analogous to the electron transport chain of green or purple photosynthetic bacteria. Exogenous electron donors such as ascorbate, glutathione,

**Table 4.** *Immobilization of hydrogenases*

| Source | Immobilization agent and conditions | Results | Reference |
|---|---|---|---|
| *Clostridium pasteurianum* | Succinyl carbodiimide glass | 6% specific activity. Improved $O_2$ stability. | Lappi *et al.* (1976) |
| | Ferredoxin bound to Sepharose 4B | Hydrogenase stabilized. No activity for ferredoxin. | Rao *et al.* (1978a) |
| | DEAE cellulose, Tris/Cl$^-$ buffer | Half-life for resistance to $O_2$ inactivation increased 25-fold. | Klibanov *et al.* (1978) |
| | PEI cellulose, phosphate buffer | Half-life for $O_2$ resistance increased 400-fold. | |
| | Glutaraldehyde(GA)-activated amino Spherosil | Half-life increased 20-fold. $H_2$ evolved from chloroplast–ferredoxin system. | Plasterk *et al.* (1981) |
| | Glutaraldehyde-activated aminopropyl glass | | |
| *Desulfovibrio gigas* | Activated aliphatic and aromatic amino Spherosil with and without cytochrome $c_3$ | $O_2$ stability improved. Immobilization yield and specific activity higher with coimmobilized $c_3$. | Hatchikian and Monsan (1980) |
| *Desulfovibrio desulfuricans* Norway | Activated amino Spherosil and aminopropyl glass with and without $c_3$ | Storage stability in $O_2$ improved. No improvement in specific activity with $c_3$ coimmobilized. | Plasterk *et al.* (1981) |
| *Desulfovibrio desulfuricans* NRC 49001 | GA-treated cells in calcium alginate | GA prevents leakage of enzyme from periplasm. Storage stability improved. | Ziomek *et al.* (1982) |
| *Alcaligenes eutrophus* soluble enzyme | GA-treated aminopropyl glass | 90% activity immobilized. Lag phase for NAD reduction abolished. Storage stability at 0°C improved. | Simon *et al.* (1978) |
| *Alcaligenes eutrophus* cells | Calcium alginate | 20% specific activity in calcium alginate. 60% activity after 1 wk at 25°C. Used for regeneration of coenzymes. | Klibanov and Puglisi (1980) |

cysteine or dithiothreitol are used in conjunction with a mediator dye (2,6-dichlorophenolindophenol or $N,N,N',N'$-tetramethyl-$p$-phenylenediamine). The electrons are donated into the photosynthetic electron transport chain prior to photosystem I (at the cytochrome $f$/plastocyanin complex). To avoid any photosynthetic oxygen evolution from the aqueous medium, the water-splitting reaction is blocked by the addition of DCMU or by controlled heat treatment of the chloroplasts, which deactivates the water-splitting enzyme complex without affecting the activities associated with photosystem I. Methyl viologen, which is thus reduced at photosystem I, is coupled to a bacterial hydrogenase for hydrogen evolution.

The conditions for optimal hydrogen production from ascorbate in a chloroplast–methyl viologen–hydrogenase catalytic system have been studied by Trebst, Krasnovski and others (see Rao and Hall, 1983). These are:

1. The presence of intact thylakoid vesicles to decrease internal cycling of electrons from reduced methyl viologen, which is produced on the outside of the vesicles, back to plastocyanin, which is located on the inside;
2. The addition of dithiothreitol or glutathione to reduce any dehydroascorbate to ascorbate and also to keep the chloroplasts, and possibly the hydrogenase, in a reducing environment; and
3. Keeping the pH of the reaction medium below neutral (Muallem and Hall, 1982).

In this system good rates of hydrogen production were only maintained for 3–5 hr. The reaction slowed down after this time mainly because the accumulated products in the reaction media gradually inhibited the coupling of electron transport from ascorbate to hydrogenase without causing permanent inactivation of any of the components of the reaction (Muallem and Hall, 1982). The fact that relatively expensive organic molecules have to be used as an electron source and that the lifetime of hydrogen production is shortened by product inhibition limits the advantage of hydrogen production with photosystem I alone compared to water photolytic systems catalyzed by photosystems II and I. However, it should be pointed out that in systems using organic electron donors there is only one gaseous product—hydrogen—and there is no need to separate oxygen from hydrogen as is the case with water photolysis.

## Photoproduction of Hydrogen by Cyanobacteria

Cyanobacteria offer advantages over chloroplasts in that their enzymes and photosystems are generally more stable. A disadvantage is that their outer walls and membranes present permeability barriers to electron carriers and hydrogenase.

*Hydrogen production by whole cells.* Hydrogen metabolism in $N_2$-fixing, heterocystous cyanobacteria is catalyzed by three types of enzymes:

1. A reversible hydrogenase occurring both in the vegetative cells and in the heterocysts;
2. An uptake hydrogenase located mainly in the heterocysts; and
3. The nitrogenase located in the heterocysts, which can catalyze hydrogen evolution with ATP consumption (see Lambert and Smith, 1981). Photosynthetic hydrogen evolution by whole cells is probably all due to this by-product of the nitrogen-fixing enzymes.

Hydrogen evolution rates of 40 $\mu$l mg$^{-1}$ dry weight hr$^{-1}$, continuing for 18 days, have been reported from *Anabaena cylindrica* cells grown in the laboratory—this represented a solar energy conversion efficiency of 1.2% (Benemann *et al.*, 1980). Outdoor systems of cyanobacteria evolving hydrogen for more than 30 days with solar radiation conversion efficiencies of 2% have also been developed. The rate of light-dependent, nitrogenase-catalyzed hydrogen evolution from *A. cylindrica* cells suspended in a nickel-free medium was higher than that from cells suspended in a normal nickel-containing medium (Daday and Smith, 1983; D.-J. Shi and D. O. Hall, unpublished).

*Hydrogen production by photosystem I.* As with chloroplasts, addition of DCMU blocks evolution of oxygen by cyanobacteria. DCMU-treated cyanobacteria can be used for photoreduction of methyl viologen with ascorbate or other photosystem I electron donors; the methyl viologen can be coupled to an exogenous hydrogenase to liberate hydrogen. Smith *et al.* (1982) demonstrated H$_2$ photoproduction from ascorbate catalyzed by photosystem I of *Mastigocladus laminosus* and *Phormidium laminosum* in the presence of methyl viologen and hydrogenase from *D. desulfuricans*. The cyanobacteria were subjected to repeated freezing and thawing to modify their cell walls to be permeable to methyl viologen. *Mastigocladus laminosus* cells retained their specific activity to reduce methyl viologen for more than 14 days. The stability and rate of hydrogen production were improved by entrapping the cells in 2% (w/v) agar.

*Hydrogen production by polyurethane-immobilized cyanobacteria.* We have used polyurethane foam pieces for immobilization of cyanobacterial cells. Cells of *Chlorogloea fritschii*, *Nostoc muscorum* and *M. laminosus* were grown into polyurethane foam pieces (polyester or polyvinyl types), harvested and used for hydrogen production with ascorbate as electron donor and methyl viologen and *D. desulfuricans* hydrogenase as catalysts. The hydrogen-evolving capacity of the reaction mixtures containing immobilized cells lasted at least 9 days, at 30°C, under continuous illumination. For comparison, the photoactivity of free cells of *M. laminosus* did not survive illumination for 24 hr (Muallem *et al.*, 1983). Thus the photosynthetic apparatus of cyanobacteria appears to be stabilized by attachment on polyurethane foam.

Attempts to couple photosynthetically reduced methyl viologen to the endogenous hydrogenase of the cyanobacteria for hydrogen production were not successful, although cell-free preparations with hydrogenase activity (hydrogen production from reduced methyl viologen) have been obtained from *A. cylindrica, N. muscorum, Oscillatoria limnetica* and *M. laminosus.*

When one compares the rate and longevity of photosystem I-mediated $H_2$ production using chloroplasts with those of cyanobacteria immobilized on polyurethane foam, it is quite evident that the latter are much more efficient and convenient for use in photoreactors. Both free and polyurethane foam-immobilized cells of *N. muscorum* and *M. laminosus* kept under continuous illumination in a reaction mixture retained their photosynthetic electron transport activity from water to methyl viologen, measured as oxygen uptake, for 14 days or longer (D. Affolter and D. O. Hall, unpublished). This suggests that both photosystems are functional in these cyanobacteria even after 2 wk of continuous illumination. However, we have not, so far, been able to observe significant hydrogen evolution with water as electron donor, using either free cells or foam-entrapped cells of cyanobacteria. Thus, most probably, in the presence of photosynthetically produced oxygen, the cyanobacteria are unable to accumulate sufficient reduced methyl viologen in the reaction medium to activate the exogenous hydrogenase and then react with the enzyme to produce hydrogen.

*Photogeneration of hydrogen by semiconductor-bound hydrogenases.* The inherent instability of isolated chloroplasts has prompted many researchers to search for more stable inorganic or synthetic materials as light-harvesting and electron transfer catalysts for use in solar energy conversion systems. Semiconductors such as titanium dioxide and cadmium sulphide are good photocatalysts (although, unlike chloroplasts, they cannot use the whole spectrum of visible light). In conjunction with hydrogenase (or Pt) they can be used for hydrogen production by water decomposition. Hydrogenases from *C. pasteurianum* and *D. desulfuricans* have been immobilized on $TiO_2$ and CdS particles by adsorption as well as by covalent coupling. On irradiation with a sunlight simulator, in the presence of an electron donor (EDTA) and methyl viologen, hydrogen was evolved continuously for a number of hours (Cuendet *et al.*, 1984). The hydrogenase bound to the semiconductor was quite stable. These types of semi-inorganic–semi-biological combinations may be forerunners of completely artificial systems which mimic natural photosynthesis.

## Acknowledgments

This work was supported by grants from the U.K. Science and Engineering Research Council and the Commission of the European Communities.

# References

Adams, M. W. W. and Hall, D. O. (1979). Purification of the membrane-bound hydrogenase of *Escherichia coli. Biochemical Journal* **183**, 11–22.

Adams, M. W. W., Mortenson, L. E. and Chen, J. S. (1981). Hydrogenase. *Biochimica et Biophysica Acta* **594**, 105–176.

Albracht, S. P. J., Graf, E. -G. and Thauer, R. K. (1982). The EPR properties of nickel in hydrogenase from *Methanobacterium thermoautotrophicum. FEBS Letters* **140**, 311–313.

Albracht, S. P. J., Kalkman, M. L. and Slater, E. C. (1983). Magnetic interaction of nickel (III) and the iron–sulphur cluster in hydrogenase from *Chromatium vinosum. Biochimica et Biophysica Acta* **724**, 309–316.

Beinert, H. and Thomson, A. J. (1983). Three-iron clusters in iron–sulfur proteins. *Archives of Biochemistry and Biophysics* **222**, 333–361.

Benemann, J. R., Berenson, J. A., Kaplan, N. O. and Kamen, M. D. (1973). Hydrogen evolution by a chloroplast–ferredoxin–hydrogenase system. *Proceedings of the National Academy of Sciences of the U.S.A.* **8**, 2317–2320.

Benemann, J. R., Miyamoto, K. and Hallenbeck, P. C. (1980). Bioengineering aspects of biophotolysis. *Enzyme and Microbial Technology* **2**, 103–111.

Bennett, M. and Weetall, H. H. (1976). Production of hydrogen using immobilized *Rhodospirillum rubrum. Journal of Solid-Phase Biochemistry* **1**, 137–142.

Berezin, I. V. and Varfolomeev, S. D. (1979). Energy-related applications of immobilized enzymes. *"Applied Biochemistry and Bioengineering* **2**, 259–290.

Berlier, Y. M., Fauque, G., Lespinat, P. A. and Le Gall, J. (1982). Activation, reduction and proton–deuterium exchange reaction of the periplasmic hydrogenase from *Desulfovibrio gigas* in relation with the role of cytochrome $c_3$. *FEBS Letters* **140**, 185–188.

Bossu, F. P. and Margerum, D. W. (1977). Electrode potentials of nickel(III,II)–peptide complexes. *Inorganic Chemistry* **16**, 1210–1214.

Braun, A. M. (ed.) (1983). "Photochemical Conversions". Presses Polytechniques Romandes, Lausanne.

Brouers, M., Collard, F., Jeanfils, J. and Loudeche, R. (1983). Long term stabilization of photobiological activities of immobilized algae. *In* "Photochemical, Photoelectrochemical and Photobiological Processes" Vol. 2 (Eds. D. O. Hall, W. Palz and D. Pirrwitz) pp. 171–178. Reidel Publishing Co., Dordrecht, Netherlands.

Cammack, R., Lalla-Maharajh, W. V. and Schneider, K. (1982a). EPR studies of some oxygen-stable hydrogenases. *In* "Electron Transport and Oxygen Utilization" (Ed. C. Ho), pp. 411–415. Elsevier/North Holland, New York.

Cammack, R., Patil, D. S., Aguirre, R. and Hatchikian, E. C. (1982b). Redox properties of the ESR-detectable nickel in hydrogenase from *Desulfovibrio gigas. FEBS Letters* **142**, 289–292.

Chen, J. -S. (1978). Structure and function of two hydrogenases from the dinitrogen-fixing bacterium *Clostridium pasteurianum* W5. *In* "Hydrogenases: Their catalytic Activity, Structure, and Function" (Eds. H. G. Schlegel and K. Schneider) pp. 57–81. Goltze KG, Göttingen.

Cocquempot, M. F., Thomas, D., Champigny, M. L. and Moyse, A. (1979). Immobilization of thylakoids in porous particles and stabilization of the photochemical processes by glutaraldehyde action at subzero temperature. *European Journal of Applied Microbiology and Biotechnology* **8**, 39–41.

Cocquempot, M. F., Thomasset, B., Barbotin, J. N., Gellf, G. and Thomas, D. (1981). Comparative stabilization of biological photosystems by several immobilization procedures. *European Journal of Applied Microbiology and Biotechnology* **11**, 193–198.

Cuendet, P. and Grätzel, M. (1982). New photosystem I electron acceptors: Improvement of hydrogen photoproduction by chloroplasts. *Photochemistry and Photobiology* **36**, 203–210.

Cuendet, P., Grätzel, M., Rao, K. K. and Hall, D. O. (1984). Immobilized enzymes on semiconductor powders: Photogeneration of $H_2$ by $TiO_2$- and Cds-bound hydrogenases. *Photobiochemistry and Photobiophysics* **7**, 331–340.

Daday, A. and Smith, G. D. (1983). The effect of nickel on the hydrogen metabolism of the cyanobacterium *Anabaena cylindrica*. *FEMS Microbiology Letters* **20**, 327–330.

Dutton, P. L. (1978). Redox potentiometry: Determination of midpoint potentials of oxidation–reduction components of biological electron-transfer systems. *In* "Methods in Enzymology" Vol. 54 (Eds. S. Fleischer and L. Packer), pp. 411–435. Academic Press, New York.

Erbes, D. L., Burris, R. H. and Orme-Johnson, W. H. (1975). On the iron–sulfur cluster in hydrogenase from *Clostridium pasteurianum* W5. *Proceedings of the National Academy of Sciences of the U.S.A.* **72**, 4795–4799.

Fernandez, V. M., Munilla, R. and Ballesteros, A. (1982). Influence of the redox potential on the activity of *Clostridium pasteurianum* and *Chromatium* hydrogenases. *Archives of Biochemistry and Biophysics* **215**, 129–135.

Fernandez, V. M., Aguirre, R. and Hatchikian, E. C. (1984). Reductive activation and redox properties of hydrogenase from *Desulfovibrio gigas*. *Biochimica et Biophysica Acta* **790**, 1–7.

Foyer, C. H. and Hall, D. O. (1980). Oxygen metabolism in the active chloroplast. *Trends in Biochemical Sciences* **5**, 188–191.

Friedrich, B., Heine, E., Fink, A. and Friedrich, C. G. (1981). Nickel requirement for active hydrogenase formation in *Alcaligenes eutrophus*. *Journal of Bacteriology* **145**, 1144–1149.

Gisby, P. E. and Hall, D. O. (1980). Biophotolytic $H_2$ production using alginate-immobilized chloroplasts, enzymes and synthetic catalysts. *Nature (London)* **287**, 251–253.

Gisby, P. E., Rao, K. K. and Hall, D. O. (1982). Hydrogen production from water using chloroplasts, enzymes and synthetic catalysts: A photobiological process. *In* "Solar World Forum" Vol. 3 (Eds. D. O. Hall and J. Morton), pp. 2242–2247. Pergamon Press, Oxford.

Graf, E. G. and Thauer, R. K. (1981). Hydrogenase from *Methanobacterium thermoautotrophicum*, a nickel-containing enzyme. *FEBS Letters* **136**, 165–169.

Grande, H. J., Dunham, W. R., Averill, B. A., Veeger. C. and Sands, R. H. (1983). Electron paramagnetic resonance and other properties of hydrogenases isolated from *Desulfovibrio vulgaris* and *Megasphaera elsdenii*. *European Journal of Biochemistry* **136**, 201–207.

Greenbaum, E. (1980). Simultaneous photoproduction of hydrogen and oxygen by photosynthesis. *In* "Biotechnology and Bioengineering Symposium No. 10" (Ed. C. D. Scott), pp. 1–13. Wiley, New York.

Greenbaum, E., Guillard, R. R. L. and Sunda, W. G. (1983). Hydrogen and oxygen photoproduction by marine algae. *Photochemistry and Photobiology* **37**, 649–655.

Gregory, D. P. and Pangborn, J. B. (1976). Hydrogen energy. *Annual Review of Energy* **1**, 279–310.

Griffith, J. S. (1971). Coupled cupric and ferric ions in cytochrome oxidase. *Molecular Physics* **21**, 141–143.

Hall, D. O., Rao, K. K., Gisby, P. E., Santos, C. P., Richoux, M. C. and Lalla-Maharajh, W. (1981). Biophotolysis of water for hydrogen production. *In* "Photosynthesis" Vol. VI (Ed. G. Akoyunoglou), pp. 639–654. Balaban International Sciences Services, Philadelphia, Pennsylvania.

Hatchikian, E. C. and Monsan, P. (1980). Highly active immobilized hydrogenase from *Desulfovibrio gigas*. *Biochemical and Biophysical Research Communications* **92**, 1091–1096.

Hatchikian, E. C., Bruschi, M. and LeGall, J. (1978). Characterization of the periplasmic hydrogenase from *Desulfovibrio gigas*. *Biochemical and Biophysical Research Communications* **82**, 451–461.

Henry, L. Adams, M. W. W., Rao, K. K. and Hall, D. O. (1980). The effect of oxygen species on the enzymatic activity of hydrogenase. *FEBS Letters* **122**, 211–214.

Ikemoto, H. and Mitsui, A. (1983). Continuous indoor and outdoor hydrogen photoproduction from sulfide by immobilized marine photosynthetic sulfur bacterium *Chromatium* sp. Miami. *International Congress on Photosynthesis, 6th, 1983*, Abstract 214-3.

Johnson, M. K., Thomson, A. J., Robinson, A. E. and Smith, B. E. (1981). Characterization of the paramagnetic centres of the molydenum–iron protein of nitrogenase from *Klebsiella pneumoniae* using low temperature magnetic circular dichroism spectroscopy. *Biochimica et Biophysica Acta* **671**, 61–70.

Johnson, M. K., Czernuszewicz, R. S., Spiro, T. G., Ramsay, R. R., Singer, T. P. and Rao, K. K. (1983). Resonance Raman studies of beef heart aconitase and a bacterial hydrogenase. *Journal of Biological Chemistry* **258**, 12771–12774.

Karube, I., Oksuka, T., Kayano, H., Matsunaga, T. and Suzuki, S. (1980). Photochemical system for regenerating NADPH from NADP with use of immobilized chloroplasts. *Biotechnology and Bioengineering* **22**, 2655–2665.

Kent, T. A., Dreyer, J. -L. Kennedy, M. C., Huynh, B. H., Emptage, M. H., Beinert, H. and Münck, E. (1982). Mössbauer studies of beef heart aconitase: Evidence for facile interconversions of iron–sulfur clusters. *Proceedings of the National Academy of Sciences of the U.S.A.* **79**, 1096–1100.

Kitajima, M. and Butler, W. L. (1976). Microencapsulation of chloroplast particles. *Plant Physiology* **57**, 746–750.

Klibanov, A. M. (1983). Immobilized enzymes and cells as practical catalysts. *Science* **219**, 722–727.

Klibanov, A. M. and Puglisi, A. V. (1980). The regeneration of coenzymes using immobilized hydrogenases. *Biotechnology Letters* **2**, 445–450.

Klibanov, A. M., Kaplan, N. O. and Kamen, M. D. (1978). A rationale for stabilization of oxygen-labile enzymes: Application to a clostridial hydrogenase. *Proceedings of the National Academy of Sciences of the U.S.A.* **75**, 3640–3643.

Kojima, N., Fox, J. A., Hausinger, R. P., Daniels, L., Orme-Johnson, W. H. and Walsh, C. (1983). Paramagnetic centers in the nickel-containing, deazaflavin-reducing hydrogenase from *Methanobacterium thermoautotrophicum*. *Proceedings of the National Academy of Sciences of the U.S.A.* **80**, 378–382.

Krasna, A. I. (1979). Hydrogenase: Properties and applications. *Enzyme and Microbial Technology* **1**, 165–172.

Krüger, H. -J., Huynh, B. H., Ljungdahl, P. O., Xavier, A. V., Dervartanian, D. V., Moura, I., Peck, H. D., Jr., Teixeira, M., Moura, J. J. G. and LeGall, J. (1982). Evidence for nickel and a three-iron center in the hydrogenase of *Desulfovibrio desulfuricans*. *Journal of Biological Chemistry* **257**, 14620–14623.

Lalla-Maharajh, W. V., Hall, D. O., Cammack, R., Rao, K. K. and LeGall, J. (1982). Purification and properties of the membrane-bound hydrogenase from *Desulfovibrio desulfuricans*. *Biochemical Journal* **209**, 445–454.

Lambert, G. R. and Smith, G. D. (1981). The hydrogen metabolism of cyanobacteria. *Biological Reviews Cambridge Philosophical Society*, **56**, 589–660.

Lancaster, J. R., Jr. (1982). Identification and detection of electron transfer components in methanogens. *In* "Methods in Enzymology" Vol. 88 (Ed. L. Packer), pp. 412–417. Academic Press, New York.

Lappi, D. A., Stolzenbach, F. E., Kaplan, N. O. and Kamen, M. D. (1976). Immobilization of hydrogenase on glass beads. *Biochemical and Biophysical Research Communications* **69**, 878–884.

Mayhew, S. G. and O'Connor, M. (1982). Structure and mechanism of bacterial hydrogenase. *Trends in Biochemical Sciences* **7**, 18–21.

Morris, P., Nash, G. V. and Hall, D. O. (1982). The stability of electron transport in *in vitro* chloroplast membranes. *Photosynthesis Research* **3**, 227–240.

Moura, J. J. G., Moura, I., Huynh, B. H., Krüger, H. -J., Teixeira, M., DuVarney, R. C., DerVartanian, D. V., Xavier, A. V., Peck, H. D., Jr. and LeGall, J. (1982). Unambiguous identification of the nickel EPR signal in $^{61}$Ni-enriched *Desulfovibrio gigas* hydrogenase. *Biochemical and Biophysical Research Communications* **108**, 1388–1393.

Muallem, A. and Hall, D. O. (1982). Ascorbate as a substrate for photoproduction of hydrogen by photosystem I of chloroplasts. *Plant Physiology* **69**, 1116–1120.

Muallem, A., Bruce, D. and Hall, D. O. (1983). Photoproduction of $H_2$ and $NADPH_2$ by polyurethane-immobilized cyanobacteria. *Biotechnology Letters* **5**, 365–368.

Musgrave, S. C., Kerby, N. W., Codd, G. A. and Stewart, W. D. P. (1983). Sustained ammonia production by immobilized filaments of the nitrogen-fixing cyanobacterium *Anabaena* 27893. *Biotechnology Letters* **4**, 647–652.

Ochiai, H., Shibata, H., Matsuo, T., Hashinokuchi, K. and Inamura, I. (1978). Immobilization of chloroplast photosystems with polyvinyl alcohols. *Agricultural and Biological Chemistry* **42**, 683–685.

Ochiai, H., Shibata, H., Sawa, Y. and Katoh, T. (1980). "Living electrode" as a long-lived photoconverter for biophotolysis of water. *Proceedings of the National Academy of Sciences of the U.S.A.* **77**, 2442–2444.

Partridge, C. D. P. and Yates, M. G. (1982). Effect of chelating agents on hydrogenase in *Azotobacter chroococcum:* Evidence that nickel is required for hydrogenase synthesis. *Biochemical Journal* **204**, 339–344.

Paul, F. and Vignais, P. M. (1980). Photophosphorylation in bacterial chromatophores entrapped in alginate gel. *Enzyme and Microbial Technology* **2**, 281–287.

Pedrosa, F. O. and Yates, M. G. (1983). Effect of chelating agents and nickel ions on hydrogenase activity in *Azospirillum brasilense, A. lipoferum* and *Derxia gummosa*. *FEMS Microbiology Letters* **17**, 101–106.

Plasterk, R. H. A., Rao, K. K. and Hall, D. O. (1981). Immobilization of hydrogenases for biophotolytic hydrogen production, stability and kinetics. *Biotechnology Letters* **3**, 99–104.

Rao, K. K. and Hall, D. O. (1979). Hydrogen production from isolated chloroplasts. *Topics in Photosynthesis* **3**, 299–329.

Rao, K. K. and Hall, D. O. (1983). Photobiological production of fuels and chemicals. *In* "Photochemical Conversions" (Ed. A. M. Braun), pp. 1–48. Presses Polytechnique Romandes, Lausanne.

Rao, K. K., Rosa, L. and Hall, D. O. (1976). Prolonged production of hydrogen gas by a chloroplast biocatalytic system. *Biochemical and Biophysical Research Communications* **68**, 21–28.

Rao, K. K., Gogotov, I. N. and Hall, D. O. (1978a). Hydrogen evolution by chloroplast–hydrogenase systems: Improvements and additional observations. *Biochimie* **60**,291–296.

Rao, K. K., Morris, P. and Hall, D. O. (1978b). Hydrogen evolution from water by a chloroplast–hydrogenase system. *In* "Hydrogenases: Their Catalytic Activity, Structure and Function" (Eds. H. G. Schlegel and K. Schneider), pp. 439–452. Goltze K. G., Göttingen.

Rao, K. K., Muallem, A., Bruce, D. L., Smith, G. D. and Hall, D. O. (1982a). Immobilization of chloroplasts, algae and hydrogenases in various solid supports for the photoproduction of hydrogen. *Biochemical Society Transactions* **10**, 527–528.

Rao, K. K., Bruce, D. L., Gisby, P. E., Muallem, A. and Hall, D. O. (1982b). Biophotolysis of water for hydrogen production via natural and artificial catalytic systems. *In* "Photochemical, Photoelectrochemical and Photobiological Processes" Vol. 1 (eds. D. O. Hall and W. Palz), pp. 195–212. Reidel Publishing Co., Dordrecht, Netherlands.

Rieder, R., Cammack, R. and Hall, D. O. (1985). Purification and properties of the cytoplasmic hydrogenase from *Desulfovibrio desulfuricans* (strain Norway 4). *European Journal of Biochemistry* **145**, 637–643.

Santos, C. P. and Hall, D. O. (1982). Thylakoid polypeptides of light and dark aged chloroplasts. *Plant Physiology* **79**, 795–802.

Schink, B. and Schlegel, H. G. (1979). The membrane-bound hydrogenase of *Alcaligenes eutrophus* 1. Solubilization, purification and biochemical properties. *Biochimica et Biophysica Acta* **567**, 315–324.

Schneider, K. and Schlegel, H. G. (1976). Purification and properties of soluble hydrogenase from *Alcaligenes eutrophus* H16. *Biochimica et Biophysica Acta* **452**, 66–80.

Schneider, K., and Schlegel, H. G. (1981). Production of superoxide radicals by soluble hydrogenase from *Alcaligenes eutrophus* H16. *Biochemical Journal* **193**, 99–107.

Schneider, K., Cammack, R., Schlegel, H. G. and Hall, D. O. (1979). The iron–sulphur centres of soluble hydrogenase from *Alcaligenes eutrophus*. *Biochimica et Biophysica Acta* **578**, 445–461.

Schneider, K., Patil, D. S. and Cammack, R. (1983). ESR properties of membrane-bound hydrogenases from aerobic hydrogen bacteria. *Biochimica et Biophysica Acta* **748**, 353–361.

Schneider, K., Jochim, K. and Schlegel, H. G. (1984a). Effect of nickel on activity and subunit composition of purified hydrogenase from *Nocardia opaca* 1b. *European Journal of Biochemistry* **138**, 533–541.

Schneider, K., Cammack, R. and Schlegel, H. G. (1984b). Content and localization of FMN, Fe–S clusters and nickel in the NAD-linked hydrogenase of *Nocardia opaca* 1b. *European Journal of Biochemistry* **142**, 75–84.

Schoenmaker, G. S., Oltman, L. F. and Stouthamer, A. H. (1979). Purification and properties of the membrane-bound hydrogenase from *Proteus mirabilis*. *Biochimica et Biophysica Acta* **567**, 511–521.

Simon, H., Egerer, P. and Günther, H. (1978). Some mechanistic aspects and immobilization of soluble hydrogenase from *Alcaligenes eutrophus*. In ''Hydrogenases: Their Catalytic Activity, Structure and Function'' (Eds. H. G. Schlegel and K. Schneider), pp. 235–251. Eric Goltze KG, Göttingen.

Smith, G. D. and Lambert, G. R. (1981). An outdoor biophotolytic system using the cyanobacterium *Anabaena cylindrica* B629. *Biotechnology and Bioengineering* **23**, 213–220.

Smith, G. D., Muallem, A. and Hall, D. O. (1982). Hydrogenase-catalyzed photoproduction of hydrogen by photosystem I of the thermophilic blue–green algae *Mastigocladus laminosus* and *Phormidium laminosum*. *Photobiochemistry and Photobiophysics* **4**, 307–319.

Takakuwa, S. and Wall, J. D. (1981). Enhancement of hydrogenase activity in *Rhodopseudomonas capsulata* by nickel. *FEMS Microbiology Letters* **12**, 359–363.

Teixeira, M., Moura, I., Xavier, A. V., DerVartanian, D. V., LeGall, J., Peck, H. D., Huynh, B. H. and Moura, J. J. G. (1983). *Desulfovibrio gigas* hydrogenase—redox properties of the nickel and iron–sulfur centers. *European Journal of Biochemistry* **130**, 481–484.

Unden, G., Böcher, R., Knecht, J. and Kröger, A. (1982). Hydrogenase from *Vibrio succinogenes*, a nickel protein. *FEBS Letters* **145**, 230–234.

Van der Westen, H. M., Mayhew, S. G. and Veeger, C. (1980) Effect of growth conditions on the content and O$_2$-stability of hydrogenase in the anaerobic bacterium *Desulfovibrio vulgaris* (Hildenborough). *FEMS Microbiology Letters* **7**, 35–39.

Van Dijk, C. and Veeger, C. (1981). The effects of pH and redox potential on the hydrogen production activity of the hydrogenase from *Megasphaera elsdenii*. *European Journal of Biochemistry* **114**, 209–219.

Weaver, P. F., Lien, S. and Siebert, M. (1980). Photobiological production of hydrogen. *Solar Energy* **24**, 3–45.

Weiss, A. R., Schneider, K. and Schlegel, H. G. (1980). Purification and properties of the membrane-bound hydrogenase of *Pseudomonas pseudoflava*. *Current Microbiology* **3**, 317–320.

Yagi, T. (1976). Properties of purified hydrogenase from the particulate fraction of *Desulfovibrio vulgaris*, Miyazaki. *Journal of Biochemistry (Tokyo)* **79**, 661–671.

Yamazaki, S. (1982). A selenium-containing hydrogenase from *Methanococcus vannielii*. Identification of the selenium moiety as a selenocysteine residue. *Journal of Biological Chemistry* **257**, 7926–7929.

Ziomek, E., Martin, W. G., Veliky, I. A. and Williams, R. E. (1982). Immobilization of *Desulfovibrio desulfuricans*: cell-associated hydrogenase in beaded matrices. *Enzyme and Microbial Technology* **4**, 405–408.

# Discussion

*R. K. Thauer:* It presently appears that hydrogenases that catalyze the uptake of $H_2$ *in vivo* are nickel enzymes and those that mediate the formation of $H_2$ *in vivo* do not contain nickel. Both types contain iron–sulphur clusters. Is there any evidence that, in the non-nickel-containing hydrogenases there is, besides the iron–sulphur clusters, additional iron functionally equivalent to the nickel in the nickel hydrogenases?

*R. Cammack:* We haven't worked on the hydrogen-producing hydrogenases, such as *C. pasteurianum*, but as you say, there has been no evidence for involvement of nickel. The situation is less clear in the case of the hydrogenases of *D. vulgaris* and *M. elsdenii*, where some nickel has been found to be present although its function has not been demonstrated. The ESR spectrum of oxidized *C. pasteurianum* hydrogenase shows unusual features. It is a rhombic spectrum with *g* values just above 2.00, rather like the nickel signal in the uptake hydrogenases [D. L. Erbes, R. H. Burris and W. H. Orme-Johnson, *Proceedings of the National Academy of Sciences of the U.S.A.* **72**, 4795–4799 (1975); J. -S. Chen, in "Hydrogenases" (Eds. H. G. Schlegel and K. Schneider), pp. 57–81. Goltze KG, Göttingen, 1978]. The linewidth is certainly unusually narrow for an iron–sulphur centre. This suggests a novel type of centre, but a present there is no evidence as to what it might be.

*O. Meyer:* Carbon monoxide dehydrogenase from *Clostridium thermoaceticum* has been reported to contain a non-covalently bound nickel cofactor of $M_r \leqq 1,357$ [S. W. Ragsdale, J. E. Clark, L. G. Ljungdahl, L. L. Lundie and H. L. Drake, *Journal of Biological Chemistry* **258**, 2364–2369 (1983)] Could you speculate on the type of compound chelating Ni in hydrogenases and CO dehydrogenases?

*Cammack:* At present, there have been no reports of a specific nickel-containing compound, equivalent to F430 of methanogens, in hydrogenase. I presume some people have looked, but not found. From the shape of the ESR spectra of nickel that we see, we can make some tentative conclusions. The shape of the spectrum can be interpreted in terms of low-spin Ni(III) in a site of distorted octahedral coordination [J. R. Lancaster, Jr., in "Methods in Enzymology" Vol. 88 (Ed. L. Packer), pp. 412–417. Academic Press, New York, 1982]. The linewidth is narrow, and yet no hyperfine splitting is observed. There is no change in linewidth on substituting $D_2O$ for $H_2O$ [R. Cammack, D. S. Patil, R. Aguirre and E. C. Hatchikian, *FEBS Letters* **142**, 289–292 (1982)]. These observations argue against coordination to a proton or an axial nitrogen. Moreover, the redox potential is much lower than in peptide complexes in which the ligands are oxygen and nitrogen [F. P. Bossu and D. W. Margerum, *Inorganic Chemistry* **16**, 1210–1214 (1977)]. One explanation for this would be the existence of sulphur ligands.

*Comment from the Floor:* There is recent (unpublished) evidence from EXAFS that sulphur is a ligand to Ni in hydrogenase.

# 5

# Properties of Two Hydrogenases from *Escherichia coli*

STUART P. BALLANTINE AND DAVID H. BOXER

*Department of Biochemistry, Medical Sciences Institute, Dundee University,*
*Dundee, Tayside, United Kingdom*

In *E. coli,* hydrogenase participates in the metabolism of hydrogen in two distinct pathways. First, it is involved in the formate–hydrogenlyase pathway present in the organism when grown anaerobically in the absence of an exogenous electron acceptor, which converts formate to $CO_2$ and $H_2$ (Gray and Gest, 1965). It also participates in the energy-conserving oxidation of hydrogen, which allows the organism to grow anaerobically on non-fermentable carbon sources such as fumarate or malate under a hydrogen-containing atmosphere (Macy *et al.*, 1976; Berhard and Gottschalk, 1978). A single species of hydrogenase has been isolated from aerobically grown *E. coli* (Adams and Hall, 1979), and previous work from our laboratory had also suggested that there was a single hydrogenase present in anaerobically grown cells (Graham *et al.*, 1980). This paper demonstrates that there are two hydrogenases in *E. coli* grown under anaerobic conditions.

A representative set of data for hydrogenase activity after cell fractionation is shown in Table 1. Hydrogenase activity is found in both the membranes and the high-speed supernatant.

Triton X-100-dispersed membranes were analysed by crossed-immunoelectrophoresis and non-dissociating polyacrylamide gel electrophoresis. Crossed-immunoelectrophoresis was performed with polyspecific antisera raised to membrane vesicles (Graham *et al.*, 1980), and at least 30 protein-staining precipitin arcs were found on the immunoplates. Activity staining of immunoplates for hydrogenase prior to staining for protein revealed two distinct arcs (Fig. 1). It should be noted that if activity staining is not carried out for prolonged periods, only one activity-stained arc may be seen. The two arcs clearly overlap and are independent, indicating their distinct identities. The isoenzymes giving rise to the observed arcs are referred to as hydrogenase 1 and hydrogenase 2. The

MICROBIAL GAS METABOLISM:
MECHANISTIC, METABOLIC
AND BIOTECHNOLOGICAL ASPECTS

**Table 1.**  *Hydrogenase activity after subcellular fractionation of
E. coli grown anaerobically on glucose*[a]

| Fraction | Total activity (%) | Specific activity (units/mg protein) |
|---|---|---|
| Crude extract | 100 | 0.42 |
| Membranes | 44 | 0.49 |
| High-speed supernatant | 25 | 0.19 |

[a]*Escherichia coli* (strain P4X) was grown anaerobically on glucose
minimal media pH 6.4 (Yamamoto and Ishimoto, 1978) supplemented with peptone (0.5%) and harvested in the late exponential
phase of growth. Cells were washed with 50 m$M$ Tris–HCl pH 7.5,
and broken by using a French pressure cell (Graham *et al.*, 1980).
Cellular debris was removed by centrifugation at 6000 $g$ for 15 min
and the membrane fraction was recovered by further centrifugation
of the supernatant at 200,000 $g$ for 2 hr. The membrane fraction was
suspended in 50 m$M$ Tris–HCl pH 7.5. Activity was measured as
$H_2$-dependent reduction of oxidised benzyl viologen (BV); 1 unit
= 1 μmol BV reduced min$^{-1}$ (Jones, 1980).

electrophoretically more mobile hydrogenase 1 corresponded to a major protein-
staining arc, in contrast to that for hydrogenase 2, which was barely visible.
Hydrogenase 2 is therefore a minor protein component of the membrane.

Electrophoresis of Triton X-100-dispersed membranes on neutrally buffered
(Hames, 1981) non-denaturing polyacrylamide gels revealed three activity-staining bands. Whether the activity bands on the gel represent distinct hydrogenases
or a single species in aggregated states was investigated immunologically. The
bands were excised, the activity eluted and analysed by crossed-immunoelectrophoresis. This allowed the gel activities to be related to the immu-
noprecipitin arcs. The two slowest migrating bands were found to correspond to
arc 2 (hydrogenase 2) and the other band to arc 1 (hydrogenase 1). Thus,
crossed-immunoelectrophoresis and polyacrylamide gel electrophoresis both
demonstrate that there are two hydrogenase isoenzymes. Both isoenzymes are
membrane-bound rather than membrane-associated, since neither could be released from the membrane by washing with high-salt or EDTA-containing
buffers.

The supernatant (soluble) cell fraction was also analysed by crossed-immunoelectrophoresis, using anti-membrane antisera. The same two activity-staining
arcs were found on the immunoplates. Similarly, analysis of the supernatant
fraction by native polyacrylamide gel electrophoresis did not reveal any activity
bands other than those found for the membrane fraction. It appears, therefore,

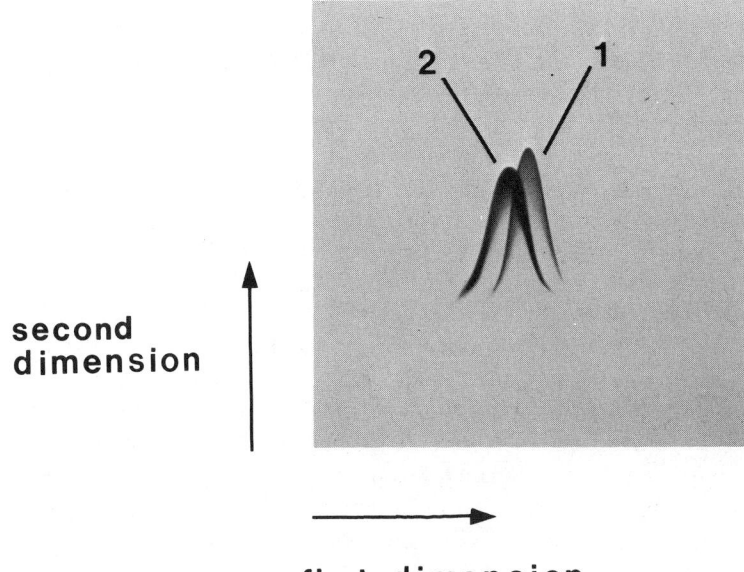

**second dimension**

**first dimension**

**Fig. 1.** Membrane-bound hydrogenases analysed by crossed-immunoelectrophoresis. A Triton X-100-dispersed membrane fraction (20–40 μg protein) was electrophoresed on immunoplates as described in Graham *et al.* (1980). Plates were stained for hydrogenase activity by immersion in 100 m$M$ PO$_4^{3-}$ pH 7.0, 0.5 m$M$ benzyl viologen and 1 m$M$ 2,3,5-triphenyltetrazolium chloride and incubation in a gas jar for 4–5 hr under an H$_2$-containing atmosphere. The activity-stained precipitin arcs marked 1 and 2 correspond to the hydrogenase isoenzymes 1 and 2 described in the text. Polyspecific antiserum raised in rabbits, with membrane vesicles as the immunogen, was employed.

that the majority of the hydrogenase activity found in the supernatant fraction arises from membrane contamination. We have found no evidence to suggest the presence of a distinct soluble hydrogenase.

Although three activity-staining bands were found by native polyacrylamide gel electrophoresis at pH 7.5 a single activity-stained band appeared when a higher pH system (Ornstein and Davies, 1963) was used. Crossed-immunoelectrophoresis of material eluted from this band identified it as hydrogenase 1. The loss of the other activity-stained bands following gel electrophoresis on the higher pH system suggested that hydrogenase 2 may be alkali-sensitive. The effect of pH variation on hydrogenase activity was investigated by incubating samples of Triton X-100-dispersed membranes for 4 hr at room temperature over the pH range 7 to 10. A three-component buffer system consisting of $N$-(2-acetamido)-2-aminoethanesulfonic acid, Tris and ethanolamine was used for this purpose (Ellis and Morrison, 1982). Most (75%) of the hydrogenase activity present at pH 7 was lost between pH 8 and pH 10. Samples incubated as described above prior to polyacrylamide gel analysis with the neutral gel system were found to have lost, with increasing pH, the activity in the two slower-migrating activity-stained bands (hydrogenase 2 activity). There was no apparent loss of activity in the band due to hydrogenase 1.

The subunit composition of hydrogenase 1 was investigated. It was isolated from $^{35}$S-labelled cultures. This was effected by electrophoresis of the Triton X-100-dispersed membrane fraction on native polyacrylamide gels and elution of the material corresponding to the band stained for hydrogenase activity. The excised material was subjected to line immunoelectrophoresis in order to remove the remaining contaminants. The activity-stained bands cut from these immuno-plates were then analysed by sodium dodecyl sulphate (SDS)–polyacrylamide gel electrophoresis (Norrild et al., 1977). Subsequent autoradiography of the gel revealed a single radioactive band of approximately $M_r = 64,000$. The subunit $M_r$ of hydrogenase 1 and its comparatively high abundance indicate that hydro-genase 1 is the anaerobic hydrogenase previously described by our laboratory (Graham et al., 1980), and for which a transmembranous location was subse-quently established (Graham, 1981). The molecular properties of hydrogenase 2 are presently under study.

These results demonstrate that there are two hydrogenases present in E. coli when grown anaerobically in the absence of exogenous electron acceptors. How-ever, both enzymes are not present under all growth conditions. Analysis of cells grown anaerobically in the presence of fumarate or nitrate as terminal electron acceptors or aerobically revealed that predominantly only hydrogenase 2 is pre-sent. Since hydrogenase 1 is present only in the absence of electron acceptors (other than the proton), conditions under which the formate hydrogenlyase path-way is active, we speculate that this hydrogenase functions in this pathway with an $H_2$-evolving role. Further support for this proposal comes from the observa-tion that growth in the presence of 0.2% formate leads to an increase in the amount of hydrogenase 1 present.

In summary, we find that there are two membrane-bound hydrogenases in E. coli. The presence of one of the hydrogenases appears to correlate with the expression of the formate hydrogenlyase pathway. This enzyme is a major mem-brane component (during growth without exogenous electron acceptors) and consists of a single polypeptide of $M_r$ about 64,000.

## Acknowledgments

The work described is supported by a research grant (GR/B/7322 4) from the Science and Engineering Research Council. D. H. B. was a Nuffield Foundation Science Research Fellow.

# References

Adams, M. W. and Hall, D. O. (1979). Purification of the membrane-bound hydrogenases of *Escherichia coli*. *Biochemical Journal* **183**, 11–22.

Bernhard, T. H. and Gottschalk, G. (1978). Cell yields of *Escherichia coli* during anaerobic growth on fumarate and molecular $H_2$. *Archives of Microbiology* **116**, 235–238.

Ellis, K. J. and Morrison, J. F. (1982). Buffers of constant ionic strength for studying pH-dependent processes. *In* "Methods in Enzymology" Vol. 87 (Ed. D. L. Purich), pp. 405–426. Academic Press, New York.

Graham, A. (1981). The organization of hydrogenase in the cytoplasmic membrane of *Escherichia coli*. *Biochemical Journal* **197**, 283–291.

Graham, A., Boxer, D. H., Haddock, B. A., Mandrand-Berthelot, M. -A. and Jones, R. W. (1980). Immunochemical analysis of the membrane-bound hydrogenase of *Escherichia coli*. *FEBS Letters* **113**, 167–172.

Gray, C. T. and Gest, H. (1965). Biological formation of molecular hydrogen. *Science* **148**, 186–192.

Hames, B. D. (1981). *In* "Gel Electrophoresis of Proteins: A Practical Approach" (Eds. B. D. Hames and D. Rickwood), p. 30. IRL Press Ltd., Oxford.

Jones, R. W. (1980). The role of the membrane-bound hydrogenase in the energy-conserving oxidation of molecular hydrogen by *Escherichia coli*. *Biochemical Journal* **188**, 345–350.

Macy, J. Kulla, H. and Gottschalk, G. (1976). $H_2$-dependent anaerobic growth of *Escherichia coli* on L-malate: Succinate formation. *Journal of Bacteriology* **125**, 423–428.

Norrild, B., Bjerrum, O. J. and Vestergaard, B. F. (1977). Polypeptide analysis of individual immunoprecipitates from crossed immunoelectrophoresis. *Analytical Biochemistry* **81**, 432–441.

Ornstein, L. and Davies, B. J. (1963). "Disc Gel Electrophoresis". Distillation Products Industries, Rochester, New York.

Yamamoto, I. and Ishimoto, M. (1978). Hydrogen-dependent growth of *Escherichia coli* in anaerobic respiration and the presence of hydrogenases with different functions. *Journal of Biochemistry (Tokyo)* **84**, 673–679.

# 6

# Genetic Analysis of Hydrogenases in *Escherichia coli*

ROBERT WAUGH,* MARIE-ANDREÉ MANDRAND-BERTHELOT† AND DAVID H. BOXER*

*Department of Biochemistry, Medical Sciences Institute, Dundee University, Dundee, Tayside, United Kingdom and †Laboratoire de Microbiologie, INSA, Villeurbanne, France*

Two distinct hydrogenase isoenzymes have recently been identified in *Escherichia coli* which are present when the organism is grown anaerobically in the absence of exogenous electron acceptors (Ballantine and Boxer, Chapter 5, this volume). These isoenzymes are immunologically and electrophoretically distinct. Mutant strains of *E. coli* deficient in hydrogenase activity have been isolated and the lesions mapped close to minute 59 on the chromosome (Pascal *et al.*, 1975; Graham *et al.*, 1980). Hyd⁻ mutants in *Salmonella* have also been isolated and shown to map in the corresponding region of the chromosome (Chippaux *et al.*, 1972). Previous evidence pointed to the presence of a single hydrogenase in *E. coli* (Adams and Hall, 1979; Graham *et al.*, 1980) and it was suggested that the structural gene for hydrogenase was located at 59 min. We have re-examined mutants at this locus to determine whether one or both hydrogenase isoenzymes are deficient.

Mutant FD12 (Graham *et al.*, 1980) was grown anaerobically in the absence of an electron acceptor and the membrane fraction prepared as described previously (Graham *et al.*, 1980). Triton X-100-dispersed membrane preparations were analysed by crossed-immunoelectrophoresis, employing antiserum prepared with membrane vesicles as the immunogen, and non-denaturing polyacrylamide gel electrophoresis at neutral pH. The hydrogenase isoenzymes were localised by staining for activity (Ballantine and Boxer, Chapter 5, this volume). Both isoenzymes were virtually absent under all conditions studied. However, occasionally when very large amounts of the membranes were analysed, small weakly stained regions on the immunoplates were found in positions that corresponded, in the free electrophoretic dimension of the immunoplate, to the expected positions of the isoenzymes. The loss of two distinct isoenzymes by a single point

MICROBIAL GAS METABOLISM:
MECHANISTIC, METABOLIC
AND BIOTECHNOLOGICAL ASPECTS

mutation is inconsistent with that mutation being present in the structural gene for either isoenzyme.

An RP4 : : Mu cointegrate plasmid, carrying the region of the *E. coli* chromosome flanked by *cys*C and *srl*, complemented the Hyd⁻ mutants confirming the original mapping of the mutation (*srl-hyd-cys*C). Both isoenzymes were restored to the wild-type levels when this plasmid was introduced by conjugation into *rec*A derivatives of the mutant strains. The cointegrate plasmid was a gift from Professor K. Stacey (University of Kent, U.K.). A 4.8-kb, pst1 fragment was subcloned from the RP4 cointegrate plasmid into the multicopy plasmid, pBR325. The resulting plasmid, pRW1, restored both isoenzymes to the wild-type levels when introduced into the mutant strains by transformation. No overproduction of either of the hydrogenase isoenzymes was observed in such strains, which would contain 20 or more functional copies of the gene. This is in further agreement with the conclusion that the mutated gene is structural for neither isoenzyme.

The above mutants were obtained by screening for colonies which are unable to catalyse the formate-dependent reduction of oxidised benzyl viologen, employing a dye-overlay technique (Graham *et al.*, 1980). Mutants defective in formate dehydrogenase (formate:benzyl viologen oxidoreductase) also possess this phenotype and must be identified in the subsequent isolation of Hyd⁻ strains. However, all the Hyd⁻ mutants isolated by this procedure have also lost formate:benzyl viologen oxidoreductase activity. This has been interpreted by assuming that the formate dehydrogenase donates electrons to oxidised benzyl viologen via hydrogenase (Chippaux *et al.*, 1977), but other explanations are possible.

We devised a new procedure for the isolation of strains defective in hydrogenase. We noted that formate dehydrogenase mutants, *fdh*A, *fdh*B, *fdh*C etc., were indistinguishable from wild-type strains when grown anaerobically on rich plates in the presence of 1 m*M* benzyl viologen (Tait *et al.*, 1981). Growth of wild-type and Fdh⁻ strains is much reduced, apparently due to the toxicity of reduced benzyl viologen, and the colonies are blue in colour. Mutants of the FD12 type grew normally as white colonies. This indicates that there is present in the organism a formate dehydrogenase-independent pathway for the reduction of benzyl viologen, which probably involves hydrogenase. Spontaneous mutants of an *fdh*B deletion strain (BW9101; Dr. B. Weiss) appearing as white colonies or papillae on such plates, were isolated. These mutants were assayed spectrophotometrically for hydrogenase activity. Many of the mutants that lacked hydrogenase activity mapped near minute 59 on the chromosome and were complemented by pRW1. These were, therefore, of the same class previously identified. Other mutants possessed hydrogenase activity. These were analysed by crossed-immunoelectrophoresis and activity staining for hydrogenase isoenzymes as described above. Some of these mutants possessed both isoenzymes, but a small proportion lacked activity specifically in one or other of the hydrogenase isoenzymes. These classes of mutants give genetic support for the existence of two hydrogenase isoenzymes and are good candidates for possessing lesions in hydrogenase structural genes. The genetic and biochemical characterisation of these mutants is now under study.

## Acknowledgments

This work is supported by a research grant (GR/B/73224) from the Science and Engineering Research Council. D.H.B. was a Nuffield Foundation Science Research Fellow.

## References

Adams, M. W. and Hall, D. O. (1979). Purification of the membrane-bound hydrogenase of *Escherichia coli*. *Biochemical Journal* **183**, 11–22.

Chippaux, M., Casse, F. and Pascal, M. -C. (1972). Isolation and phenotype of mutants from *S. typhimurium* defective in formate hydrogenylase activity. *Journal of Bacteriology* **110**, 766–768.

Chippaux, M., Pascal, M. -C. and Casse, F. (1977). Formate hydrogenylase system in *Salmonella typhimurium* LT2. *European Journal of Biochemistry* **72**, 149–155.

Graham, A., Boxer, D. H., Haddock, B. A., Mandrand-Berthelot. M. -A. and Jones, R. W. (1980). Immunochemical analysis of the membrane-bound hydrogenase of *Escherichia coli*. *FEBS Letters* **113**, 167–172.

Pascal, M. -C., Casse, F., Chippaux, M. and Lepelletier, M. (1975). Genetic analysis of mutants of *Escherichia coli* K12 and *Salmonella typhimurium* LT2 deficient in hydrogenase activity. *Molecular and General Genetics* **141**, 173–179.

Tait, R. C., Anderson, K., Cangelosi, G. and Shanmugam, K. T. (1981). Hydrogenase genes. *In* "Trends in the Biology of Fermentations for Fuels and Chemicals" (Eds. A. Hollaender *et al.*), pp. 279–300. Plenum Press, New York.

# Part IV
# Carbon Monoxide

# 7

# Properties and Function of Carbon Monoxide Dehydrogenase from Anaerobic Bacteria

GABRIELE DIEKERT,* GEORG FUCHS† AND RUDOLF K. THAUER*

*Mikrobiologie, Fachbereich Biologie, Philipps-Universität Marburg, Marburg, Federal Republic of Germany and †Universität Ulm, Abteilung Angewandte Mikrobiologie, Ulm, Federal Republic of Germany

## Introduction

In 1958 Yagi discovered that cell extracts of *Desulfovibrio desulfuricans* catalyzed the anaerobic oxidation of carbon monoxide to $CO_2$ with methyl viologen (MV) as electron acceptor (Yagi, 1958, 1959).

$$CO + H_2O + 2MV \rightleftarrows CO_2 + 2H^+ + 2MV^-$$

Four years later he provided evidence for the reversibility of the reaction (Yagi and Tamiya, 1962). A carbon monoxide:methyl viologen oxidoreductase activity (from here on designated as anaerobic carbon monoxide dehydrogenase) was subsequently also found in many other anaerobic bacteria: *Clostridium pasteurianum* (Thauer *et al.*, 1974; Fuchs *et al.*, 1974), methanogenic bacteria (Daniels *et al.*, 1977; Kluyver and Schnellen, 1947), acetogenic bacteria (Diekert and Thauer, 1978; Diekert and Ritter, 1982; Clark *et al.*, 1982; Lynd *et al.*, 1982; Sharak Genthner and Bryant, 1982) and *Rhodopseudomonas gelatinosa*, anaerobically grown on CO in the dark (Uffen, 1983; Wakim and Uffen, 1983). But not all strict anaerobes contain a carbon monoxide dehydrogenase. For example *Clostridium butyricum*, *Clostridium kluyveri*, *Clostridium acidiurici*, and *Clostridium cylindrosporum* have been shown to be devoid of this activity (Diekert and Thauer, 1978; Andress, 1976).

The bacteria that contain an anaerobic carbon monoxide dehydrogenase do not form a distinct phylogenetic group. Amongst them are archaebacteria and both gram-negative and -positive eubacteria. In spite of this fact, the properties of the enzyme appear to be very similar, suggesting a similar function.

MICROBIAL GAS METABOLISM:
MECHANISTIC, METABOLIC
AND BIOTECHNOLOGICAL ASPECTS

Many aerobic bacteria can grow on CO and $O_2$ as sole carbon and energy source (Hegemann, 1980; Uffen, 1981; Meyer and Schlegel, 1983). The organisms contain a membrane-associated carbon monoxide dehydrogenase which catalyzes the irreversible oxidation of carbon monoxide to $CO_2$ with methylene blue as artificial electron acceptor (methyl viologen is not reduced) (Meyer and Schlegel, 1980). The aerobic enzyme contains molybdenum and is stable in the presence of molecular oxygen. For a review on the properties and function of carbon monoxide dehydrogenase from aerobic bacteria the reader is referred to Meyer (Chapter 8, this volume).

In the following, the kinetic and molecular properties and the function of the carbon monoxide dehydrogenase found in anaerobic bacteria are discussed. It is shown that the anaerobic enzyme, which is rapidly inactivated by $O_2$, is a nickel protein (at least in those enzymes studied) and that the function in some of the organisms is probably to reduce $CO_2$ to CO rather than to oxidize CO to $CO_2$.

## Kinetic Properties

Most of the properties discussed in this section were studied with cell extracts of the respective organisms rather than with the purified enzyme. Some of the properties are deduced from studies of only one or two organisms. Generalization of the results may turn out not to be justifiable.

The carbon monoxide dehydrogenases catalyze the reduction of methyl viologen with carbon monoxide linearly with time in the range measurable photometrically. The dependence of the rate on substrate concentration follows simple Michaelis/Menten kinetics: plots of $1/v$ versus $1/S$ are linear. Apparent $K_m$ values for CO and for MV obtained from such reciprocal plots as well as the specific activity of the enzyme in crude extracts of various anaerobic bacteria are summarized in Table 1.

The rate of methyl viologen reduction with CO increases with increasing pH. A pH optimum is not observed. At pH 9 the rate is approximately three times higher than at pH 7 (Thauer *et al.*, 1974; Diekert and Thauer, 1978).

The catalytic mechanism of carbon monoxide oxidation with MV was found to be "Ping-Pong", indicating that the carbon monoxide dehydrogenase (E) can be present in an oxidized and a reduced form (Diekert and Thauer, 1978).

$$E_{ox} + CO + H_2O \rightleftharpoons E_{red}^{2-} + CO_2 + 2\ H^+$$

$$E_{red}^{2-} + 2\ MV \rightleftharpoons E_{ox} + 2\ MV^-$$

The enzyme also catalyzes the reduction of triquat ($E^{0'} = -548$ mV) with CO ($CO_2/CO$: $E_0' = -560$ mV), suggesting that the electron-accepting group of the enzyme has a very negative redox potential.

Incubation of the enzyme with cyanide results in a time-dependent loss of

**Table 1.**  *Carbon monoxide:methyl viologen oxidoreductase activity in cell extracts of anaerobic bacteria*[a]

| Organism (with growth substrate) | Specific activity [U (mg protein)$^{-1}$] | $[CO]_{0.5v}$ ($\mu M$) | $[MV]_{0.5v}$ (m$M$) | Reference |
|---|---|---|---|---|
| *Clostridium pasteurianum* (glucose) | 0.010 | 5 | 0.4 | Thauer *et al.* (1974) |
| *Clostridium thermoaceticum* (glucose) | 7 | 150 | 1.3 | Diekert and Thauer (1978) |
| *Clostridium formicoaceticum* (fructose) | 15 | 2.2 | 3.0 | Diekert and Thauer (1978) |
| *Clostridium thermoautotrophicum* (H$_2$ and CO$_2$) | 10 | n.d.[b] | n.d. | Clark *et al.* (1982) |
| *Acetobacterium woodii* (H$_2$ and CO$_2$) | 10 | 15 | 1.0 | Diekert (1980) |
| *Butyribacterium methylotrophicum* (methanol plus acetate) | 47 | n.d. | n.d. | Lynd *et al.* (1982) |
| *Methanobacterium thermoautotrophicum* (H$_2$ and CO$_2$) | 0.4 | >1000 | 0.02 | Daniels *et al.* (1977) |
| *Methanobrevibacter arboriphilicus* (H$_2$ and CO$_2$) | 0.1 | n.d. | n.d. | Hammel *et al.* (1984) |
| *Methanosarcina barkeri* (acetate) | 5 | n.d. | n.d. | Krzycki *et al.* (1982) |
| *Desulfovibrio baarsii* (formate plus sulphate) | 2.5 | 18 | 2 | Jansen *et al.* (1984) |
| *Rhodopseudomonas gelatinosa* (anaerobically on CO in the dark) | 3.5 | 12.5 | 0.1 | Uffen (1983) |

[a]1 unit = 1 $\mu$mol CO oxidized min$^{-1}$ (mg protein)$^{-1}$.
[b]n.d., not determined.

activity. The rate and extent of inactivation are decreased in the presence of carbon monoxide. Cyanide-inactivated enzyme is reactivated by incubation with CO. These findings are interpreted as indicating that the oxidized form ($E_{ox}$) reversibly reacts with cyanide (Thauer *et al.*, 1974; Diekert and Thauer, 1978; Daniels *et al.*, 1977; Drake *et al.*, 1980; Ragsdale *et al.*, 1983a,b).

$$E_{ox} + CN^- \xrightleftharpoons{slow} E\text{-}CN^- \text{ (inactive)}$$

Incubation of extracts of *Clostridium pasteurianum, Clostridium thermoaceticum* or *Clostridium formicoaceticum* with methyl iodide results in loss of CO oxidation activity. The rate of inactivation is increased in the presence of CO. Extracts inactivated by the alkyl halide can be reactivated by photolysis (Thauer *et al.,* 1974; Diekert and Thauer, 1978). The latter finding could, until now, not be reproduced with purified enzyme (Drake *et al.,* 1980; Ragsdale *et al.,* 1983a,b), making interpretation of the results difficult.


## Molecular Properties


The carbon monoxide dehydrogenase activity in anaerobic bacteria is mainly associated with the soluble cell fraction except for the enzyme of *R. gelatinosa,* which appears to be membrane-bound (Wakim and Uffen, 1983). It is difficult to purify since the enzyme is very labile and is rapidly inactivated by molecular oxygen. Purification to apparent homogeneity has presently been achieved only for the enzymes of *C. thermoaceticum* (Ragsdale *et al.,* 1983a; Diekert and Ritter, 1983a), *Acetobacterium woodii* (Ragsdale *et al.,* 1983b) and *Methanosarcina barkeri* (Krzycki and Zeikus, 1984). Partial purification of the enzyme from *Methanobrevibacter arboriphilicus* has been described (Hammel *et al.,* 1984).

The first evidence that the carbon monoxide dehydrogenase from anaerobic bacteria contains nickel was provided by the finding that the synthesis of active carbon monoxide dehydrogenase in *C. pasteurianum* (Diekert *et al.,* 1979a) and in acetogenic bacteria (Diekert and Thauer, 1980; Diekert and Ritter, 1982) is dependent on the presence of nickel in the growth medium and that the uptake of nickel by the growing cells and the synthesis of carbon monoxide dehydrogenase proceed in parallel. Drake *et al.* (1980) reported that the carbon monoxide dehydrogenase of *C. thermoaceticum* copurifies with nickel. The enzyme was purified 14-fold and, based on the specific activity, was less than 10% pure (see also Drake, 1982). Only 2 years later purification to apparent homogeneity was described for the enzyme from *C. thermoaceticum* by Ragsdale *et al.* (1983a) and by Diekert and Ritter (1983a). The purified enzyme contains 2 mol nickel per mol enzyme with an $\alpha\beta$ structure. Also, 11 iron, 14 acid-labile sulphur, and 1 zinc were found (Ragsdale *et al.,* 1983a,b). Electron paramagnetic resonance (EPR) studies indicate that nickel is involved in the catalytic mechanism of CO oxidation (Ragsdale *et al.,* 1982, 1983c).

From the finding that anaerobic carbon monoxide dehydrogenase in the ox-idized form reversibly reacts with cyanide and in the reduced form reversibly reacts with alkyl halides it was deduced that the prosthetic group of the enzyme could be a corrinoid (Diekert *et al.*, 1979b). A nickel porphinoid was recently found to be the prosthetic group of methyl–CoM reductase in methanogenic bacteria (Diekert *et al.*, 1980; Jaenchen *et al.*, 1981; Pfaltz *et al.*, 1982; Ellefson *et al.*, 1982; Livingston *et al.*, 1984). Attempts to demonstrate the presence of a tetrapyrrole in anaerobic carbon monoxide dehydrogenase failed until recently. When cells of *C. thermoaceticum* were grown in the presence of $^{14}$C-labelled δ-aminolevulinic acid (δ-ALA) and the enzyme was isolated, the purified protein contained very little radioactivity. δ-ALA was taken up by the cells, as was indicated by the finding that vitamin $B_{12}$ became labelled. From the specific radioactivity of the corrinoid and the dpm/mg enzyme protein it was calculated that the purified enzyme (specific activity 350 units/mg and nickel content 10 nmol/mg protein) contained at most 0.2 mol tetrapyrrole per mol nickel (G. Diekert and M. Hansch, unpublished results).

## Physiological Electron Acceptor

The physiological electron acceptor of carbon monoxide dehydrogenase in anaerobic bacteria is not definitively known. In cell extracts of *Methanobacterium thermoautotrophicum* and of *Methanobrevibacter arboriphilicus*, coenzyme $F_{420}$ is reduced with carbon monoxide (Daniels *et al.*, 1977). The ability of the enzyme to couple with $F_{420}$ is, however, gradually lost during purification of the carbon monoxide:methyl viologen oxidoreductase activity (unpublished results). The purified carbon monoxide dehydrogenase of *C. thermoaceticum* and of *A. woodii* mediates the reduction of rubredoxin (Ragsdale *et al.*, 1983a,b). Cell extracts of *C. pasteurianum* mediate the reduction of FMN and FAD with carbon monoxide. Free flavins are, however, not considered to be physiological electron acceptors (Thauer *et al.*, 1974).

## Function

*Rhodopseudomonas gelatinosa* (Hirsch, 1968; Uffen, 1976), *Eubacterium limosum* (Sharak Genthner and Bryant, 1982), *Butyribacterium methylotrophicum* (Kerby *et al.*, 1983), *C. thermoaceticum* (Kerby and Zeikus, 1983) and *M. thermoautotrophicum* (Daniels *et al.*, 1977) were shown to grow on CO as electron donor. Growth, however, is generally slow and, in all cases documented, considerably slower than on other substrates. The physiological function of carbon monoxide dehydrogenase in most anaerobes is thus probably not to enable growth on carbon monoxide. Only in the case of *R. gelatinosa* does the enzyme appear to be induced when the bacterium is grown on CO (Uffen, 1983).

   Evidence has recently accumulated that CO could an intermediate rather than a substrate in the metabolism of chemotrophic anaerobes (Thauer *et al.*, 1983). In the following the physiological function of carbon monoxide dehydrogenase in acetogenic bacteria, in autotrophic methanogens, in acetoclastic methanogens and in mini-methane producers is discussed.

*Acetogenic Bacteria*

Acetogenic bacteria synthesize acetate from $2CO_2$ and couple this reaction with the synthesis of ATP.

$$2CO_2 + 8[H] \rightarrow CH_3COOH + 2H_2O$$
$$ATP$$

Two pathways for the total synthesis of acetate from $CO_2$ exist (Fig. 1). In *C. thermoaceticum*, *C. formicoaceticum* and *A. woodii* the methyl tetrahydrofolate pathway is operative (Wood *et al.*, 1982; Ljungdahl and Wood, 1982); in *C. acidiurici* and in *C. cylindrosporum* growing on hypoxanthine the serine pathway is functional (Vogels and van der Drift, 1976; Dürre and Andreesen, 1983). The two pathways are identical in the reactions leading from $CO_2$ to methylene tetrahydrofolate.

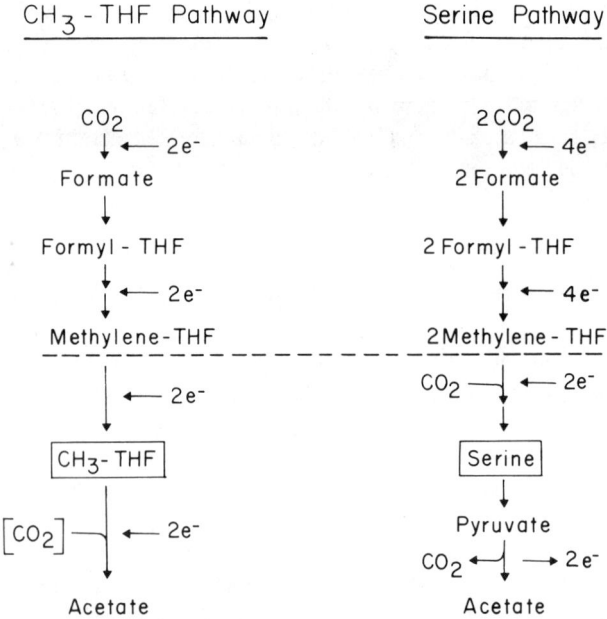

**Fig. 1.** Total synthesis of acetate from $CO_2$ via the $CH_3$-tetrahydrofolate and serine pathways. The scheme is simplified. For details see the review by Vogels and van der Drift (1977b), Dürre and Andreesen (1983) and Ljungdahl and Wood (1982).

A clue to understanding the function of carbon monoxide dehydrogenase was the finding that only those acetogenic bacteria with the methyl tetrahydrofolate ($CH_3$-THF) pathway contain carbon monoxide dehydrogenase activity and that carbon monoxide oxidation by cell suspensions of these organisms is stimulated by pyruvate (Diekert and Thauer, 1978). Since acetate formation from methyl tetrahydrofolate is pyruvate-dependent (Schulman et al., 1973), it was concluded that carbon monoxide dehydrogenase is somehow involved in the catalysis of acetate formation from $CH_3$-THF and pyruvate. The multi-enzyme system mediating the reaction was subsequently shown to contain the carbon monoxide dehydrogenase (Drake et al., 1980, 1981). In 1982, Hu et al. published evidence that this multi-enzyme system catalyzes the formation of acetyl–CoA from $CH_3$-THF, CoA and CO at a rate of 20 nmol min$^{-1}$ (mg protein)$^{-1}$. Carbon monoxide is specifically incorporated into the carboxyl group of acetyl–CoA. An isotopic exchange between C1 of acetyl–CoA and CO was also observed.

$$CH_3\text{-THF} + CoA + CO \rightleftharpoons CH_3CO\text{-CoA} + THF$$

Diekert and Ritter (1983b) then showed that growing cultures of acetogenic bacteria specifically incorporate $^{14}CO$ into the carboxyl group of acetate. Similar results were obtained by Kerby et al. (1983). These findings indicate that acetate is formed in a carbonylation reaction analogous to the chemical synthesis of acetic acid from methanol and carbon monoxide.

$$CH_3OH + CO \rightarrow CH_3COO^- + H^+ \qquad \Delta G^{0\prime} = -97 \text{ kJ/mol}$$

Cobalt and rhodium complexes are required as catalysts (Mullen, 1980). The rhodium-catalyzed reaction is commonly known as the Monsanto process. In the catalysis the methyl–metal complex reacts with carbon monoxide in an insertion reaction leading to the formation of an acetyl–metal compound (Fig. 2). Hydrolysis of the latter leads to the irreversible formation of acetic acid, the carboxyl group of which is derived from CO. One can easily envisage that a thiolytic cleavage of an acetyl–metal enzyme could lead to the reversible formation of acetyl–CoA via the same mechanism.

Methyl-$B_{12}$ is converted to acetic acid by cell extracts of *C. thermoaceticum*. Trideuteromethyl-$B_{12}$ leads to the formation of trideuteroacetic acid (Parker *et*

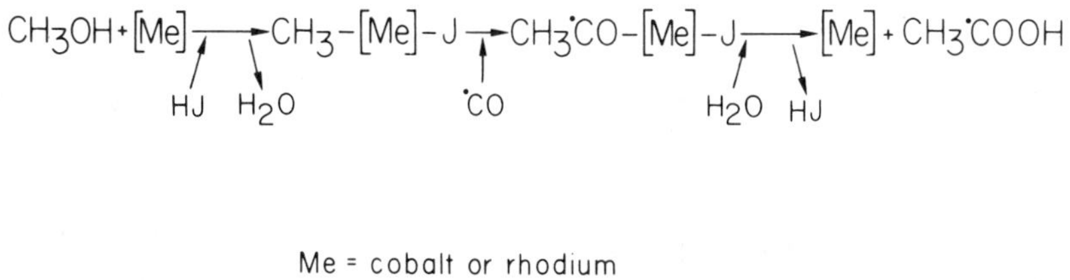

$$CH_3OH + [Me] \longrightarrow CH_3 - [Me] - J \longrightarrow CH_3CO - [Me] - J \longrightarrow [Me] + CH_3COOH$$

Me = cobalt or rhodium

**Fig. 2.** Acetic acid formation from methanol and CO. The scheme is simplified. For details see Mullen (1980).

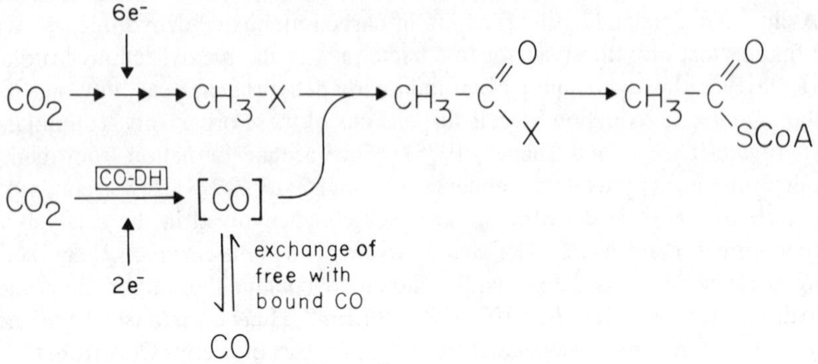

**Fig. 3.** Proposed mechanism of acetyl–CoA formation from $2CO_2$ in acetogenic bacteria, in autotrophic methanogens and in autotrophic sulphate-reducing bacteria.

*al.*, 1972). This finding is consistent with the proposed carbonylation reaction, which predicts that the deuterium in the methyl group should be retained.

Diekert *et al.* (1984) recently showed that growing cells of *C. thermoaceticum* and of *A. woodii* produce CO from $CO_2$. They found that CO formation from $CO_2$ rather than incorporation of CO into acetate was inhibited by cyanide, which inactivates carbon monoxide dehydrogenase. From the results they concluded that carbon monoxide dehydrogenase is not directly involved in the carbonylation reaction. The findings are consistent with the involvement of carbon monoxide dehydrogenase in $CO_2$ reduction to CO as depicted in Fig. 3 (see also Pezacka and Wood, 1984).

The scheme in Fig. 3 is still speculative. It does not explain why carbon monoxide dehydrogenase appears to be required as one of four proteins for the synthesis of acetyl-CoA from $CH_3$-THF, CO and CoA (Hu *et al.*, 1982). It also does not readily explain why—during growth in the presence of $^{14}CO$—only a few percent of the acetate formed is synthesized from $CO_2$ and CO and the rest from $2CO_2$ (Diekert and Ritter, 1983b).

It is of historical interest that Fischer *et al.* (1932) demonstrated that acetate is formed from CO by anaerobic sewage sludge. One of the authors (F. Fischer), together with H. Tropsch, developed the procedure for the synthesis of gasoline from CO and $H_2$. Carbonylation reactions are involved in the formation of the alkanes (Masters, 1979).

*Autotrophic Methanogens*

Most methanogenic bacteria can grow on $H_2$ and $CO_2$ as sole energy source (Balch *et al.*, 1979).

$$4H_2 + CO_2 \rightarrow CH_4 + 2H_2O$$
$$\curvearrowright ATP$$

These archaebacteria contain a carbon monoxide dehydrogenase, which can be inactivated by cyanide (Daniels et al., 1977). The specific activity in methanogens is, however, much lower than in acetogenic bacteria growing on $H_2$ and $CO_2$ (Table 1), indicating an anabolic rather than a catabolic function. This was substantiated by the observation that methane formation from $CO_2$ and $H_2$ is not affected by cyanide (Stupperich and Fuchs, 1983).

When growing on $H_2$ and $CO_2$ as energy source, the methanogens generally use $CO_2$ as the sole or main carbon source. The bacteria thus grow autotrophically. Exceptions are *Methanobrevibacter ruminantium*, *Methanobrevibacter smithii* (Balch et al., 1979) and *Methanospirillum hungatei* (strain GP1) (Patel et al., 1976), which require acetate and $CO_2$ for cell carbon synthesis.

The autotrophic methanogens do not assimilate $CO_2$ via the Calvin cycle (for a review, see Fuchs and Stupperich, 1982). In a series of labelling experiments backed up by enzymatic studies it was shown that acetyl–CoA (formed from $2CO_2$ in a non-cyclic process) is an early intermediate in $CO_2$ fixation. From acetyl–CoA pyruvate is formed by reductive carboxylation. Until recently the mechanism of acetyl–CoA formation from $2CO_2$ remained obscure.

A breakthrough in understanding was the finding that [14]CO is specifically incorporated into C2 of pyruvate (which is derived from the carboxyl group of acetyl–CoA) when *M. thermoautotrophicum* grows on $H_2$, $CO_2$ and [14]CO (Stupperich et al., 1983). It had been shown earlier that methanol is specifically incorporated into the methyl group of pyruvate (which is derived from the methyl group of acetyl–CoA) in methanol-growing *M. barkeri* (Kenealy and Zeikus, 1982). These observations indicate that—as in acetogens—a carbonylation reaction is involved in acetyl–CoA synthesis from $2CO_2$ and that the function of carbon monoxide dehydrogenase in autotrophic methanogens could be to provide the CO required.

The hypothesis (Fig. 3) is supported by the following findings. (1) CO is formed from $CO_2$ and $H_2$ by growing cells and by cell suspensions of *M. thermoautotrophicum* (Conrad and Thauer, 1983). CO formation is inhibited by cyanide (Eikmanns et al., 1984). (2) Cyanide affects acetyl–CoA formation from $2CO_2$ but not the formation of the thioester from $CO_2$ and CO (Stupperich and Fuchs, 1983). In the presence of cyanide the carboxyl group of acetyl–CoA is exclusively derived from added CO (Stupperich and Fuchs, 1984a, b).

Some sulphate-reducing bacteria can grow autotrophically (Pfennig and Widdel, 1981). These anaerobic eubacteria also contain a carbon monoxide dehydrogenase, assimilate $CO_2$ via acetyl–CoA and incorporate [14]CO specifically into its carboxyl group (Jansen et al., 1984). The mechanism of acetyl–CoA formation from $2CO_2$ via CO (in a free or bound form as shown in Fig. 3) thus appears to be operative in acetogenic bacteria, autotrophic methanogens and autotrophic sulphate-reducing bacteria. For details the reader is referred to a recent review on autotrophic $CO_2$ fixation by chemotrophic anaerobes (Fuchs and Stupperich, 1982).

*Acetoclastic Methanogens*

*Methanosarcina barkeri* (Balch *et al.*, 1979; Smith and Mah, 1980; Mah *et al.*, 1981), *Methanothrix soehngenii* (Huser *et al.*, 1982) and *Methanococcus mazei* (Mah, 1980; Touzel and Albagnac, 1983; Fathepure, 1983) ferment acetate to $CH_4$ and $CO_2$ and couple this reaction with the synthesis of ATP:

$$CH_3\text{-}COO^- + H^+ \rightarrow CH_4 + CO_2$$
$$\curvearrowright ATP$$

Trideuteroacetate yields trideuteromethane, excluding the possibility that acetate is first oxidized to $CO_2$ and then $CO_2$ reduced to methane (Pine and Barker, 1956; Blaut and Gottschalk, 1982).

*Methanosarcina barkeri* can also grow on $H_2$ and $CO_2$. The specific activity of carbon monoxide dehydrogenase in acetate-grown cells is five times higher than in cells grown on $H_2$ plus $CO_2$ (Krzycki *et al.*, 1982). This finding indicates that carbon monoxide dehydrogenase is somehow involved in the acetoclastic reaction. Acetate-grown cells also contain an active acetate kinase and phosphotransacetylase (Kenealy and Zeikus, 1982), indicating that acetyl–CoA is an intermediate. Hu *et al.* (1982) showed that in acetogenic bacteria the formation of acetyl–CoA from methyl tetrahydrofolate and CO is principally a reversible reaction. Based on these observations, the mechanism depicted in Fig. 4 has been proposed. It involves a decarbonylation reaction leading to the formation of CO and methyl-X. For thermodynamic reasons the CO must be tightly bound to a

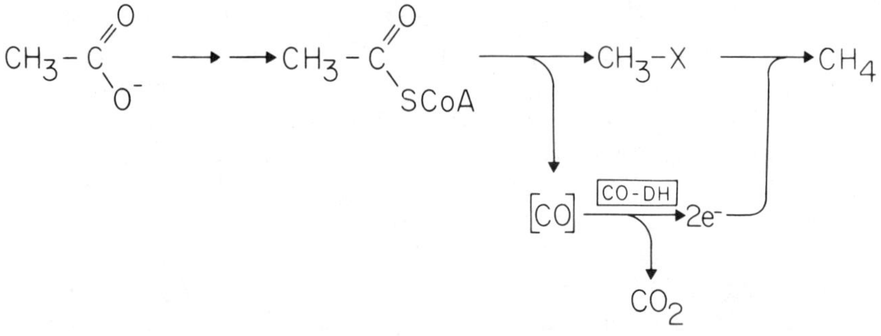

**Fig. 4.** Proposed mechanism of methane and $CO_2$ formation from acetate in *Methanosarcina barkeri*.

metal (Eikmanns and Thauer, 1984). According to the scheme, the carbon monoxide dehydrogenase in acetoclastic methanogens has the function of oxidizing bound CO to $CO_2$. The electrons thus generated serve to reduce methyl-X to $CH_4$.

The proposed mechanism is supported by several observations, the most important of which is that methane formation from acetate is inhibited by low concentrations of cyanide, which inactivates the carbon monoxide dehydrogenase of the acetoclastic methanogens (Eikmanns and Thauer, 1984). However, the finding that acetate-grown cells mediate an isotopic exchange between $CO_2$ and the carboxyl group of acetate but not (or only at very low rates) an isotopic exchange between CO and the carboxyl group of acetate is difficult to explain, assuming CO to be an intermediate. The proposed function of carbon monoxide dehydrogenase in acetoclastic methanogens is thus not proven.

*"Mini-Methane Producers"*

*Desulfovibrio desulfuricans* (Yagi, 1958; Postgate, 1970) and *C. pasteurianum* (Thauer *et al.*, 1974) contain carbon monoxide dehydrogenase (Table 1). The sulphate reducer requires acetate as a carbon source; *C. pasteurianum* produces acetate from pyruvate rather than from $2CO_2$. A function of CO dehydrogenase similar to that in acetogenic bacteria or autotrophic methanogens can therefore be excluded.

Postgate (1969) reported that cell extracts of *C. pasteurianum* and of *D. desulfuricans* produce, in small amounts, $CH_4$ from the methyl group of pyruvate. Acetyl–CoA is probably an intermediate. A role of CO dehydrogenase similar to that proposed for the acetoclastic methanogens can therefore be envisaged.

The enzymes mediating "mini-methane" formation and their function remain to be elucidated, since growing cultures of *C. pasteurianum* and of *D. desulfuricans* do not produce methane, even in minute amounts.

## Acknowledgments

This work was supported by grants from the Deutsche Forschungsgemeinschaft and by the Fonds der Chemischen Industrie.

## References

Andress, G. (1976). "Repression" der Kohlenmonoxid-Dehydrogenase-Aktivität durch Cystein in wachsenden Kulturen von *Clostridium pasteurianum*. Diploma Thesis, Ruhr-Universität, Bochum.

Balch, W. E., Fox, G. E., Magrum, L. J., Woese, C. R. and Wolfe,R. S. (1979). Methanogens: Reevaluation of a unique biological group. *Microbiological Reviews* **43**, 260–296.

Blaut, M. and Gottschalk, G. (1982). Effect of trimethylamine on acetate utilization by *Methanosarcina barkeri. Archives of Microbiology* **133**, 230–235.

Clark, J. E., Ragsdale, S. W., Ljungdahl, L. G. and Wiegel, J. (1982). Levels of enzymes involved in the synthesis of acetate from $CO_2$ in *Clostridium thermoautotrophicum. Journal of Bacteriology* **151**, 507–509.

Conrad, R. and Thauer, R. K. (1983). Carbon monoxide production by *Methanobacterium thermoautotrophicum. FEMS Microbiology Letters* **20**, 229–232.

Daniels, L., Fuchs, G., Thauer, R. K. and Zeikus, J. G. (1977). Carbon monoxide oxidation by methanogenic bacteria. *Journal of Bacteriology* **132**, 118–126.

Diekert, G. (1980). Kohlenmonoxid-Oxidation und die Reduktion von $CO_2$ zu Acetat in Clostridien. Ph.D. Thesis, Philipps-Universität, Marburg.

Diekert, G. and Ritter, M. (1982). Nickel requirement of *Acetobacterium woodii. Journal of Bacteriology* **151**, 1043–1045.

Diekert, G. and Ritter, M. (1983a). Purification of the nickel protein carbon monoxide dehydrogenase of *Clostridium thermoaceticum. FEBS Letters* **151**, 41–44.

Diekert, G. and Ritter, M. (1983b). Carbon monoxide fixation into the carboxyl group of acetate during growth of *Acetobacterium woodii* on $H_2$ and $CO_2$. *FEMS Microbiology Letters* **17**, 299–302.

Diekert, G. and Thauer, R. K. (1978). Carbon monoxide oxidation by *Clostridium thermoaceticum* and *Clostridium formicoaceticum. Journal of Bacteriology* **136**, 597–606.

Diekert, G. and Thauer, R. K. (1980). The effect of nickel on carbon monoxide dehydrogenase formation in *Clostridium thermoaceticum* and *Clostridium formicoaceticum. FEMS Microbiology Letters* **7**, 187–189.

Diekert, G., Graf, E. G. and Thauer, R. K. (1979a). Nickel requirement for carbon monoxide dehydrogenase formation in *Clostridium pasteurianum. Archives of Microbiology* **122**, 117–120.

Diekert, G. B., Graf, E. G. and Thauer, R. K. (1979b). Carbon monoxide oxidation by clostridia: Evidence for the involvement of a corrinoid-like compound. *In* ''Vitamin $B_{12}$'' (Eds. B. Zagalak and W. Friedrich), pp. 1033–1036. de Gruyter, Berlin.

Diekert, G., Jaenchen, R. and Thauer, R. K. (1980). Biosynthetic evidence for a nickel tetrapyrrole structure of factor $F_{430}$ from *Methanobacterium thermoautotrophicum. FEBS Letters* **119**, 118–120.

Diekert, G., Hansch, M. and Conrad, R. (1984). Acetate synthesis from 2 $CO_2$ in acetogenic bacteria: Is carbon monoxide an intermediate? *Archives of Microbiology* **138**, 224–228.

Drake, H. L. (1982). Occurrence of nickel in carbon monoxide dehydrogenase from *Clostridium pasteurianum* and *Clostridium thermoaceticum. Journal of Bacteriology* **149**, 561–566.

Drake, H. L., Hu, S.-I. and Wood, H. G. (1980). Purification of carbon monoxide dehydrogenase, a nickel enzyme from *Clostridium thermoaceticum. Journal of Biological Chemistry* **255**, 7174–7180.

Drake, H. L., Hu, S.-I. and Wood, H. G. (1981). Purification of five components from *Clostridium thermoaceticum* which catalyze synthesis of acetate from pyruvate and methyltetrahydrofolate. *Journal of Biological Chemistry* **10**, 11137–11144.

Dürre, P. and Andreesen, J R. (1983). Purine and glycine metabolism by purinolytic clostridia. *Journal of Bacteriology* **154**, 192–199.

Eikmanns, B. and Thauer, R. K. (1984). Catalysis of an isotopic exchange between $CO_2$ and the carboxyl group of acetate by *Methanosarcina barkeri* grown on acetate. *Archives of Microbiology* **138**, 365–370.

Eikmanns, B., Fuchs, G. and Thauer, R. K. (1984). Formation of carbon monoxide from $CO_2$ and $H_2$ by *Methanobacterium thermoautotrophicum. Eur. J. Biochem.* (in press).

Ellefson, W. L., Whitman, W. B. and Wolfe, R. S. (1982). Nickel-containing factor $F_{430}$: Chromophore of the methylreductase of *Methanobacterium. Proceedings of the National Academy of Sciences of the U.S.A.* **79**, 3707–3710.

Fathepure, B. Z. (1983). Isolation and characterization of an aceticlastic methanogen from a biogas digester. *FEMS Microbiology Letters* **19**, 151–156.

Fischer, F., Lieske, R. and Winzer, K. (1932). Biologische Gasreaktionen. *Biochemische Zeitschrift* **245**, 2–12.

Fuchs, G. and Stupperich, E. (1982). Autotrophic $CO_2$ fixation pathway in *Mlethanobacterium thermoautotrophicum. Zentralblatt für Bakteriologie, Mikrobiologie und Hygiene, Abteilung I, Originale C* **3**, 277–288.

Fuchs, G., Schnitker, U. and Thauer, R. K. (1974). Carbon monoxide oxidation by growing cultures of *Clostridium pasteurianum. European Journal of Biochemistry* **49**, 111–115.

Hammel, K. E., Cornwell, K. L., Diekert, G. B. and Thauer, R. K. (1984). Evidence for a nickel-containing carbon monoxide dehydrogenase in *Methanobrevibacter arboriphilicus. Journal of Bacteriology* **157**, 975–978.

Hegemann, G. (1980). Oxidation of carbon monoxide by bacteria. *Trends in Biochemical Sciences* **5**, 256–259.

Hirsch, P. (1968). Photosynthetic bacterium growing under carbon monoxide. *Nature (London)* **217**, 555–556.

Hu, S.-I., Drake, H. L. and Wood, H. G. (1982). Synthesis of acetyl coenzyme A from carbon monoxide, methyltetrahydrofolate, and coenzyme A by enzymes from *Clostridium thermoaceticum. Journal of Bacteriology* **149**, 440–448.

Huser, B. A., Wuhrmann, K. and Zehnder, A. J. B. (1982). *Methanothrix soehngenii* gen. nov. sp. nov., a new acetotrophic non-hydrogen-oxidizing methane bacterium. *Archives of Microbiology* **132**, 1–9.

Jaenchen, R., Diekert, G. and Thauer, R. K. (1981). Incorporation of methionine-derived methyl groups into factor $F_{430}$ by *Methanobacterium thermoautotrophicum. FEBS Letters* **130**, 133–136.

Jansen, K., Thauer, R. K., Widdel, F. and Fuchs, G. (1984). Formate, $CO_2$, and carbon monoxide assimilation by *Desulfovibrio baarsii. Archives of Microbiology* **138**, 257–262.

Kenealy, W. R. and Zeikus, J. G. (1982). One-carbon metabolism in methanogens: Evidence for synthesis of a two-carbon cellular intermediate and unification of catabolism and anabolism in *Methanosarcina barkeri. Journal of Bacteriology* **151**, 932–941.

Kerby, R. and Zeikus, J. G. (1983). Growth of *Clostridium thermoaceticum* on $H_2/CO_2$ or CO as energy source. *Current Microbiology* **8**, 27–30.

Kerby, R., Niemczura, W. and Zeikus, J. G. (1983). Single-carbon catabolism in acetogens: Analysis of carbon flow in *Acetobacterium woodii* and *Butyribacterium methylotrophicum* by fermentation and $^{13}C$ nuclear magnetic resonance measurement. *Journal of Bacteriology* **155**, 1208–1218.

Kluyver, A. J. and Schnellen, C. G. T. P. (1947). On the fermentation of carbon monoxide by pure cultures of methane bacteria. *Archives of Biochemistry* **14**, 57–78.

Krzycki, J. A. and Zeikus, J. G. (1984). Characterization and purification of carbon monoxide dehydrogenase from *Methanosarcina barkeri. Journal of Bacteriology* **158**, 231–237.

Krzycki, J. A., Wolkin, R. H. and Zeikus, J. G. (1982). Comparison of unitrophic and mixotrophic substrate metabolism by an acetate-adapted strain of *Methanosarcina barkeri. Journal of Bacteriology* **149**, 247–254.

Livingston, D. A., Pfaltz, A., Schreiber, J., Eschenmoser, A., Ankel-Fuchs, D., Moll, J., Jaenchen, R. and Thauer, R. K. (1984). Zur Kenntnis des Faktors F 430 aus methanogenen Bakterien: Struktur des proteinfreien Faktors. *Helvetica Chimica Acta* **67**, 334–351.

Ljungdahl, L. G. and Wood, H. G. (1982). Acetate biosynthesis. *In* "$B_{12}$" Vol. 2 (Ed. D. Dolphin), pp. 165–202. Wiley, New York.

Lynd, L., Kerby, R. and Zeikus, J. G. (1982). Carbon monoxide metabolism of the methylotrophic acidogen *Butyribacterium methylotrophicum. Journal of Bacteriology* **149**, 255–263.

Mah, R. A. (1980). Isolation and characterization of *Methanococcus mazei. Current Microbiology* **3**, 321–326.

Mah, R. A., Smith, M. R., Ferguson, T. and Zinder, S. (1981). Methanogenesis from $H_2$–$CO_2$, methanol, and acetate by *Methanosarcina. In* "Microbial Growth on $C_1$ Compounds" (Ed. H. Dalton), pp. 131–142. Heyden, London.

Masters, C. (1979). The Fischer-Tropsch reaction. *In* "Advances in Organometallic Chemistry" Vol. 17 (Eds. F. G. A. Stone and R. West), pp. 61–103. Academic Press, New York.

Meyer, O. and Schlegel, H. G. (1980). Carbon monoxide:methylene blue oxidoreductase from *Pseudomonas carboxydovorans. Journal of Bacteriology* **141**, 74–80.

Meyer, O. and Schlegel, H. G. (1983). Biology of aerobic carbon monoxide-oxidizing bacteria. *Annual Reviews of Microbiology* **37**, 277–310.

Mullen, A. (1980). Carbonylations catalyzed by metal carbonyls—Reppe reactions. *In* "New Syntheses with Carbon Monoxide" (Ed. J. Falbe), pp. 286–290. Springer-Verlag, Berlin and New York.

Parker, D. J., Wood, H. G., Ghambeer, R. K. and Ljungdahl, L. G. (1972). Total synthesis of acetate from carbon dioxide. Retention of deuterium during carboxylation of trideuteriomethyltetrahydrofolate or trideuteriomethylcobalamin. *Journal of Biological Chemistry* **11**, 3074–3080.

Patel, G. B., Roth, L. A., van den Berg, L. and Clark, D. S. (1976). Characterization of a strain of *Methanospirillum hungatei. Canadian Journal of Microbiology* **22**, 1404–1410.

Pezacka, E. and Wood, H. G. (1984). The synthesis of acetyl-CoA by *Clostridium thermoaceticum* from carbon dioxide, hydrogen, coenzyme A and methyltetrahydrofolate. *Archives of Microbiology* **137**, 63–69.

Pfaltz, A., Jaun, B., Fässler, A., Eschenmoser, A., Jaenchen, R., Gilles, H. H., Diekert, G. and Thauer, R. K. (1982). 81. Zur Kenntnis des Faktors $F_{430}$ aus methanogenen Bakterien: Struktur des porphinoiden Ligandsystems. *Helvetica Chimica Acta* **65**, 828–865.

Pfennig, N. and Widdel, F. (1981). Ecology and physiology of some anaerobic bacteria from the microbial sulfur cycle. *In* "Biology of Inorganic Nitrogen and Sulfur" (Eds. H. Bothe and A. Trebst), pp. 169–177. Springer-Verlag, Berlin and New York.

Pine, M. J. and Barker, H. A. (1956). Studies on the methane fermentation. XII. The pathway of hydrogen in the acetate fermentation. *Journal of Bacteriology* **71**, 644–648.

Postgate, J. R. (1969). Methane as a minor product of pyruvate metabolism by sulphate-reducing and other bacteria. *Journal of General Microbiology* **57**, 293–302.

Postgate, J. R. (1970). Carbon monoxide as a basis for primitive life on other planets: A comment. *Nature (London)* **226**, 978.

Ragsdale, S. W., Ljungdahl, L. G. and DerVartanian, D. V. (1982). EPR evidence for nickel–substrate interaction in carbon monoxide dehydrogenase from *Clostridium thermoaceticum. Biochemical and Biophysical Research Communications* **108**, 658–663.

Ragsdale, S. W., Clark, J. E., Ljungdahl, L. G., Lundie, L. L. and Drake, H. L. (1983a). Properties of purified carbon monoxide dehydrogenase from *Clostridium thermoaceticum*, a nickel, iron–sulfur protein. *Journal of Biological Chemistry* **258**, 2364–2369.

Ragsdale, S. W., Ljungdahl, L. G. and DerVartanian, D. V. (1983b). Isolation of carbon monoxide dehydrogenase from *Acetobacterium woodii* and comparison of its properties with those of the *Clostridium thermoaceticum* enzyme. *Journal of Bacteriology* **155**, 1224–1237.

Ragsdale, S. W., Ljungdahl, L. G. and DerVertanian, D. V. (1983c). $^{13}C$ and $^{61}Ni$ isotope substitutions confirm the presence of a nickel(III)–carbon species in acetogenic CO dehydrogenases. *Biochemical and Biophysical Research Communications* **115**, 658–665.

Schulman, M., Ghambeer, R. K., Ljungdahl, L. G. and Wood, H. G. (1973). Total synthesis of acetate from $CO_2$. VII. Evidence with *Clostridium thermoaceticum* that the carboxyl of acetate is derived from the carboxyl of pyruvate by transcarboxylation and not by fixation of $CO_2$. *Journal of Biological Chemistry* **248**, 6255–6261.

Sharak Genthner, B. R. and Bryant, M. P. (1982). Growth of *Eubacterium limosum* with carbon monoxide as the energy source. *Applied and Environmental Microbiology* **43**, 70–74.

Smith, M. R. and Mah, R. A. (1980). Acetate as sole carbon and energy source for growth of *Methanosarcina* strain 227. *Applied and Environmental Microbiology* **39**, 993–999.

Stupperich, E. and Fuchs, G. (1983). Autotrophic acetyl coenzyme A synthesis *in vitro* from two $CO_2$ in *Methanobacterium*. *FEBS Letters* **156**, 345–348.

Stupperich, E. and Fuchs, G. (1984a). Autotrophic synthesis of activated acetic acid from two $CO_2$ in *Methanobacterium thermoautotrophicum*. I. Properties of in vitro system. *Archives of Microbiology* **139**, 8–13.

Stupperich, E. and Fuchs, G. (1984b). Autotrophic synthesis of activated acetic acid from two $CO_2$ in *Methanobacterium thermoautotrophicum*. II. Evidence for different origins of acetate carbon atoms. *Archives of Microbiology* **139**, 14–20.

Stupperich, E., Hammel, K. E., Fuchs, G. and Thauer, R. K. (1983). Carbon monoxide fixation into the carboxyl group of acetyl coenzyme A during autotrophic growth of *Methanobacterium*. *FEBS Letters* **152**, 21–23.

Thauer, R. K., Fuchs, G., Käufer, B. and Schnitker, U. (1974). Carbon-monoxide oxidation in cell-free extracts of *Clostridium pasteurianum*. *European Journal of Biochemistry* **45**, 343–349.

Thauer, R. K., Brandis-Heep, A., Diekert, G., Gilles, H.-H., Graf, E. G., Jaenchen, R. and Schönheit, P. (1983). Drei neue Nickelenzyme aus anaeroben Bakterien. *Naturwissenschaften* **70**, 60–64.

Touzel, J. P. and Albagnac, G. (1983). Isolation and characterization of *Methanococcus mazei* strain MC$_3$. *FEMS Microbiology Letters* **16**, 241–245.

Uffen, R. L. (1976). Anaerobic growth of a *Rhodopseudomonas* species in the dark with carbon monoxide as sole carbon and energy substrate. *Proceedings of the National Academy of Sciences of the U.S.A.* **73**, 3298–3302.

Uffen, R. L. (1981). Metabolism of carbon monoxide. *Enzyme and Microbial Technology* **3**, 197–206.

Uffen, R. L. (1983). Metabolism of carbon monoxide by *Rhodopseudomonas gelatinosa*: Cell growth and properties of the oxidation system. *Journal of Bacteriology* **155**, 956–965.

Vogels, G. D. and van der Drift, C. (1976). Degradation of purines and pyrimidines by microorganisms. *Bacteriological Reviews* **40**, 403–468.

Wakim, B. T. and Uffen, R. L. (1983). Membrane association of the carbon monoxide oxidation system in *Rhodopseudomonas gelatinosa*. *Journal of Bacteriology* **153**, 571–573.

Wood, H. G., Drake, H. L. and Hu, S.-I. (1982). Studies with *Clostridium thermoaceticum* and the resolution of the pathway used by acetogenic bacteria that grow on carbon monoxide or carbon dioxide and hydrogen. *In* "Amino Acids, Fermentations, and Nucleic Acids: A Symposium" (Ed. E. E. Snell), pp. 29–56. Annual Reviews Inc., Palo Alto, California.

Yagi, T. (1958). Enzymic oxidation of carbon monoxide. *Biochimica et Biophysica Acta* **30**, 194–195.

Yagi, T. (1959). Enzymic oxidation of carbon monoxide. II. *Journal of Biochemistry (Tokyo)* **46**, 949–955.

Yagi, T. and Tamiya, N. (1962). Enzymic oxidation of carbon monoxide. III. Reversibility. *Biochimica et Biophysica Acta* **65**, 508–509.

# Discussion

*G. Taylor:* Would Professor Thauer like to speculate on the nature of the carriers of the $C_1$ unit in the methanogens?

**R. K. Thauer:** Three carriers of $C_1$ units are presently known in methanogenic bacteria: (1) the carbon dioxide-reducing factor (CDR factor), methanofurane, (2) methanopterin and (3) coenzyme M. There are still some disputes with respect to the exact structure and function of methanopterin. Whether factor F430 is a $C_1$ carrier is not known.

**D. Kell:** Since you use ATP to activate acetate in acetoclastic methanogens, and probably only get one ATP back in the reduction of $CH_3$-S-CoM to methane, how do acetate-grown methanogens make ATP in this scheme?

**Thauer:** The scheme of methane and $CO_2$ formation from acetate (Fig. 4) with acetyl–CoA as intermediate implies that the reduction of methyl-X to methane with a bound CO species as electron donor is coupled with the synthesis of more than 1 mol of ATP. We have no idea how this is achieved.

**Thauer:** After rupture of the cells, the carbon monoxide dehydrogenase is found in the soluble cell fraction. There is no evidence that the enzyme is a peripheral membrane protein located on the outer aspect of the cytoplasmic membrane. A simple vectorial organization of the electron transport chain from CO to methyl CoM can, therefore, not readily be envisaged.

# 8

# Metabolism of Aerobic Carbon Monoxide–Utilizing Bacteria

ORTWIN MEYER

*Institut für Mikrobiologie der Georg-August-Universität Göttingen, Göttingen, Federal Republic of Germany*

## The Natural Carbon Monoxide Cycle

Carbon monoxide is a highly poisonous, odorless, colorless, tasteless and explosive gas. The concentration of CO in the atmosphere is low (0.11 ppmv) but can reach a couple of hundred parts per million in urban areas with heavy traffic (Höschele, 1969) and up to 10% (v/v) in the gas vacuoles of some brown algae occurring free in kelp (Langdon, 1917). Carbon monoxide also vents from submarine hydrothermal systems, and its formation seems to be at least partly due to the activity of thermophilic bacteria (Baross *et al.*, 1982). The residence time of CO in the atmosphere is 0.3 to 0.4 years and the gas undergoes a natural cycle of formation and destruction (Fig. 1). As reported by Seiler (1974, 1978), CO is produced by various anthropogenic processes, mainly by the incomplete combustion of carbonaceous fossil fuels used in motor vehicles, aircrafts and home heating, as well as by some industrial processing ($6.4 \times 10^{14}$ g yr$^{-1}$). Major amounts of CO are also produced by reaction of OH radicals with methane in the troposphere ($4 \times 10^{14}$ g yr$^{-1}$). These figures show that both sources each contribute about 50% of the total annual CO production ($13.1–14.1 \times 10^{14}$ g). Minor sources are the oceans ($0.1–1.2 \times 10^{14}$ g yr$^{-1}$), bush fires ($0.6 \times 10^{14}$ g yr$^{-1}$), plants ($1.3 \times 10^{14}$ g yr$^{-1}$) and the oxidation of hydrocarbons ($0.6 \times 10^{14}$ g yr$^{-1}$). The oxidation of CO by OH radicals in the troposphere ($6 \times 10^{14}$ g yr$^{-1}$) and the uptake of CO by the soil surface ($5 \times 10^{14}$ g yr$^{-1}$) are considered major CO sinks and each amounts to approximately 50% of the total sink capacity ($13.1 \times 10^{14}$ g yr$^{-1}$).

Aerobic CO-oxidizing bacteria have been isolated from different soil and water samples taken all over the world (Kim and Hegeman, 1983; Nozhevnikova and Yurganov, 1978; Meyer and Schlegel, 1978, 1983; Zavarzin and Nozhevnikova, 1977). Some isolates were shown to oxidize CO even at ambient CO

131

MICROBIAL GAS METABOLISM:
MECHANISTIC, METABOLIC
AND BIOTECHNOLOGICAL ASPECTS

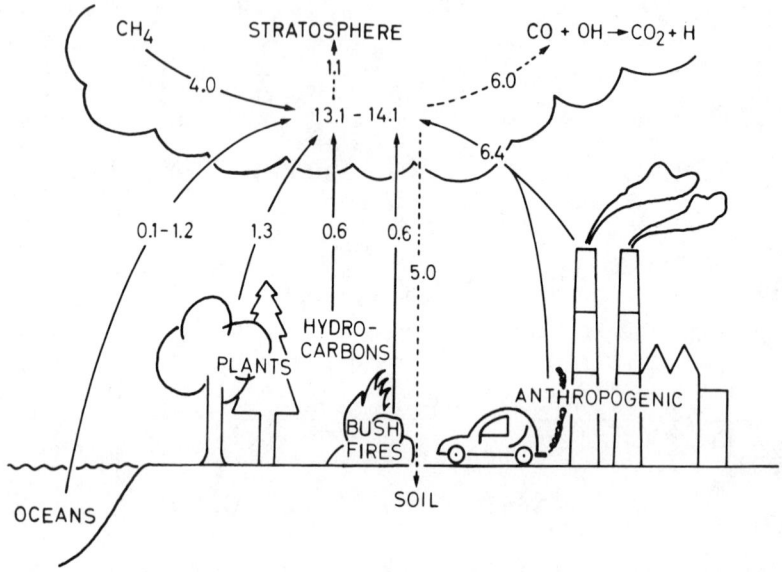

**Fig. 1.** Cycle of atmospheric carbon monoxide. Emission and uptake rates are given in $10^{14}$ g $yr^{-1}$. The data are taken from Seiler (1978).

concentrations (Conrad *et al.*, 1981). Therefore, the capacity of soil and water to decompose CO has been ascribed to these bacteria. On the other hand, the kinetic properties of natural soil and pure cultures of some CO-oxidizing bacteria are different (Conrad, 1984; Conrad *et al.*, 1981), and there are other indications that CO-autotrophic metabolism is not responsible for the removal of atmospheric CO by soil (Bartholomew and Alexander, 1979; Conrad and Seiler, 1982). As speculated by Conrad (1984), the aerobic CO-oxidizing bacteria represent a low-affinity CO uptake system, exhibiting a high $K_m$ of 500 $\mu$l CO litre$^{-1}$ and a high $V_{max}$. On the other hand, natural soil displayed a low $K_m$ of only 5 $\mu$l CO litre$^{-1}$ and low $V_{max}$ values (Conrad, 1984; Conrad *et al.*, 1981). These considerations suggest the coexistence of two different CO-oxidizing bacterial communities in natural soil, a low-$K_m$ system of high affinity responsible for the uptake of low concentrations of atmospheric CO (Conrad, 1984; Conrad and Seiler, 1982; Conrad *et al.*, 1981), and a high-$K_m$ system of low affinity acting in ecological niches where elevated CO concentrations are supplied, for example by fermentation, or as an emergency CO uptake system in areas of temporarily high CO concentrations, such as roadside or city soils (Spratt and Hubbard, 1981). However, it must be noted that recent isolates of carboxydotrophic bacteria (*Pseudomonas thermocarboxydovorans* strain C2 and the unidentified strain G26) exhibited $K_m$ values of 10 to 30 $\mu$l CO litre$^{-1}$ (J. Colby, personal communication;

Lyons *et al.*, 1982) that closely approach those observed for natural soils. These data suggest that carboxydotrophic bacteria may represent or contribute to the natural sink for CO. Whether methane-oxidizing bacteria contribute to the natural CO sink is not known.

## Groups of CO-Oxidizing Bacteria

The bacterial groupings that oxidize CO under aerobic or anaerobic conditions are listed in Fig. 2. They may be subdivided according to whether they are able to use CO as an energy source for growth (utilitarian CO oxidation) or whether the oxidation process is a gratuitous one (non-utilitarian CO oxidation). Some acetogens have been shown to grow anaerobically with CO, forming acetate and $Co_2$ (Diekert *et al.*, Chapter 7, this volume; Genthner and Bryant, 1982; Kerby and Zeikus, 1983; Kerby *et al.*, 1983; Lynd *et al.*, 1982; Wood *et al.*, 1982). Methanogens grow very slowly with CO by reducing it with $H_2$ to form methane (Daniels *et al.*, 1977; Kluyver and Schnellen, 1947). Some phototrophic bacteria can grow with CO anaerobically in the dark, forming $CO_2$ and $H_2$ (Hirsch, 1968; Uffen, 1976, 1981, 1983; Wakim and Uffen, 1983) and *Rhizobium japonicum* can use CO as an electron donor for anaerobic growth with nitrate as terminal electron acceptor (Gunatilaka *et al.*, 1984). The acetogenic *Clostridium acidiurici* and *C. cylindrosporum* (Diekert and Thauer, 1978) and *C. purinolyticum* and *Peptococcus glycinophilus* (Dürre and Andreesen, 1982; Dürre *et al.*, 1983) are devoid of CO-oxidizing activity, and some other *Clostridium* species containing CO dehydrogenase activity are non-utilitarian (Fuchs *et al.*, 1974, 1975). Under anaerobic conditions, acetogenic clostridia such as *C. thermoaceticum, C. pasteurianum* and *C. formicoaceticum* contain active CO dehydrogenase, even in the absence of CO (Thauer *et al.*, 1974, Fuchs *et al.*, 1974). This clostridial CO dehydrogenase and the enzyme from *Acetobacterium woodii* are new nickel-containing iron–sulphur proteins (Drake, 1982; Drake *et al.*, 1980; Diekert and Ritter, 1983; Diekert *et al.*, 1979; Ragsdale *et al.*, 1983; Thauer *et al.*, 1983) and it seems that they function as $CO_2$ reductases rather than CO dehydrogenases (Fuchs and Stupperich, 1984; Diekert *et al.*, Chapter 7, this volume). Sulphate-reducing bacteria, such as *Desulfovibrio desulfuricans*, contain a soluble, viologen-dependent CO dehydrogenase (Meyer and Rohde, 1984; Yagi, 1958, 1959; Yagi and Tamiya, 1962). The visible absorption spectrum of the enzyme is very similar to that of the clostridial enzyme, and electron paramagnetic resonance measurements suggest the presence of $Fe_2S_2$ centers (K. Fiebig, O. Meyer and R. Cammack, unpublished data). It has been concluded from theoretical considerations that sulphate-reducing bacteria should be capable of growing with CO (Meyer and Rohde, 1984), but that has not yet been substantiated. CO is co-oxidized but not used for growth by the methanotrophic bacteria, owing to broad

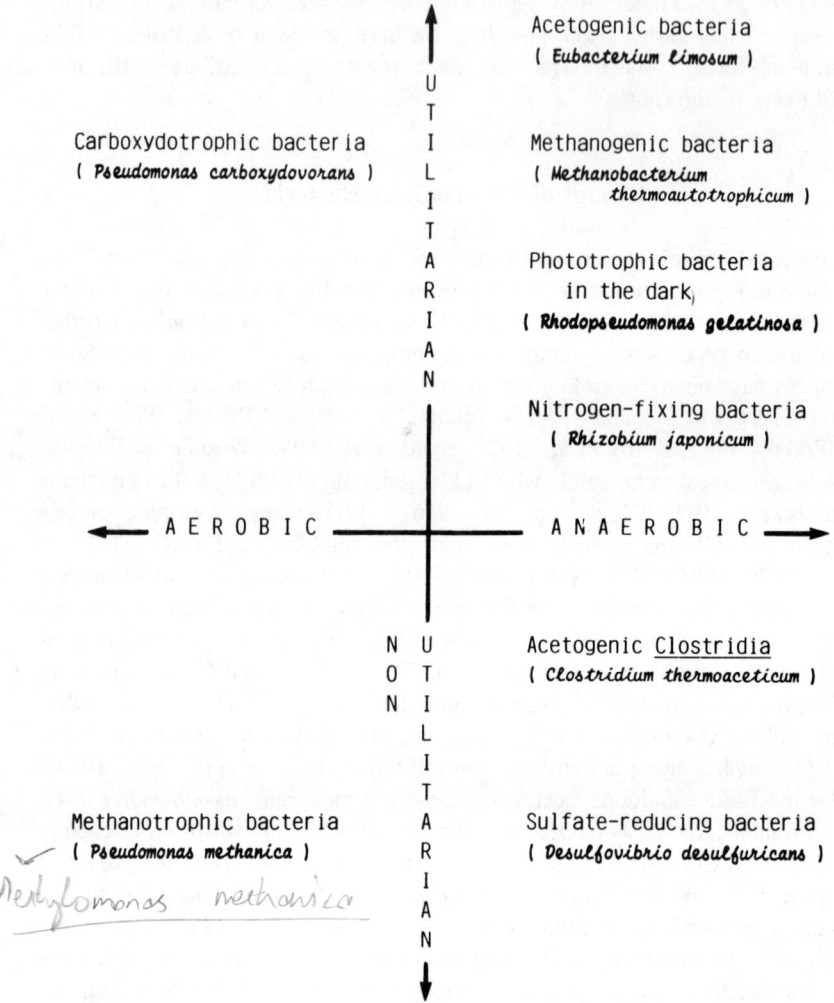

Fig. 2. Bacteria that oxidize CO under aerobic and anaerobic conditions.

specificity of the methane monooxygenase complex (for a review, see Colby *et al.*, 1979). This reaction requires oxygen and reducing equivalents.

Carbon monoxide serves as a sole carbon and energy source for the aerobic carboxydotrophic bacteria (Kim and Hegeman, 1983; Hegeman, 1980; Meyer, 1980, 1981; Meyer and Schlegel, 1983; Nozhevnikova and Yurganov, 1978; Zavarzin and Nozhevnikova, 1977). Some of them are also capable of growing mixotrophically with CO and a heterotrophic substrate (Kiessling and Meyer, 1982). Carboxydotrophs represent the sole group of aerobic, utilitarian CO-oxidizing bacteria and will be the main topic of the present contribution.

## Strains of Carboxydotrophic Bacteria

Growing cells of carboxydotrophic bacteria oxidize CO according to the following equation (Meyer and Rohde, 1984; Meyer and Schlegel, 1978, 1983)

$$O_2 + 2.19 \, CO \rightarrow 1.83 \, CO_2 + 0.36 \text{ cell carbon} \tag{1}$$

Carboxydotrophs represent a new type of chemolithoautotrophic metabolism in being able to (1) catalyze the oxidation of CO to $CO_2$, (2) use the electrons derived from this reaction for growth, (3) assimilate parts of the $CO_2$ formed via the ribulosebisphosphate cycle and (4) withstand CO inhibition. They are exceptional among the chemolithotrophic bacteria (Fig. 3) in being independent of exogenous or separately added $CO_2$. Besides CO, carboxydotrophic bacteria can grow with hydrogen plus $CO_2$, indicating that they also possess the metabolism for hydrogen oxidation. Like the hydrogen bacteria, but unlike the majority of nitrifying, sulphur- and iron-oxidizing bacteria, carboxydotrophic bacteria are facultative lithotrophs, able to use organic substrates for heterotrophic growth.

Defined by their ability to grow aerobically with CO as a sole carbon and energy source, carboxydotrophic bacteria are taxonomically diverse (for a review, see Meyer and Schlegel, 1983). As evident from Table 1, the group comprises mostly gram-negative species such as the different strains of *Pseudomonas carboxydovorans, Ps. carboxydohydrogena, Ps. carboxydoflava, Ps. compransoris, Ps. gazotropha, Alcaligenes carboxydus, Arthrobacter* 11/x

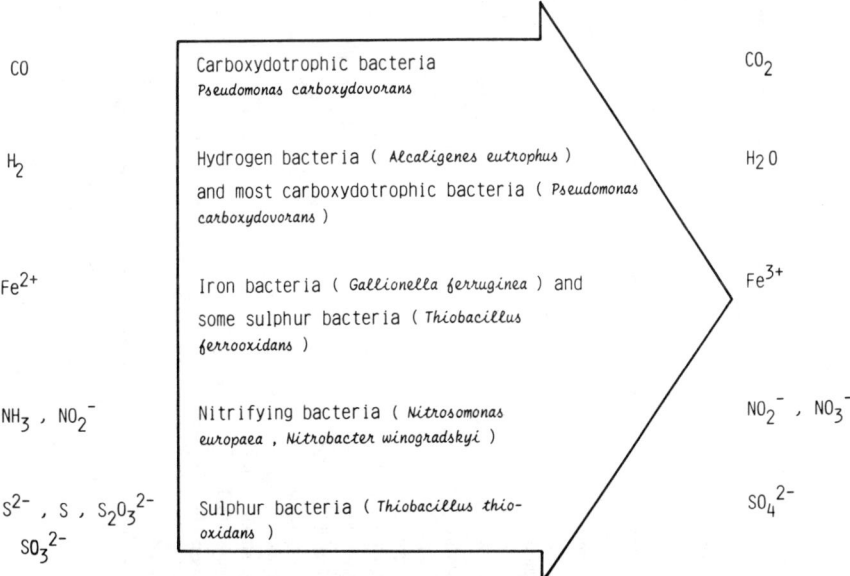

**Fig. 3.** Principal groups of aerobic chemolithoautotrophic bacteria.

**Table 1.** Properties of taxonomically described carboxydotrophic species (January 1984)[a]

| Species name | Gram type | Length (μm) | Width (μm) | Flagellation | G + C of the DNA (mol %) | Pigmentation | Growth temperature (°C) |
|---|---|---|---|---|---|---|---|
| *Ps. carboxydovorans* | | | | | | | |
| Strain OM5 | − | 1.0 to 3.0 | 0.4 to 0.7 | Subpolarly monotrichous | 60.7 to 62.2 | None | 30 |
| Strain OM2 | − | 0.9 to 1.8 | 0.6 to 0.7 | Subpolarly monotrichous | 63.2 | None | 30 |
| Strain OM3 | − | 1.0 to 2.0 | 0.4 to 0.6 | Non-flagellated | 63.2 | None | 30 |
| Strain OM4 | − | 0.9 to 1.8 | 0.5 to 0.7 | Subpolarly monotrichous | 63.8 | None | 30 |
| *Ps. carboxydohydrogena* | − | 1.0 to 2.0 | 0.3 to 0.7 | Subpolarly monotrichous | 58.2 | None | 30 |
| *Ps. carboxydoflava* | − | 1.1 to 2.4 | 0.6 to 0.9 | Non-flagellated | 68.0 | Yellow | 30 |
| *Ps. compransoris* | − | 1.5 to 2.1 | 0.7 to 0.9 | Polarly monotrichous | 63.3 | None | 30 |
| *P. gazotropha* | − | 1.1 | 0.7 | Polarly monotrichous | 66.9 | None | 30 |
| *A. carboxydus* | − | 0.9 to 1.7 | 0.7 to 0.9 | Peritrichous | 63.8 | None | 30 |
| Arthrobacter strain 11/x | + | Distinctive pleomorphism | | Non-flagellated | 70.2 | None | 37 |
| *Ps. thermocarboxydovorans* | − | n.d. | n.d. | Polarly monotrichous | 72.0 | None | 50 |
| *B. schlegelii* | | | | | | | |
| Strains MA48 and MA51 | + | 2.5 to 5 | 0.6 | Peritrichous | 67.7; 67 to 68 | None | 70 |
| Strain OMT 1 | + | 3.2 to 4.3 | 0.6 to 0.8 | Peritrichous | 64.9 | None | 65 |
| Strain OMT 4 | + | 2.5 to 3.4 | 0.6 to 0.7 | Peritrichous | 65.4 | None | 65 |
| Strain OMT 7 | + | 4.1 to 5.2 | 0.6 to 0.7 | Peritrichous | 63.9 | None | 65 |
| *Bacillus* | | | | | | | |
| OMT 2 | + | 3.6 to 5.8 | 0.6 to 0.8 | Peritrichous | 68.6 | None | 65 |

[a]For further details see Meyer and Rohde (1984), Meyer and Schlegel (1983), Nozhevnikova and Yurganov (1978) and Zavarzin and Nozhevnikova (1977).

and *Ps. thermocarboxydovorans.* *Bacillus schlegelii* and *Arthrobacter* 11/x are the only gram-positive carboxydotrophs known to date. Most carboxydotrophic bacteria are mesophilic. However, *Pseudomonas thermocarboxydovorans,* which was recently isolated in Colby's group, is thermophilic and grows best at 50°C (Lyons *et al.,* 1982, 1984). Like *A. carboxydus,* this bacterium is one of the very few exceptions among carboxydotrophs in being unable to grow with H$_2$ plus CO$_2$. The *B. schlegelii* strains MA48, MA51, OMT1, OMT4 and OMT7 and *Bacillus* strain OMT2 are spore-forming, obligate thermophilic bacteria (Krüger, 1983). Their optimal growth temperature with CO was 70°C; no growth occurred at temperatures below 50°C. CO is the best substrate for *B. schlegelii* and *Ps. thermocarboxydovorans;* these two strains grow on CO with a doubling time of 3 hr, compared to approximately 20 hr for the mesophilic carboxydotrophs. The former are the fastest growing carboxydotrophic bacteria known to date.

## Carbon Monoxide Oxidase

Carbon monoxide:acceptor oxidoreductase (CO oxidase) is the key enzyme in CO metabolism of carboxydotrophic bacteria. It serves three functions (Fig. 4): (1) to feed electrons into a special CO-insensitive respiratory chain for electron transport-linked phosphorylation, (2) to provide the carbon source CO$_2$ and (3) to provide ATP and, by inverse electron transfer, NADH for CO$_2$ assimilation via the reductive pentosephosphate cycle.

Some properties of CO oxidase purified from *P. carboxydovorans* have been summarized by Meyer and Rohde (1984) and Meyer and Schlegel (1983); most of them also apply to the enzymes from *Ps. carboxydohydrogena* and *Ps. carboxydoflava.* Purified CO oxidase catalyzes the oxidation of CO according to the following equation (Meyer and Schlegel, 1979, 1980):

$$CO + H_2O + X(ox) \rightarrow CO_2 + XH_2 \text{ (red)} \qquad (2)$$

The redox potential of this reaction is very negative [$E^{\circ\prime} = -540$ mV (Fuchs *et al.,* 1975)]. Although the metabolism of carboxydotrophs is strictly aerobic, water is the source of the second oxygen atom in the CO$_2$ formed. The two electrons released can be transferred to unphysiological electron acceptors with a redox potential around 0 mV. Enzyme activity is routinely tested photometrically by following the reduction of methylene blue or thionine (Kim and Hegeman, 1981a; Meyer and Schlegel, 1979, 1980). Artificial electron acceptors with a negative redox potential, such as methyl or benzyl viologen, or soluble physiological electron acceptors, such as NAD, NADP, FAD, FMN, spinach ferredoxin or cytochrome *c,* are not reduced. CO-grown *Ps. carboxydovorans* contains

## Cytoplasm   Cytoplasmic   Outside
### |membrane|

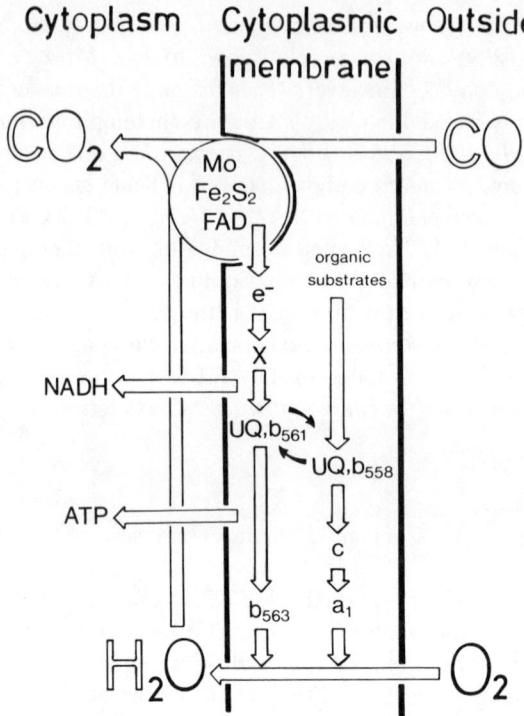

**Fig. 4.** Diagrammatic representation of the function and localization of CO oxidase in *Ps. carboxydovorans* and its interaction with the CO-insensitive branch of the respiratory chain.

ubiquinone ($UQ_{10}$) as sole quinone [about 4 $\mu$mol $UQ_{10}$ (g particulate protein)$^{-1}$ (Meyer and Schlegel, 1980)]. Restoration of the CO-oxidizing activity in ultraviolet-treated extracts of *Ps. carboxydohydrogena* with $UQ_{10}$ suggested that a quinone was necessary and might serve as a physiological electron acceptor (Kim and Hegeman, 1981b).

The visible absorption spectra of CO oxidases from different carboxydotrophic bacteria are identical (Meyer and Rajagopalan, 1984; Meyer and Schlegel, 1983). They exhibit a striking similarity to the spectra of bovine milk xanthine oxidase and chicken liver xanthine dehydrogenase. The spectra of CO oxidase, xanthine oxidase and xanthine dehydrogenase are characterized by the typically broad iron–sulphur absorption between 400 and 600 nm and absorption due to FAD at 450 nm (Meyer, 1982; Meyer and Rajagopalan, 1984; Meyer and Schlegel, 1983). Unlike clostridial CO dehydrogenase, which contains nickel and iron–sulphur centers of the $Fe_4S_4$ type, CO oxidase from carboxydotrophic bacteria is a molybdenum-containing iron–sulphur flavoprotein, containing $Fe_2S_2$ centers and molybdenum bound to a molybdopterin-containing molybdenum cofactor (Meyer and Rajagopalan, 1984; Meyer and Schlegel, 1983). Nickel is not contained in the carboxydobacterial CO oxidase (Meyer, 1982). The molecular weight of CO oxidase depends somewhat on the method employed but, when applied to a high-performance liquid chromatographic (HPLC) sizing column, the enzymes from *Ps. carboxydovorans, Ps. carboxydoflava* and *Ps. carboxydohydrogena* comigrate with chicken liver xanthine dehydrogenase, suggesting an $M_r$ of 300,000. For CO oxidase from *Ps. carboxydohydrogena*, an $M_r$ of 400,000 has been reported by Kim and Hegeman (1981a).

In agreement with the subunit composition reported for CO oxidase from *Ps. carboxydohydrogena* and other carboxydotrophs (Kim and Hegeman, 1981a; Kim *et al.*, 1982), the completely dissociated enzyme from *Ps. carboxydovorans* displayed a large subunit of $M_r$ 86,000, a medium one of $M_r$ 34,000 and a small one of $M_r$ 17,000 (Meyer and Rohde, 1984). However, only a single type of subunit is formed when the enzyme is dissociated under conditions which do not destroy the iron–sulphur centers.

The presence of 2 mol molybdenum per mol CO oxidase is evident from (1) chemical analysis (Meyer, 1982), (2) the presence of the molybdenum cofactor and molybdopterin (Meyer and Rajagopalan, 1984), (3) inhibition of the enzyme activity by methanol, which traps Mo in the pentavalent state (Meyer, 1982), (4) a visible absorption spectrum very similar to that of other molybdenum enzymes (Meyer and Rajagopalan, 1984; Meyer and Schlegel, 1983), (5) the appearance of an electron paramagnetic resonance (EPR) signal of the Mo(V)-resting I type (Bray *et al.*, 1983), (6) dependence on molybdenum of CO-autotrophic growth exclusively (Kalnowski, 1980; Meyer and Schlegel, 1983) and (7) inhibition of CO-autotrophic growth by the molybdate antagonist tungstate (Kalnowski, 1980).

Chemical analysis revealed the presence of eight iron and eight acid-labile sulphides (Meyer, 1982). The CO oxidases from *Ps. carboxydovorans* and *Ps. carboxydohydrogena* have been examined by EPR spectroscopy (Bray *et al.*, 1983). The spectra from the two enzymes were very similar and showed signals from two distinct FeS centers, characteristic of $Fe_2S_2$ clusters. As with milk xanthine oxidase, two different types of FeS centers could be resolved, one appearing at liquid nitrogen temperatures (FeS I) and the other at liquid helium temperatures (FeS II). A distance of 0.8 to 1.5 nm between the Mo and FeS centers was calculated from the spin–spin interaction of Mo(V) with the FeS I center (Bray *et al.*, 1983).

CO oxidases from different strains of carboxydotrophic bacteria seem to be rather similar (Rohde *et al.*, 1982). This is indicated by identical $M_r$, electrophoretic mobility and the range of electron acceptors reduced (Cypionka *et al.*, 1980); the oxidases revealed reactions of partial identity when subjected to Ouchterlony double immunodiffusion (Kim *et al.*, 1982; Meyer and Rohde, 1984; Rohde *et al.*, 1982).

## Carbon Monoxide–Insensitive Respiration

The metabolism of carboxydotrophic bacteria is strictly respiratory and they contain $b$-, $c$- and $a$-type cytochromes (Cypionka and Meyer, 1982a,b, 1983a,b; Kim and Hegeman, 1981b; Meyer and Schlegel, 1983, Meyer and Rohde, 1984; Zavarzin and Nozhevnikova, 1977). Growth and respiration of most carboxydotrophs is not impaired by CO (Cypionka and Meyer, 1982a,b), in contrast to the majority of non-carboxydotrophic aerobic organisms. Experiments employing electron transport inhibitors, room- and low-temperature spectroscopy and photochemical CO action spectra have led to a model for the respiratory chain of Ps. carboxydovorans (Cypionka and Meyer, 1983b). The chain is branched at the level of $b$-type cytochromes or ubiquinone. One branch contains cytochromes $b_{558}$, $c$ and $a_1$ and has been termed the heterotrophic branch. The second branch allows growth in the presence of CO and has been termed the autotrophic branch; it contains cytochromes $b_{561}$ and $b_{563}$. The heterotrophic branch was found to be sensitive to antimycin A (which inhibits electron transport from cytochrome $b$ to cytochrome $a$), CO and micromolar concentrations of cyanide. The autotrophic branch was sensitive to the quinone analogue HQNO (2-$n$-heptyl-4-hydroxyquinoline-$N$-oxide) and to only millimolar concentrations of cyanide; it was absolutely insensitive to CO. Cytochrome $a_1$ was inhibited by CO; therefore, its function as terminal oxidase of the heterotrophic branch could be established by photochemical CO action spectra. When extracts were reduced and cytochrome $a_1$ was inhibited by CO, cytochrome $b_{563}$ was the first to be reoxidized upon addition of small amounts of oxygen. That cytochrome $b_{563}$ is the terminal oxidase of the autotrophic branch and that CO insensitivity of Ps. carboxydovorans is due to the function of this unusual cytochrome as alternative, CO-insensitive terminal oxidase of the autotrophic branch has been concluded from its reluctance to become reduced, the absence of $H_2O_2$ as an end product and the absence of cytochrome $d$ and of the salicylhydroxamic acid-sensitive alternative terminal oxidase present in many plants. Cytochrome $b_{563}$ is constitutive and is also employed in other carboxydotrophic bacteria.

## Localization of CO Oxidase

The specific CO $\rightarrow$ methylene blue and $H_2$ $\rightarrow$ methylene blue activities were examined in the cytoplasmic and in the particulate fractions of CO-grown carboxydotrophs (Cypionka et al., 1980). With all strains, $H_2$-oxidizing activity was high in the membrane fraction and very low in the cytoplasmic fraction, characterizing hydrogenase of carboxydotrophic bacteria as a membrane-bound enzyme. It could not reduce pyridine nucleotides. Alcaligenes carboxydus contained very low levels of hydrogenase activity, although this bacterium could not grow with $H_2$ plus $CO_2$. CO-oxidizing activity was about equally distributed

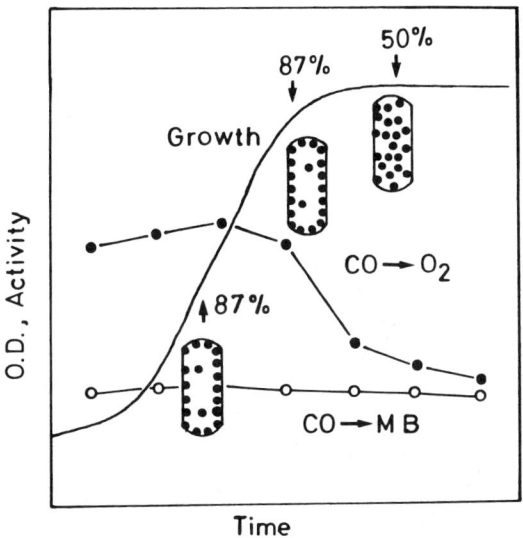

**Fig. 5.** Schematic representation of the dependence of the reduction of oxygen or methylene blue (MB) on the growth phase by whole cell suspensions of *Ps. carboxydovorans* growing auto-trophically with CO. The figures indicate the percentages of CO oxidase attached to the inner aspect of the cytoplasmic membrane. Redrawn from Rohde (1983).

between the cytoplasmic and the membrane fraction. This was difficult to explain in view of the failure of CO oxidase to reduce soluble physiological electron carriers. Thus, we speculated that, *in vivo,* CO oxidase might be associated with the cytoplasmic membrane (Schlegel and Meyer, 1981). Immunolocalization studies using the ferritin and protein A–gold method have established that CO oxidase of *Ps. carboxydovorans* is loosely attached to the inner aspect of the cytoplasmic membrane *in vivo* (Meyer and Rohde, 1984; Rohde, 1983) but is released into the cytoplasm when the cells enter the stationary growth phase (Fig. 5). Migration of CO oxidase into the cytoplasm explains the observed decrease of CO $\rightarrow$ $O_2$ activity in stationary-phase cells and the independence of growth phase of the CO $\rightarrow$ methylene blue activity; contact of the enzyme with the respiratory chain components is not a prerequisite for the latter activity. These results suggest that, *in vivo,* CO oxidase transfers the electrons to a membrane-bound electron acceptor, such as a cytochrome.

## Molybdenum Cofactor and Molybdopterin

The existence of a molybdenum cofactor (MoCo) common to nitrate reductase and xanthine dehydrogenase was first postulated by Pateman (Pateman *et al.,* 1964). MoCo seems to be universal to all molybdenum-containing enzymes, except nitrogenase, which contains an iron–molybdenum cofactor, FeMoCo (Pienkos *et al.*, 1977; Shah and Brill, 1977; Yang *et al.*, 1982). It was shown (Johnson, 1980; Johnson and Rajagopalan, 1982; Johnson *et al.*, 1980; Rajagopalan *et al.*, 1982) and later confirmed by others that MoCo is composed of molybdenum and a novel pterin (Claassen *et al.*, 1982a,b; Claassen, 1983; Ishizuka *et al.*, 1983). The pterin has been termed molybdopterin (MPT) and a structure has been proposed (Johnson and Rajagopalan, 1982). The pterin nucleus is substituted in the 6-position by a side chain of four carbon atoms, containing a hydroxyl group, a terminal monophosphate ester, susceptible to cleavage with alkaline phosphatase, and two sulphydryls attached to a double bond (Fig. 6). Additional evidence for the presence of sulphydryl groups in MPT stems from the observation that MoCo released from xanthine oxidase by heat treatment under anaerobic conditions binds to Thiol-Sepharose 4B and can be eluted with reduced sulphydryl compounds (Mendel and Alikulov, 1983). MPT is tightly, but non-covalently, attached to molybdoenzymes and can be released by treatment with sodium dodecyl sulphate, guanidine, acid or heat. It has been concluded from the absence of fluorescence in native molybdoenzymes that enzyme-bound MPT is in its tetrahydro state. Free MPT is unstable and autoxidizes to different fluorescent degradation products (Fig. 6). The latter can no longer complement aponitrate reductase in extracts of the *Neurospora crassa nit-1* mutant. Most of the information on native MPT has been derived from structural studies of oxidized, fluorescent derivatives (Fig. 6) and from the

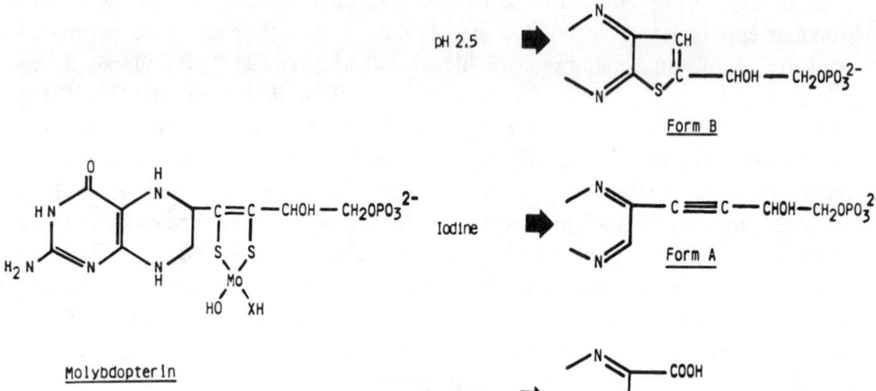

**Fig. 6.** Molybdopterin and some of its oxidation products. X stands for oxygen (sulphite oxidase, nitrate reductase and presumably CO oxidase) or for the cyanolysable sulphur (xanthine oxidase, xanthine dehydrogenase, aldehyde oxidase). For details see Ishizuka *et al.* (1983), Johnson and Rajagopalan (1982), Johnson *et al.* (1980, 1983, 1984), Meyer and Rajagopalan (1984) and Rajagopalan *et al.* (1982).

structural relationship of form A with urothione (Johnson and Rajagopalan, 1982). Urothione is a sulphur-containing pterin that occurs in human urine but is not found in patients deficient in MoCo.

The ability of material released from CO oxidase to complement aponitrate reductase in extracts of the *N. crassa nit-1* mutant indicated the presence of the molybdenum cofactor in CO oxidase also (R. V. Hageman and K. V. Rajagopalan, personal communication; Meyer and Schlegel, 1983). That MPT is also a constituent of CO oxidases from carboxydotrophic bacteria and that CO oxidase, xanthine oxidase, xanthine dehydrogenase and sulphite oxidase each contain two MPTs is indicated by the formation of the same amounts of identical fluorescent cofactor derivatives, by the presence of two organic phosphates additional to FAD (Meyer and Rajagopalan, 1984) and by similar complementation activities of the enzymes in the *N. crassa nit-1* assay (R. V. Hageman and K. V. Rajagopalan, personal communication).

The molybdenum cofactor in *Ps. carboxydovorans* and other carboxydotrophic bacteria does not occur free within the cell, but is found either incorporated into CO oxidase (or any other molybdoenzyme) or associated with a stabilizing

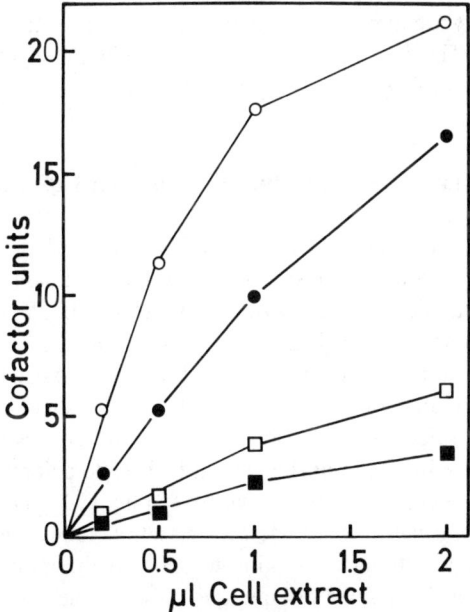

**Fig. 7.** Effect of increasing amounts of extracts of pyruvate-grown *Ps. carboxydovorans* mutant strain OM5-15 (see Table 2) on complementation activity, using a 17-hr complementation incubation on ice. Incubation of the cofactor source with *nit-1* extract in the presence (○, ●) or absence (□, ■) of molybdate (13.3 m*M* final concentration in the complementation assay) was done with the untreated extracts (●, ■) or after incubation for 90 s at 70°C (○, □). Further details of the complementation procedure are given in the legend to Table 2.

protein. The former cofactor is inducible, tightly bound and complements aponitrate reductase in *N. crassa nit-1* only after treatment with acid, defolding agents
(sodium dodecyl sulphate, guanidine hydrochloride) or heat (Fig. 7); the free
cofactor seems to be constitutive, is very loosely bound and complements nitrate
reductase in *nit-1* without any further treatment of the extract (Fig. 7). This
situation is similar to that found in liver, *Escherichia coli, Rhodospirillum
rubrum, N. crassa* and *Drosophila melanogaster,* where the loosely bound
MoCo existing in the soluble cell fraction is considered attached to a molybdenum cofactor carrier protein (for a review, see Johnson, 1980). Addition of 10
m$M$ molybdate to the assays increased the observed complementation activities
about five-fold and may be indicative of molybdenum-free cofactor (Fig. 7).
Marked stimulation of complementation activity by molybdate has also been
observed in *N. crassa* (Lee *et al.,* 1974), *E. coli* (Amy and Rajagopalan, 1979),
*Proteus mirabilis* (Claassen *et al.,* 1982b) and tobacco plants (Mendel, 1983).
When the soluble fraction of *Ps. carboxydovorans* grown in the presence of
[99]Mo-labelled molybdate was subjected to gel filtration, material capable of
complementing nitrate reductase in *nit-1* without any pretreatment coincided
with radioactivity and, after iodine oxidation, with fluorescence typical of form
A of molybdopterin; the $M_r$ of this material was 45,000, and it presumably
represents the molybdenum cofactor carrier protein (M. Rieth and O. Meyer,
unpublished data). This value compares favorably with the $M_r$ of 40,000 reported
for the carrier protein in *E. coli* (Amy and Rajagopalan, 1979).

## Mutants Devoid of the Molybdenum Cofactor

*Pseudomonas carboxydovorans* grows with CO and is able to reduce nitrate to
nitrite when growing with organic substrates. Both CO oxidase and nitrate reductase are molybdenum-containing enzymes, and growth of *Ps. carboxydovorans*
with poor nitrogen sources, such as leucine, is inhibited by 100 m$M$ chlorate.
Our strategy for isolation of mutants devoid of MoCo followed the idea that such
mutants should lack all molybdoenzyme activities and should be insensitive to
chlorate. Thus, we selected for mutants of *Ps. carboxydovorans* unable to grow
with CO or to reduce nitrate to nitrite and insensitive to chlorate. These mutants
were then grown heterotrophically or with $H_2$ plus $CO_2$ and checked for the
presence of material capable of complementing nitrate reductase in extracts of *N.
crassa nit-1,* with or without attempting to split off the molybdenum cofactor
(Table 2). The mutants obtained fell into two classes. The overwhelming majority of them retained both forms of MoCo. One could speculate that among this
group of mutants some may have lost the ability to express CO oxidase apoprotein or to synthesize enzyme of complete cofactor composition and others might
have lost their insensitivity towards CO. The second class of mutants was of the
desired type and is represented by the mutant OM5-41. The latter is incapable of

growing with CO and has lost both forms of MoCo. Both groups of mutants retained the capability of growing with $H_2$ plus $CO_2$, indicating that the mutation had not affected $CO_2$ assimilation via the Calvin cycle. Whether CO oxidase apoprotein devoid of the molybdenum cofactor is expressed in the mutant

**Table 2.** *Description of mutants*

| | Growth[a] with | | Nitrate reduction[b] | Presence of MoCo in extracts after treatment with | |
|---|---|---|---|---|---|
| Mutant strain | CO | $H_2$ | | Heat[c] | None[c] |
| *Pseudomonas carboxy-dohydrogena* strain Z-1062, wild type | + | + | + | + | + |
| *Alcaligenes carboxydus* strain Z-1171, wild type | + | − | + | + | + |
| *Pseudomonas carboxy-dovorans* strain OM5, wild type | + | + | + | + | + |
| OM5-3 | − | + | − | + | + |
| OM5-6 | − | ± | − | + | + |
| OM5-11 | − | ± | − | + | + |
| OM5-12 | − | + | − | + | + |
| OM5-14 | − | + | − | + | +. |
| OM5-15 | − | + | − | + | + |
| OM5-16 | − | − | − | + | + |
| OM5-17 | − | − | − | + | + |
| OM5-25 | − | + | ± | + | + |
| OM5-34 | − | + | − | + | + |
| OM5-41 | − | + | − | − | − |

[a]Autotrophic growth with CO or $H_2$ plus $CO_2$ was tested in liquid batch culture.
[b]The formation of nitrite from nitrate was tested in a medium containing 0.2% (w/v) pyruvate and 0.1% (w/v) nitrate.
[c]Bacteria were grown heterotrophically in liquid batch culture, harvested by centrifugation and disrupted with the French pressure cell. The *Neurospora crassa nit-1* mutant grown with $NH_4Cl$ as nitrogen source, was induced with nitrate for 2 hr at 30°C. Extracts were prepared by grinding the mycelia together with sea sand and several crystals of phenylmethylsulphonyl fluoride in a mortar. Complementation of aponitrate reductase was done by mixing 100 μl of *nit-1* extract with 200 μl of the cofactor source, diluted in complementation buffer (10 m$M$ phosphate buffer, pH 6.8, containing 20 m$M$ sodium molybdate), and incubation on ice overnight (17 hr). Complemented NADPH : nitrate reductase activity was determined in the presence of sulphite under the conditions described by Amy and Rajagopalan (1979). Heat treatment was done by incubating 100 μl of the cofactor source for 90 s at 70°C. Cofactor activity is defined as the amount of cofactor that will reconstitute 1 unit of nitrate reductase activity per 17 hr of complementation incubation on ice. One unit of nitrate reductase activity produces 1 nmol of nitrite per minute.

OM5-41, and whether extracts of this mutant induced for CO oxidase can be reactivated upon incubation with a cofactor source, must await further studies. The nonappearance of any mutant devoid of MoCo, complementing *nit-1* without pretreatment but retaining the enzyme-bound cofactor, is consistent with the hypothesis of a molybdenum cofactor carrier protein which protects MoCo against autoxidation and makes it available for molybdoenzyme synthesis.

These and other mutants are presently under investigation. They will allow us to take a genetic approach to the problems of CO-insensitive electron transport in carboxydotrophs, the nature of the physiological electron acceptor, the structure and function of CO oxidase and its cofactors and, in particular, the molybdenum cofactor.

## Note Added in Proof

Recently, *Desulfovibrio vulgaris* strain Madison has been shown to grow with CO as the sole energy source with sulphate as the electron acceptor and acetate as an additional carbon source. However, CO concentrations exceeding 4.5% (v/v) in the culture headspace were inhibitory [Lupton, F. S., Conrad, R. and Zeikus, J. G. (1984). CO metabolism of *Desulfovibrio vulgaris* strain Madison: Physiological function in the absence or presence of exogeneous substrates. *FEMS Microbiology Letters* **23**, 263–268].

## Acknowledgments

I wish to thank R. Conrad, H. Cypionka, K. Fiebig, J. L. Johnson, B. Krüger, T. Malchow, F. Mayer, K. V. Rajagopalan, M. Rieth, M. Rohde and H. G. Schlegel for their contribution to the work and/or ideas incorporated in this article. I would also like to thank the Deutsche Forschungsgemeinschaft for grants that have supported many aspects of the work described here.

## References

Amy, N. K. and Rajagopalan, K. V. (1979). Characterization of the molybdenum cofactor from *Escherichia coli*. *Journal of Bacteriology* **140**, 114–124.

Baross, J. A., Lilley, M. D. and Gordon, L. I. (1982). Is the $CH_4$, $H_2$ and CO venting from submarine hydrothermal systems produced by thermophilic bacteria? *Nature (London)* **298**, 366–368.

Bartholomew, G. W. and Alexander, M. (1979). Microbial metabolism of carbon monoxide in culture and in soil. *Applied and Environmental Microbiology* **37**, 932–937.

Bray, R. C., George, G. N., Lange, R. and Meyer, O. (1983). Studies by e.p.r. spectroscopy of carbon monoxide oxidase from *Pseudomonas carboxydovorans* and *Pseudomonas carboxydohydrogena*. *Biochemical Journal* 211, 687–694.

Claassen, V. P. (1983). Studies on the molybdenum cofactor common to several enzymes. Ph.D. thesis, Vrije Universiteit, Amsterdam, The Netherlands.

Claassen, V. P., Oltmann, L. F., Van't Riet, J., Brinkman, U. A. T. and Stouthamer, A. H. (1982a). Purification of molybdenum cofactor and its fluorescent oxidation products. *FEBS Letters* 142, 133–137.

Claassen, V. P., Oltmann, L. F., Vader, C. E. M., Van't Riet, J. and Stouthamer, A. H. (1982b). Molybdenum cofactor from the cytoplasmic membrane of *Proteus mirabilis*. *Archives of Microbiology* 133, 283–288.

Colby, J., Dalton, H. and Whittenbury, R. (1979). Biological and biochemical aspects of microbial growth on C$_1$ compounds. *Annual Review of Microbiology* 33, 481–517.

Conrad, R. (1984). Capacity of aerobic microorganisms to utilize and grow on atmospheric trace gases (H$_2$, CO, CH$_4$). *In* "Current Perspectives in Microbial Ecology" (Eds. M. J. Klug and C. A. Reddy), pp. 461–467. American Society for Microbiology, Washington, D.C.

Conrad, R. and Seiler, W. (1982). Utilization of traces of carbon monoxide by aerobic oligotrophic microorganisms in ocean, lake and soil. *Archives of Microbiology* 132, 41–46.

Conrad, R., Meyer, O. and Seiler, W. (1981). Role of carboxydobacteria in consumption of atmospheric carbon monoxide by soil. *Applied and Environmental Microbiology* 42, 211–215.

Cypionka, H. and Meyer, O. (1982a). Why carboxydobacteria are insensitive to carbon monoxide. *Zentralblatt für Bakteriologie, Mikrobiologie und Hygiene, Abteilung 1, Originale C* 3, 534.

Cypionka, H. and Meyer, O. (1982b). Influence of carbon monoxide on growth and respiration of carboxydobacteria and other aerobic organisms. *FEMS Microbiology Letters* 15, 209–214.

Cypionka, H. and Meyer, O. (1983a). Carbon monoxide-insensitive respiratory chain of *Pseudomonas carboxydovorans*. *Journal of Bacteriology* 156, 1178–1187.

Cypionka, H. and Meyer, O. (1983b). The cytochrome composition of carboxydotrophic bacteria. *Archives of Microbiology* 135, 293–298.

Cypionka, H., Meyer, O. and Schlegel, H. G. (1980). Physiological characteristics of various species of strains of carboxydobacteria. *Archives of Microbiology* 127, 301–307.

Daniels, L., Fuchs, G., Thauer, R. K. and Zeikus, J. G. (1977). Carbon monoxide oxidation by methanogenic bacteria. *Journal of Bacteriology* 132, 118–126.

Diekert, G. B. and Ritter, M. (1983). Purification of the nickel protein carbon monoxide dehydrogenase of *Clostridium thermoaceticum*. *FEBS Letters* 151, 41–43.

Diekert, G. B. and Thauer, R. K. (1978). Carbon monoxide oxidation by *Clostridium thermoaceticum* and *Clostridium formicoaceticum*. *Journal of Bacteriology* 136, 597–606.

Diekert, G. B., Graf, E. G. and Thauer, R. K. (1979). Nickel requirement for carbon monoxide dehydrogenase formation in *Clostridium pasteurianum*. *Archives of Microbiology* 122, 117–120.

Drake, H. L. (1982). Occurrence of nickel in carbon monoxide dehydrogenase from *Clostridium pasteurianum* and *Clostridium thermoaceticum*. *Journal of Bacteriology* 149, 561–566.

Drake, H. L., Hu, S. I. and Wood, H. G. (1980). Purification of carbon monoxide dehydrogenase, a nickel enzyme from *Clostridium thermoaceticum*. *Journal of Biological Chemistry* 255, 7174–7180.

Dürre, P. and Andreesen, J. R. (1982). Pathway of carbon dioxide reduction to acetate without energy requirement in *Clostridium purinolyticum*. *FEMS Microbiology Letters* 15, 51–56.

Dürre, P., Spahr, R. and Andreesen, J. (1983). Glycine fermentation via a glycine reductase in *Peptococcus glycinophilus* and *Peptococcus magnus*. *Archives of Microbiology* 134, 127–135.

Fuchs, G. and Stupperich, E. (1984). CO$_2$ reduction to cell carbon in methanogens. *In* "Microbial Growth on C$_1$ Compounds" (Eds. R. L. Crawford and R. S. Hanson), pp. 199–202. American Society for Microbiology, Washington, D.C.

Fuchs, G., Schnitker, U. and Thauer, R. K. (1974). Carbon monoxide oxidation by growing cultures of *Clostridium pasteurianum*. *European Journal of Biochemistry* **49**, 111–115.

Fuchs, G., Andress, G. and Thauer, R. K. (1975). Co-oxidation by anaerobic bacteria: Indications for the involvement of a vitamin $B_{12}$ compound. *In* "Microbial Production and Utilization of Gases $CH_2$, $CH_4$, CO" (Eds. H. G. Schlegel, G. Gottschalk and N. Pfennig), pp. 231–236. Akademie der Wissenschaften, Göttingen.

Genthner, B. R. S. and Bryant, M. P. (1982). Growth of *Eubacterium limosum* with carbon monoxide as the energy source. *Applied and Environmental Microbiology* **43**, 70–74.

Gunatilaka, M. K., Allen, G. C. and Neal, J. L. (1984). Poster presented at the 4th International Symposium on Microbial Growth on One Carbon Compounds, University of Minnesota, Minneapolis, September 6–10, 1983.

Hegeman, G. (1980). Oxidation of carbon monoxide by bacteria. *Trends in Biochemical Sciences* **5**, 256–259.

Hirsch, P. (1968). Photosynthetic bacterium growing under carbon monoxide. *Nature (London)* **217**, 555–556.

Höschele, K. (1969). Luftverunreinigung durch Autoabgase. *Umschau* **69**, 852.

Ishizuka, M., Ushio, K., Toraya, T. and Fukui, S. (1983). Formation of thieno(3,2-g)pteridines from the molybdenum cofactor. *Biochemical and Biophysical Research Communications* **111**, 537–543.

Johnson, J. L. (1980). The molybdenum cofactor common to nitrate reductase, xanthine dehydrogenase and sulfite oxidase. *In* "Molybdenum and Molybdenum-containing Enzymes" (Ed. M. Coughlan), pp. 347–383. Pergamon Press, Oxford.

Johnson, J. L. and Rajagopalan, K. V. (1982). Structural and metabolic relationship between the molybdenum cofactor and urothione. *Proceedings of the National Academy of Sciences of the U.S.A.* **79**, 6856–6860.

Johnson, J. L., Hainline, B. E. and Rajagopalan, K. V. (1980). Characterization of the molybdenum cofactor of sulfite oxidase, xanthine oxidase, and nitrate reductase. *Journal of Biological Chemistry* **255**, 1783–1786.

Johnson, J. L., Hainline, B. E. and Rajagopalan, K. V. (1983). The pterin component of the molybdenum cofactor (molybdopterin): Relationship to urothione. *In* "Chemistry and Biology of Pteridines" (Ed. J. A. Blair), pp. 1043–1047. de Gruyter, Berlin.

Johnson, J. L., Hainline, B. E., Rajagopalan, K. V. and Arison, B. H. (1984). The pterin component of the molybdenum cofactor. Structural characterization of two fluorescent derivatives. *Journal of Biological Chemistry* **259**, 5414–5422.

Kalnowski, G. (1980). Physiologische Untersuchung der Kohlenmonoxid-Oxidation und des chemolithotrophen Wachstums an zwei neu isolierten Carboxidobakterien. Ph.D. thesis, Technische Universität, Braunschweig, Federal Republic of Germany.

Kerby, R. and Zeikus, J. G. (1983). Growth of *Clostridium thermoaceticum* on $H_2/CO_2$ or CO as energy source. *Current Microbiology* **8**, 27–30.

Kerby, R., Niemczura, W. and Zeikus, J. G. (1983). Single-carbon catabolism in acetogens: Analysis of carbon flow in *Acetobacterium woodii* and *Butyribacterium methylotrophicum* by fermentation and $^{13}C$ nuclear magnetic resonance measurement. *Journal of Bacteriology* **155**, 1208–1218.

Kiessling, M. and Meyer, O. (1982). Profitable oxidation of carbon monoxide or hydrogen during heterotrophic growth of *Pseudomonas carboxydoflava*. *FEMS Microbiology Letters* **13**, 333–338.

Kim, Y. M. and Hegeman, G. D. (1981a). Purification and some properties of carbon monoxide dehydrogenase from *Pseudomonas carboxydohydrogena*. *Journal of Bacteriology* **148**, 904–911.

Kim, Y. M. and Hegeman, G. D. (1981b). Electron transport system of an aerobic carbon monoxide-oxidizing bacterium. *Journal of Bacteriology* **148**, 991–994.

Kim, Y. M. and Hegeman, G. D. (1983). Oxidation of carbon monoxide by bacteria. *International Review of Cytology* **81**, 1–32.

Kim, Y. M., Kirkconnell, S. and Hegeman, G. (1982). Immunological relationships among carbon monoxide dehydrogenases of carboxydobacteria. *FEMS Microbiology Letters* **13**, 219–223.

Kluyver, A. J. and Schnellen, C. G. (1947). Fermentation of carbon monoxide by pure cultures of methane bacteria. *Archives of Biochemistry* **14**, 57–70.

Krüger, B. (1983). Extrem thermophile carboxydotrophe Bakterien. Diplom Thesis, Universität Göttingen, Göttingen, Federal Republic of Germany.

Langdon, S. E. (1917). Carbon monoxide occurrence free in kelp. *Journal of the American Chemical Society* **39**, 149–156.

Lee, K. Y., Pan, S. S., Erickson, R. and Nason, A. (1974) Involvement of molybdenum and iron in the *in vitro* assembly of assimilatory nitrate reductase utilizing *Neurospora* mutant *nit-1*. *Journal of Biological Chemistry* **249**, 3941–3952.

Lynd, L., Kerby, R. and Zeikus, J. G. (1982). Carbon monoxide metabolism of the methylotrophic acidogen *Butyribacterium methylotrophicum*. *Journal of Bacteriology* **149**, 255–263.

Lyons, C. M., Williams, E. and Colby, J. (1982). Characterization of novel strains of carboxydobacteria. *Society for General Microbiology Quarterly* **9**, M7.

Lyons, C. M., Justin, P., Colby, J., and Williams, E. (1984). Isolation, characterization and autotrophic metabolism of a moderately thermophilic carboxydobacterium, *Pseudomonas thermocarboxydovorans* sp. nov. *Journal of General Microbiology* **130**, 1097–1105.

Mendel, R. R. (1983). Release of molybdenum co-factor from nitrate reductase and xanthine oxidase by heat treatment. *Phytochemistry* **22**, 817–819.

Mendel, R. R. and Alikulov, Z. A. (1983). Reversible immobilization of molybdenum cofactor on a gel matrix via sulphydryl groups. *Journal of Chromatography* **267**, 409–413.

Meyer, O. (1980). Using carbon monoxide to produce single-cell protein. *BioScience* **30**, 405–407.

Meyer, O. (1981). Growth of carbon monoxide oxidizing bacteria with industrial gas mixtures, automobile exhaust gas and other unconventional CO-containing gases. *Studies in Environmental Science* **9**, 79–86.

Meyer, O. (1982). Chemical and spectral properties of carbon monoxide : methylene blue oxidoreductase. The molybdenum-containing iron–sulfur flavoprotein from *Pseudomonas carboxydovorans*. *Journal of Biological Chemistry* **257**, 1333–1341.

Meyer, O. and Rajagopalan, K. V. (1984). Molybdopterin in carbon monoxide oxidase from carboxydotrophic bacteria. *Journal of Bacteriology* **157**, 643–648.

Meyer, O. and Rohde, M. (1984). Enzymology and bioenergetics of carbon monoxide-oxidizing bacteria. *In* "Microbial Growth on C₁ Compounds" (Eds. R. L. Crawford and R. S. Hanson), pp. 26–33. American Society for Microbiology, Washington, D.C.

Meyer, O. and Schlegel, H. G. (1978). Reisolation of the carbon monoxide utilizing hydrogen bacterium *Pseudomonas carboxydovorans* (Kistner) comb. nov. *Archives of Microbiology* **118**, 35–43.

Meyer, O. and Schlegel, H. G. (1979). Oxidation of carbon monoxide in cell extracts of *Pseudomonas carboxydovorans*. *Journal of Bacteriology* **137**, 811–817.

Meyer, O. and Schlegel, H. G. (1980). Carbon monoxide : methylene blue oxidoreductase from *Pseudomonas carboxydovorans*. *Journal of Bacteriology* **141**, 74–80.

Meyer, O. and Schlegel, H. G. (1983). Biology of aerobic carbon monoxide-oxidizing bacteria. *Annual Review of Microbiology* **37**, 277–310.

Nozhevnikova, A. N. and Yurganov, L. N. (1978). Microbiological aspects of regulating the carbon monoxide content in the earth's atmosphere. *Advances in Microbial Ecology* **2**, 203–238.

Pateman, J. A., Cove, D. J., Rever, B. M. and Roberts, D. B. (1964). A common co-factor for nitrate reductase and xanthine dehydrogenase which also regulates the synthesis of nitrate reductase. *Nature (London)* **201**, 58–60.

Pienkos, P. T., Shah, V. K. and Brill, W. J. (1977). Molybdenum cofactors from molybdoenzymes and *in vitro* reconstitution of nitrogenase and nitrate reductase. *Proceedings of the National Academy of Sciences of the U.S.A.* **74**, 5468–5471.

150                                ORTWIN MEYER

Ragsdale, S. W., Ljungdahl, L. G. and DerVartanian, D. V. (1983). Isolation of carbon monoxide dehydrogenase from *Acetobacterium woodii* and comparison of its properties with those of the *Clostridium thermoaceticum* enzyme. *Journal of Bacteriology* 115, 1224–1237.

Rajagopalan, K. V., Johnson, J. L. and Hainline, B. E. (1982). The pterin of the molybdenum cofactor. *Federation Proceedings, Federation of the American Societies for Experimental Biology* 41, 2608–2612.

Rohde, M. (1983). Immunelektronenmikroskopische Untersuchung der in situ Lokalisierung von Kohlenmonoxid-Oxidase aus *Pseudomonas carboxydovorans* mit Hilfe der Immunferritin- und der Protein A-Gold-Technik. Ph.D. Thesis, Universität Göttingen, Göttingen, Federal Republic of Germany.

Rohde, M., Cypionka, H. and Meyer, O. (1982). Carbon monoxide oxidation in different carboxydobacteria is catalyzed by identical enzymes. *Zentralblatt für Bakteriologie, Mikrobiologie und Hygiene, Abteilung 1, Originale C* 3, 543–544.

Schlegel, H. G. and Meyer, O. (1981). Microbial growth on carbon monoxide, formate and hydrogen: A biochemical assessment. In "Microbial Growth on C$_1$ Compounds" (Ed. H. Dalton), pp. 543–544. Heyden, London.

Seiler, W. (1974). The cycle of atmospheric CO. *Tellus* 26, 116–135.

Seiler, W. (1978). The influence of the biosphere on the atmospheric CO and H$_2$ cycles. In "Environmental Biogeochemistry and Geomicrobiology" Vol. 3 (Ed. W. E. Krumbein), pp. 773–810. Ann Arbor Science Publishers, Ann Arbor, Michigan.

Shah, V. K. and Brill, W. J. (1977). Isolation of an iron–molybdenum cofactor from nitrogenase. *Proceedings of the National Academy of Sciences of the U.S.A.* 74, 3249–3253.

Spratt, H. G. and Hubbard, J. S. (1981). Carbon monoxide metabolism in roadside soils. *Applied and Environmental Microbiology* 41, 1192–1201.

Thauer, R. K., Fuchs, G., Käufer, B. and Schnitker, U. (1974). Carbon monoxide oxidation in cell-free extracts of *Clostridium pasteurianum*. *European Journal of Biochemistry* 45, 343–349.

Thauer, R. K., Brandis-Heep, A., Diekert, G., Gilles, H.-H., Graf, E. G., Jaenchen, R. and Schönheit, P. (1983). Drei neue Nickelenzyme aus anaeroben Bakterien. *Naturwissenschaften* 70, 60–64.

Uffen, R. L. (1976). Anaerobic growth of a *Rhodopseudomonas* species in the dark with carbon monoxide as sole carbon and energy substrate. *Proceedings of the National Academy of Sciences of the U.S.A.* 73, 3298–3302.

Uffen, R. L. (1981). Metabolism of carbon monoxide. *Enzyme and Microbial Technology* 3, 197–206.

Uffen, R. L. (1983). Metabolism of carbon monoxide by *Rhodopseudomonas gelatinosa:* Cell growth and properties of the oxidation system. *Journal of Bacteriology* 155, 956–965.

Wakim, B. T. and Uffen, R. L. (1983). Membrane association of the carbon monoxide oxidation system in *Rhodopseudomonas gelatinosa*. *Journal of Bacteriology* 153, 571–573.

Wood, H. G., Drake, H. L. and Hu, S.-I. (1982). Studies with *Clostridium thermoaceticum* and the resolution of the pathway used by acetogenic bacteria that grow on carbon monoxide or carbon dioxide and hydrogen. *Proceedings of the Biochemistry Symposium*, pp. 29–56.

Yagi, T. (1958). Enzymic oxidation of carbon monoxide. *Biochimica et Biophysica Acta* 30, 194–195.

Yagi, T. (1959). Enzymic oxidation of carbon monoxide. II. *Journal of Biochemistry (Tokyo)* 46, 949–955.

Yagi, T. and Tamiya, N. (1962). Enzymic oxidation of carbon monoxide. III. Reversibility. *Biochimica et Biophysica Acta* 56, 508–509.

Yang, S. S., Pan, W. H., Delabert Friesen, G., Burgess, D. K., Corbin, J. L., Stiefel, E. I. and Newton, W. E. (1982). Iron–molybdenum cofactor from nitrogenase. Modified extraction methods as probes for composition. *Journal of Biological Chemistry* 257, 8042–8048.

Zavarzin, G. A. and Nozhevnikova, A. N. (1977). Aerobic carboxydobacteria. *Microbial Ecology* **3**, 305–326.

## Discussion

*J. Colby:*  Bearing in mind the ability of gram-positive hydrogen bacteria from a culture collection to grow on CO, why have published isolations always yielded a relatively few types of gram-positive organisms?

*O. Meyer:*  This is a tough question, and obviously the same situation applies to the hydrogen bacteria. Carboxydotrophs and hydrogenotrophs have a strictly aerobic metabolism and CO uptake in natural soil is confined to the upper surface layer. Therefore, samples taken from the surface layer of soil or aqueous environments have been favoured for the enrichment and isolation of these organisms. These habitats are not only aerobic, but are also exposed to sunlight. An organism which could minimize ultraviolet killing by reducing the amount of thymine in its DNA should gain a significant selective advantage [C. E. Singer and B. N. Ames, *Science* **170**, 822–826 (1970)]. Following these considerations, carboxydotrophic bacteria should reveal relatively high G + C values, which is actually the case. If the assumption is correct that most gram-negative bacteria display elevated G + C contents compared to gram-positive ones, this would explain why the former are prevailing in enrichment cultures of carboxydotrophic bacteria.

*R. K. Poole:*  Could you summarize the evidence, at which you hinted in your talk, for the role of cytochrome $b_{563}$ as the terminal oxidase?

*Meyer:*  Cytochrome $b_{563}$ ($o$) was the least reducible cytochrome, and it was the first to react with oxygen during reoxidation of reduced extracts. Cytochrome $d$ and the salicylhydroxamic acid-sensitive alternative terminal oxidase of many eukaryotes were absent. Hydrogen peroxide was not formed by the autotrophic branch and, in CO difference spectra, a maximum at 416 nm and troughs at 433 and 563 nm were indicative of cytochrome $o$.

*Poole:*  Could the $O_2$-reactive (and CO-unreactive) component be a non-cytochrome oxidase, e.g. a flavoprotein?

*Meyer:*  This seems to be not very likely, bearing in mind its sensitivity to inhibitors typical of terminal cytochrome oxidases, such as cyanide or azide. However, direct proof for the function of cytochrome $b_{563}$ ($o$) as the CO-insensitive terminal oxidase of the autotrophic branch has to await its isolation.

*R. K. Thauer:*  Carbon monoxide reacts with cytochrome oxidase only when in the reduced form ($Fe^{II}$). The sensitivity of the oxidase towards CO is thus dependent on the steady-state oxidation state. If the redox potential of cytochrome $aa_3$ in the CO-oxidizing bacteria is more negative than in other organisms, this would explain the relative CO insensitivity.

*Meyer:*  The midpoint potential of the terminal oxidase of the autotrophic branch (cytochrome $b_{563}$) is indeed very negative; a potential of $-105$ mV was recently determined for cytochrome $b_{563}$ in extracts of CO-grown *Pseudomonas carboxydovorans* (H. Cypionka, personal communication). In addition, the apparent $K_m$ of that cytochrome was lower than 0.5 $\mu M$ $O_2$.

# 9

# Carbon Monoxide Oxidoreductases from Thermophilic Carboxydobacteria

JUDITH M. BELL,* EDWIN WILLIAMS† AND JOHN COLBY*

*Biology Department, Sunderland Polytechnic, Sunderland, United Kingdom, and †Microbiology Department, The University, Newcastle-upon-Tyne, United Kingdom

## Isolation of Thermophilic Species

Aerobic bacteria that can grow at the expense of CO as sole carbon and energy source (carboxydotrophic bacteria; carboxydobacteria) were first isolated by Kistner (1953) and have since been isolated from a variety of water, sewage and soil samples (Nozhevnikova and Yurganev, 1978; Meyer and Schlegel, 1978). These isolations have mostly been done by conventional liquid enrichment techniques, usually at 25 to 30°C, and have yielded a relatively few well-described species. All are mesophilic and belong to the gram-negative genera *Pseudomonas* and *Achromobacter* (Meyer and Schlegel, 1983).

Our efforts have been directed at isolating novel types of carboxydotrophic bacteria with particular emphasis on thermophilic species. One of our isolates, *Pseudomonas thermocarboxydovorans* (NCIB 11893; type strain C2), is a gram-negative, obligately aerobic, thermophilic carboxydobacterium from sewage. A full description of this species has been published elsewhere (Lyons *et al.*, 1984). *Pseudomonas thermocarboxydovorans* grows well in mineral medium under an atmosphere of 25 to 80% (v/v) CO in air with a doubling time in small shake flasks of about 3 hr. The organism is a facultative autotroph able to grow on a variety of organic acids and amino acids, but unable to grow on sugars. *Pseudomonas thermocarboxydovorans* differs from 'classical' carboxydotrophic bacteria such as *Pseudomonas carboxydovorans* in its optimum growth temperature of 45 to 65°C, the high G + C ratio of its DNA (72 mole %) and its inability to grow as a hydrogen bacterium, to oxidise hydrogen or to synthesize hydrogenase.

*Streptomyces* G26 is a gram-positive mycelial bacterium isolated from compost. Like *Ps. thermocarboxydovorans*, it is a moderate thermophile with an

MICROBIAL GAS METABOLISM:
MECHANISTIC, METABOLIC
AND BIOTECHNOLOGICAL ASPECTS

optimum growth temperature range of 45 to 65°C and grows rapidly in mineral medium under an atmosphere of 25 to 80% (v/v) CO in air.


## Carbon Monoxide Oxidoreductase from *Pseudomonas thermocarboxydovorans*

CO:acceptor oxidoreductase from aerobic carboxydotrophic bacteria catalyses the following reaction:

$$CO + X + H_2O \rightarrow CO_2 + XH_2$$

where X is an artificial electron acceptor of $E^{0'}$ between $+0.011$ and $+0.220$ V, for example methylene blue or thionin (Meyer and Schlegel, 1980; Kim and Hegeman, 1981). The enzyme has been observed in several mesophilic carboxydobacteria (Cypionka *et al.*, 1980) and purified from two, *Ps. carboxydovorans* (Meyer and Schlegel, 1980; Meyer, 1982) and *Pseudomonas carboxydohydrogena* (Kim and Hegeman, 1981). Chemical and spectral analyses have shown the CO oxidoreductase from *Pseudomonas carboxydovorans* to be an iron–sulphur molybdenoflavoprotein containing two molecules of FAD, eight atoms of iron, eight of acid-labile sulphide and probably two of molybdenum (Meyer, 1982). Electron paramagnetic resonance (EPR) spectroscopy revealed the enzyme from *Ps. carboxydovorans* and *Ps. carboxydohydrogena* to contain Mo(V) atoms in two different chemical environments and two different iron–sulphur centres (Bray *et al.*, 1983). Carbon monoxide dehydrogenases from an anaerobic homoacetate-fermenting bacterium *Clostridium thermoaceticum, Clostridium pasteurianum* and the acetogenic *Acetobacterium woodii* have also been described (Thauer *et al.*, 1974; Drake *et al.*, 1980; Ragsdale *et al.*, 1983a,b), although these enzymes contain nickel and use electron acceptors of much lower potential such as viologen dyes, ferredoxin and rubredoxin.


*Purification of CO Oxidoreductase from Pseudomonas Thermocarboxydovorans*

Carbon monoxide oxidoreductase from strain C2 has been purified to near homogeneity, as shown in Table 1. Strain C2 was grown under CO in a 3-litre batch fermenter, and a washed suspension prepared as described by Lyons *et al.* (1984). For enzyme extraction, the suspension was passed once through a French pressure cell at 137 MPa. Carbon monoxide oxidoreductase was purified from the crude cell extract by ammonium sulphate fractionation, two DEAE–cellulose columns and gel filtration on Ultragel AcA34. The purified protein contains about 5% of a contaminating protein that has no enzyme activity (as determined by fast protein liquid chromatography or non-denaturing polyacrylamide gel electrophoresis).

**Table 1.**  *Purification of CO oxidoreductase from Pseudomonas thermocarboxydovorans[a]*

| Fraction | Volume (ml) | Total activity (units) | Specific activity [units (mg protein)$^{-1}$] | Yield (%) | Purification factor |
|---|---|---|---|---|---|
| $S_{35}$ extract | 38 | 1400 | 0.8 | 100 | 1 |
| DEAE–cellulose eluate I | 68 | 994 | 6.2 | 71 | 7.7 |
| 30 to 55% ammonium sulphate fraction | 4.2 | 952 | 6.4 | 68 | 8.0 |
| Ultragel AcA34 eluate | 70 | 714 | 9.3 | 51 | 10.4 |
| DEAE–cellulose eluate II | 4.7 | 434 | 11.6 | 31 | 14.5 |

[a]Bacteria were grown and a crude French press homogenate prepared as described in the text. The homogenate was centrifuged at 35,000 $g$ for 1 hr and the supernatant ($S_{35}$ extract) pumped onto a 2.5 × 15 cm column of DEAE–cellulose equilibrated with 10 m$M$ Tris–HCl buffer, pH 7.0 (buffer A). The column was eluted with a linear gradient (500 ml) of 0 to 0.3 $M$ NaCl in buffer A at a flow rate of 1.5 ml min$^{-1}$. The active fractions from the column were combined and then fractionated with ammonium sulphate, the fraction precipitating between 30 and 55% saturation being retained. This material was briefly dialysed against 1 litre of buffer A and then pumped onto a 2.5 × 90 cm column of Ultragel AcA34 and eluted with the same buffer at 0.25 ml min$^{-1}$. The active fractions from the column were combined and then pumped onto a 1.5 × 6 cm column of DEAE–cellulose equilibrated with 10 m$M$ Tris–HCl buffer, pH 8.0 (buffer B). This second DEAE–cellulose column was eluted with a linear gradient (300 ml) of 0 to 0.5 $M$ NaCl in buffer B at a flow rate of 0.5 ml min$^{-1}$. The final enzyme preparation was obtained by concentrating the active fractions from the second DEAE–cellulose column with 60% saturated ammonium sulphate followed by dialysis of the concentrated material against 1 litre of buffer A for 3 hr. Enzyme activity was determined spectrophotometrically with PES and DCPIP as electron acceptors, as described in the legend to Table 4. One unit of CO oxidoreductase activity is the amount catalysing the reduction of 1 μmol of DCPIP in 1 min at 50 °C.

## Properties of the Purified Enzyme

These are listed in Table 2. In most of its properties, the CO oxidoreductase resembles the analogous enzyme from mesophilic bacteria. Nevertheless, CO oxidoreductase from *Ps. thermocarboxydovorans* shows the following unique properties: (1) the enzyme is heat-stable with an optimum temperature in the usual assay of 80°C, (2) the enzyme has a very low apparent $K_m$ for CO (Table 3), some 100-fold lower than the published values for other CO oxidoreductases (Meyer and Schlegel, 1980; Kim and Hegeman, 1981) and (3) the enzyme can use both horse heart cytochrome $c$ and potassium ferricyanide as electron acceptors *in vitro* (Table 4). The natural electron acceptor for CO oxidoreductase in aerobic carboxydotrophic bacteria is thought to be ubiquinone (Kim and Hegeman, 1981; Cypionka and Meyer, 1983) and the electron acceptor range of the *Ps. thermocarboxydovorans* enzyme is consistent with this hypothesis. Further work is required to elucidate the electron transport pathway involved in CO respiration.

**Table 2.**  *Kinetic and molecular properties of the CO oxidoreductase from Pseudomonas thermocarboxydovorans*

| | |
|---|---|
| Optimum pH | 7.5 |
| Temperature for maximum activity | 80°C |
| $K_m$ for CO | 0.6 $\mu M$ |
| Inhibitors | Acetylene, cyanide, methanol, 8-hydroxyquinoline, iodoacetate, p-hydroxymercuribenzoate |
| Electron acceptors | Methylene blue, ferricyanide, thionin, phenazine ethosulphate/methosulphate, cytochrome $c$, 2,6-dichlorophenolindophenol |
| $K_m$ for phenazine ethosulphate | 3.8 $\mu M$ |
| Molecular weight (gel filtration) | 310,000 |
| Molecular weight (by sedimentation velocity) | 230,000 |
| Fe content | 6.9 g atom/mol[a] |
| Acid-labile sulphide | 6.9 g atom/mol[a] |
| Mo content | 0.7 g atom/mol[a] |
| Flavin | 1.8[a] |
| Absorption maxima | 340 and 420 nm |

[a]Calculated assuming a molecular weight of 270,000.

**Table 3.**  *Apparent $K_m$ values for CO and electron acceptors determined with CO oxidoreductases from Pseudomonas thermocarboxydovorans and Streptomyces-G26[a]*

| | Organism | | |
|---|---|---|---|
| Compound | Ps. thermo-carboxydovorans | Strepto-myces G26 | Ps. carboxy-dovorans[b] |
| CO | 0.5 | 1.3 | 53.0 |
| Phenazine ethosulphate | 5.1 | 3.0 | n.t.[c] |
| Benzyl viologen | n.a.[d] | 220.0 | n.a. |

[a]The $K_m$ values were determined from Lineweaver–Burk double reciprocal plots on data obtained by using the oxygen electrode assay described in the legend to Table 4. Carbon monoxide oxidoreductase was purified from *Pseudomonas* and partially purified from *Streptomyces* G26 as described in the text. All values are in micromoles per litre.
[b]Data from Meyer and Schlegel (1980).
[c]n.t., not tested.
[d]n.a., not applicable.

**Table 4.** *Activity of different bacterial CO oxidoreductases with a variety of electron acceptors*[a]

| | | Organism | | |
|---|---|---|---|---|
| Electron acceptor | $E^{0'}$ (mV) | *Ps. thermo-carboxydovorans* C2 | *Strepto-myces* G26 | *Ps. carboxy-dovorans*[b] |
| Methyl viologen | −440 | 0 | 25 | 0 |
| Benzyl viologen | −359 | 0 | 98 | 0 |
| NAD/NADP | −320 | 0 | 0 | 0 |
| FAD/FMN | −319 | 0 | 0 | 0 |
| Methylene blue | + 11 | 50 | 62 | 142 |
| Thionin | + 70 | 38 | 62 | 143 |
| PMS | + 80 | 100 | 100 | 100 |
| Toluidine blue O | +110 | n.t.[c] | 54 | 126 |
| DCPIP | +217 | 23 | 12 | 29 |
| Cytochrome *c* | +245 | 32 | 10 | 0 |
| Ferricyanide | +429 | 47 | 19 | 0 |

[a]Carbon monoxide oxidoreductase was purified from *Ps. thermocarboxydovorans* C2 and partially purified from *Streptomyces* G26 as described in the text. Capped cuvettes contained 0.15 nmol Tris–HCl buffer (pH 7), 0.26 μmol 2,6-dichlorophenolindophenol (DCPIP), 15.0 μmol phenazine ethosulphate (PES) and 0.5 ml of CO-saturated water in a final volume of 3 ml. The cuvettes were prewarmed to 50°C and then the reaction started by the addition of enzyme solution through the cap. Other electron acceptors were tested similarly by replacing DCPIP and PES in the above assay mixture. Autoxidizable acceptors were tested in a reaction mixture saturated with CO and containing catalase (0.03 μg Sigma type C100), glucose oxidase (0.01 μg Sigma type X) and glucose (0.1 mM), to remove residual oxygen, or by measuring oxygen consumption in a Rank oxygen electrode containing the same reaction mixture. Values are percentages of the rate with PMS.
[b]Data from Meyer and Schlegel (1979).
[c]n.t., not tested.

## Carbon Monoxide Oxidoreductase from *Streptomyces* G26

Carbon monoxide oxidoreductase has been partially purified from this organism by ion-exchange chromatography and ammonium sulphate fractionation. The enzyme differs markedly from the isofunctional enzymes from gram-negative carboxydobacteria (including *Ps. thermocarboxydovorans*) with respect to its range of artificial electron acceptors (Tables 3 and 4). The enzyme from *Streptomyces* G26 will reduce low-potential acceptors such as benzyl viologen ($E^{0'} = -0.359$ V) in addition to dyes such as phenazine ethosulphate (PES) and methylene blue.

In order to determine whether the activities with benzyl viologen and PES are associated with a single active site, the effects of pH, temperature and inhibitors on the two activities were investigated. Enzyme activities were measured spectrophotometrically (except that inhibitors were tested in the oxygen electrode assay) as described in the legend to Table 4. The following buffers (all at a final concentration of 50 m$M$) were used to obtain different pH values: 2-($N$-morpholino)ethanesulphonic acid–NaOH (pH 5.5 to 7.0), sodium phosphate (pH 6.0 to 7.0), Tris–HCl (pH 7.0 to 8.5), sodium pyrophosphate–HCl (pH 7.5 to 8.5). Both activities had the same optimum pH of 7.2 with identically shaped pH–activity curves. Similarly, the temperature–activity curves were identical, with maximum activity with both dyes at about 65°C. Of the 15 potential inhibitors tested, only 2,2-dipyridyl (21 to 36% inhibition at 10 m$M$), potassium cyanide (100% inhibition at 1 m$M$) and $p$-hydroxymercuribenzoate (100% inhibition at 0.25 m$M$) were effective and all three had a similar effect on both activities. The following compounds failed to inhibit enzyme activity measured with either acceptor: semicarbazide–HCl (1 m$M$), hydroxylamine–HCl (1 m$M$), EDTA (1 m$M$), 8-hydroxyquinoline (1 m$M$), sodium fluoride (10 m$M$), sodium chlorate (10 m$M$), sodium azide (10 m$M$), neocuproine (1 m$M$), sodium iodoacetate (10 m$m$), dithio-bis-dinitrobenzoic acid (1 m$M$), dithio-bis-dinitrobenzoic acid (1 m$M$), $N$-ethylmaleimide (1 m$M$) and superoxide dismutase (1 unit).

The available evidence therefore suggests that a single active site is responsible for reactivity with both PES and benzyl viologen, although this will require confirmation by purifying the enzyme to homogeneity. This ability to use viologen dyes is shared with the nickel-containing CO dehydrogenases from anaerobic bacteria (Ragsdale *et al.*, 1983a,b). The latter enzymes, however, are extremely oxygen-sensitive, requiring the careful exclusion of oxygen during their extraction and purification (Ragsdale *et al.*, 1983b). No such precautions are necessary with the CO oxidoreductase from G26.

The observation of viologen reduction by CO in *Streptomyces* G26 suggests the intriguing possibility of electron transfer from CO to NAD(P) in this organism without the need to invoke reversed electron flow. Ragsdale *et al.* (1983b) have suggested that CO dehydrogenase from anaerobic bacteria is coupled to ferredoxin or rubredoxin; the presence of these iron–sulphur proteins in *Streptomyces* G26 has yet to be demonstrated.

## Biotechnological Applications of Carbon Monoxide Oxidoreductase

Carbon monoxide is a colourless, odourless and very toxic gas because of its ability to bind to metalloproteins. As a major component of industrial gases, such as synthesis gas, and as a universal by-product of the combustion of fossil fuels, it is a potential hazard both in the home and at the workplace. A cheap enzyme-based sensor able to detect toxic levels of CO (1 to 50 ppm) might make a significant contribution to protecting the quality of breathable air. The construction of an enzyme-based CO sensor is discussed in Turner *et al.* (Chapter 10, this volume).

## References

Bray, R. C., George, G. N., Lange, R. and Meyer, O. (1983). Studies by epr spectroscopy of carbon monoxide oxidases from *Pseudomonas carboxydovorans* and *Pseudomonas carboxydohydrogena*. *Biochemical Journal* **211**, 687–694.

Cypionka, H. and Meyer, O. (1983). CO-insensitive respiratory chain of *Pseudomonas carboxydovorans*. *Journal of Bacteriology* **156**, 1178–1187.

Cypionka, H., Meyer, O. and Schlegel, H. G. (1980). Physiological characteristics of various species of strains of carboxydobacteria. *Archives of Microbiology* **127**, 301–307.

Drake, H. L., Hu, S.-I. and Wood, H. G. (1980). Purification of carbon monoxide dehydrogenase, a nickel enzyme from *Clostridium thermoaceticum*. *Journal of Biological Chemistry* **255**, 7174–7180.

Kim, Y. M. and Hegeman, G. D. (1981). Purification and some properties of carbon monoxide dehydrogenase from *Pseudomonas carboxydohydrogena*. *Journal of Bacteriology* **148**, 904–911.

Kistner, A. (1953). On a bacterium oxidising carbon monoxide. *Proceedings of the Koninklijike Nederlands e Akademie van Wetenschappen, Series C* **56**, 43.

Lyons, C. M., Justin, P., Colby, J. and Williams, E. (1984). Isolation, characterisation and autotrophic metabolism of a moderately thermophilic carboxydobacterium, *Pseudomonas thermocarboxydovorans* sp. nov. *Journal of General Microbiology* **130**, 1097–1105.

Meyer, O. (1982). Chemical and spectral properties of carbon monoxidemethylene blue oxidoreductase. The molybdenum containing iron sulphur flavoprotein from *Pseudomonas carboxydovorans*. *Journal of Biological Chemistry* **257**, 1333–1341.

Meyer, O. and Schlegel, H. G. (1978). Reisolation of the carbon monoxide utilizing bacterium *Pseudomonas carboxydovorans* (Kistner) comb. nov. *Archives of Microbiology* **118**, 35–43.

Meyer, O. and Schlegel, H. G. (1979). Oxidation of carbon monoxide in cell extracts of *Pseudomonas carboxydovorans*. *Journal of Bacteriology* **137**, 811–817.

Meyer, O. and Schlegel, H. G. (1980). Carbon monoxide:methylene blue oxidoreductase from *Pseudomonas carboxydovorans*. *Journal of Bacteriology* **141**, 74–80.

Meyer, O. and Schlegel, H. G. (1983). Biology of aerobic carbon monoxide utilizing bacteria. *Annual Reviews of Microbiology* **37**, 277–310.

Nozhevnikova, A. N. and Yurganev, L. N. (1978). Microbiological aspects of regulating the CO content of the earth's atmosphere. *In* "Advances in Microbial Ecology" Vol. 2 (Ed. M. Alexander), pp. 203–244. Plenum Press, New York.

Ragsdale, S. W., Clark, J. E., Ljungdahl, L. G., Lundie, L. L. and Drake, H. L. (1983a). Properties of purified carbon monoxide dehydrogenase from *Clostridium thermoaceticum*, a nickel, iron sulphur protein. *Journal of Biological Chemistry* **258**, 2364–2369.

Ragsdale, S. W., Ljungdahl, L. G. and DerVartanian, D. V. (1983b). Isolation of carbon monoxide dehydrogenase from *Acetobacterium woodii* and comparison of its properties with those of the *Clostridium thermoaceticum* enzyme. *Journal of Bacteriology* **155**, 1224–1237.

Thauer, R. K., Fuchs, G., Kaufer, B. and Schnitker, U. (1974). Carbon monoxide oxidation in cell free extracts of *Clostridium pasteurianum*. *European Journal of Biochemistry* **45**, 343–349.

# 10

# Enzyme-Based Carbon Monoxide Sensors

ANTHONY P. F. TURNER*, WILLIAM J. ASTON*, GRAHAM DAVIS*,
I. JOHN HIGGINS*, H. ALLEN O. HILL† AND JOHN COLBY‡

*Biotechnology Centre, Cranfield Institute of Technology, Cranfield,
Bedfordshire, United Kingdom, †Inorganic Chemistry Laboratory, University of
Oxford, Oxford, United Kingdom, and ‡Biology Department, Sunderland
Polytechnic, Sunderland, United Kingdom

## Introduction

The affinity and specificity of biological systems for certain molecules have secured an increasingly significant role for biosensors in analytical chemistry. Biosensor development to date has exploited tissues, cells, antibodies and enzymes under relatively mild physical and chemical conditions, as exemplified by the detection of a wide variety of substances in dilute aqueous solution (Aston and Turner, 1984). As the art develops, however, there will be incentives to extend the advantages offered by biosensors to more hostile environments and to overcome the stability problems inherent in such applications.

## Carbon Monoxide Sensors

There is considerable commercial interest in the potential of biosensors for the specific measurement of gases, with one prime target for such development programmes being CO. Detection of this toxic gas is required in mines, underground car parks, road tunnels and various industrial environments. Carbon monoxide is a product of incomplete combustion, and quantification of the gas is important in combustion control systems for furnaces and engines; it also provides an effective basis for a fire alarm. The implication of the gas in the emotive area of cigarette smoking is part of a broad clinical and biological interest which frequently requires accurate measurement of CO both in solution and in the gas phase.

The diversity of situations in which CO measurement is required is mirrored by the range of conventional techniques adopted. Where expense and size are no

MICROBIAL GAS METABOLISM:
MECHANISTIC, METABOLIC
AND BIOTECHNOLOGICAL ASPECTS

object, gas chromatography and infrared absorption have found acceptance. The ubiquitous mass spectrometer, however, experiences problems in the presence of dinitrogen, which has the same mass number as CO. A limited range of low-cost miniature CO gas sensors are commercially available which operate either by the direct electrochemical oxidation of CO or by adsorption at coated semi-conductor devices. When analysing authentic samples, however, the most important limitation of these devices is lack of selectivity and interference from, for example, $N_2O$, $NO_2$ and $H_2S$.

## Gas Biosensors

Biosensors for gas analysis have already achieved some commercial success; an immobilised-cholinesterase reactor system capable of detecting gaseous nerve poisons, such as pesticides, in the ppm range has been developed (Goodson and Jacobs, 1974). The device consists of a pad of immobilised cholinesterase held between a pair of platinum electrodes. A stream of sample air together with the enzyme substrate butyrylthiocholine iodide is passed through the pad and the hydrolysis product thiocholine iodide detected electrochemically. In the presence of enzyme inhibitors no easily oxidisable thiol is formed and the cell voltage rises.

Biosensors incorporating intact microorganisms have been described for the determination of methane (Okada *et al.*, 1981; Karube *et al.*, 1982), ammonia (Hikuma *et al.*, 1980) and nitrogen dioxide (Suzuki and Karube, 1982). These systems require the respective gases to be dissolved prior to delivery to the immobilised cells. The metabolism of the substrates is reflected by the consumption of oxygen, which is monitored by using a Clark oxygen electrode. The use of intact microorganisms rather than the relevant purified enzymes facilitates the construction of stable devices, with outputs reported to remain steady for 10 to 24 days (Suzuki and Karube, 1982)

In order to exploit fully the specificity of biological components in sensors, it is often preferable to monitor a single interaction rather than the range of metabolic events occurring in tissues and cells. The use of a purified enzyme for assay of its substrate directly in the gas phase has been reported by Guilbault (1983) and represents a significant departure from conventional approaches. The principle of the detector is that the frequency of vibration of an oscillating piezoelectric quartz crystal is decreased when its mass is increased by the adsorption of material. Guilbault (1983) has reported that the adsorption may be made specific for formaldehyde by coating the crystal with a mixture of formaldehyde dehydrogenase, reduced glutathione and NAD. The implication that the specific binding capacity of an enzyme may be exploited under dry conditions has important connotations, particularly for gas analysis.

## Carbon Monoxide–Binding Proteins

A prerequisite for a successful biosensor is the identification of an appropriate biological component. The ability of CO to bind the terminal oxidases of aerobic organisms and to inhibit the respiration of most of them provides the basis for one approach. Intact microorganisms have been incorporated into biosensors based on secondary transducers, such as the oxygen electrode (Diviès, 1975), and on mediated electron transfer (Turner et al., 1983). In addition, cytochrome oxidase has been coupled to an electrode via cytochrome c (Hill et al., 1981). Inhibitor electrodes based on the irreversible binding of CO to these systems may be envisaged.

A more flexible approach to CO quantification requires the specific metabolism of the gas. Carbon monoxide–oxidising bacteria are well known and CO oxidoreductases have been purified from several Pseudomonads (Meyer and Schlegel, 1983). The characteristics of the CO oxidoreductase from *Pseudomonas thermocarboxydovorans,* however, are particularly well suited for use in an enzyme-based sensor, offering greater stability (half-life of several days at room temperature) and a higher affinity for CO (apparent $K_m$ for CO = 0.6 $\mu M$) than the enzyme from similar but mesophilic carboxydobacteria.

*Pseudomonas thermocarboxydovorans* was isolated from sewage (Lyons et al., 1983) and grown on mineral salts medium supplemented with p-aminobenzoic acid under 33% (v/v) CO in air. Carbon monoxide oxidoreductase was purified by column chromatography, further details of which are given elsewhere in this volume (Bell et al., Chapter 9, this volume). The enzyme utilised a number of artificial electron acceptors including methylene blue, phenazine ethosulphate (PES), ferricyanide, cytochrome c, thionin, $N,N,N',N'$-tetramethyl-4-phenylenediamine (TMPD), ferrocene monocarboxylic acid and 1,1'-dimethylferrocene, all of which could be reoxidised at either platinum or carbon electrodes.

## Amperometric Biosensors

Most biosensors depend on a secondary transducer, quite discrete from the main reaction, which is capable of detecting the consumption of substrates or formation of products (Aston and Turner, 1984). Such an approach is prone to interference by environmental fluctuations (for example, $O_2$ tension, humidity, pH) which may interact directly with the transducer. In addition, the complexity of such systems is liable to increase costs, reduce reliability and impair the theoretical resolution and limits of detection. Biological redox catalysts may be coupled more simply and effectively to electronic systems by the use of promoters and/or mediators facilitating electron transfer between the catalyst and an

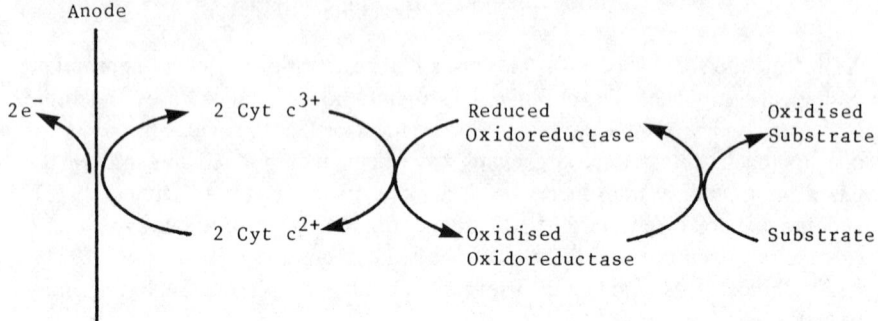

**Fig. 1.** Scheme representing biosensor construction with redox enzymes such as CO oxidoreductase.

electrode. Rapid reversible electrochemistry of horse heart cytochrome *c* at a 4,4′-bipyridyl-modified gold electrode has been described (Albery *et al.*, 1981). This reaction has been used to couple efficiently electron transport at an electrode to the reduction of dioxygen to water, via a terminal oxidase, thus providing the basis for an oxygen sensor (Hill *et al.*, 1981). Horse heart cytochrome *c* will act as an electron acceptor for several redox enzymes including CO oxidoreductase, allowing the construction of biosensors according to the general scheme shown in Fig. 1.

Alternatively, low-molecular-weight mediators may be used in place of cytochrome *c* to couple the oxidation reaction to an electrode. Initial work with CO oxidoreductase used a fuel cell to demonstrate the feasibility of this approach. The CO fuel cell was similar in construction to that previously described for an enzyme-based methanol fuel cell (Turner *et al.*, 1982; Davis *et al.*, 1983). The reactions are summarised in Fig. 2.

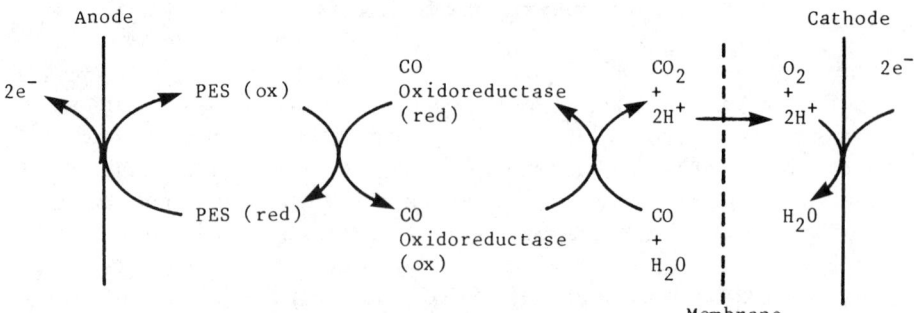

**Fig. 2.** Reactions in a CO fuel cell.

When the anode compartment (3 ml total volume, containing 1 mg enzyme) was continuously sparged with CO the fuel cell produced a current of approximately 1 mA for several hours. This principle may be readily applied to the quantitative analysis of CO; by integrating the current–time curve obtained in response to the addition of aliquots of CO, the number of coulombs passed may be calculated. The CO fuel cell described, however, was a relatively inefficient device, due principally to the reaction of the reduced mediator with oxygen and loss of CO from the anode to the cathode compartment. By restricting the apparatus to a single compartment and poising the potential of the electrode, the charge obtained reflected the stoichiometry of the two-electron reaction. Figure 3 shows the charge passed in response to addition of CO, using TMPD as a mediator with the working electrode poised at a potential of +110 mV versus a saturated calomel electrode (SCE). Clearly, this configuration provided a highly sensitive detector for CO.

A probe configuration is preferred for most practical applications of sensors and so we sought a more convenient version of this enzyme-based amperometric system. A glucose biosensor was recently described which uses a substituted ferricinium ion as a mediator of electron transfer between glucose oxidase and a graphite electrode (Cass et al., 1984). This class of mediator combines well-behaved electrochemistry and insensitivity to oxygen with a choice of physical

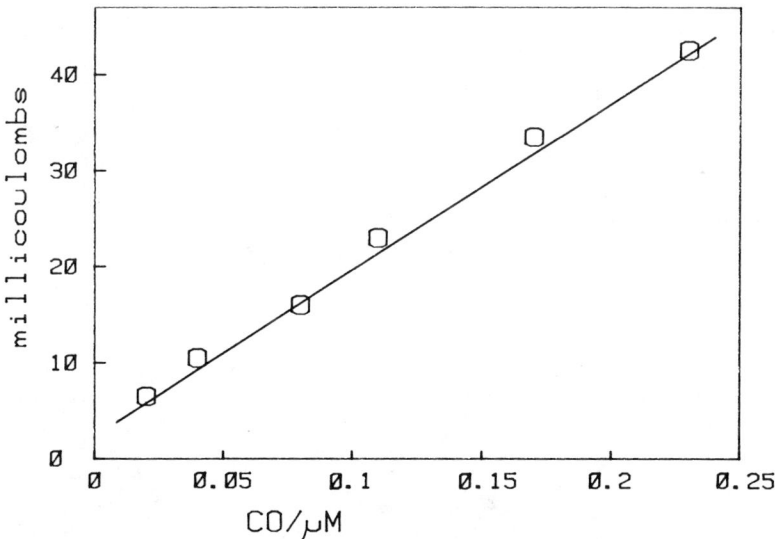

**Fig. 3.** Calibration curve for a CO sensor. Aliquots of CO-saturated buffer were added to a stirred solution of TMPD and CO oxidoreductase. The oxidised form of the mediator was regenerated with a platinum electrode poised at +110 mV (SCE) and the charge passed was calculated.

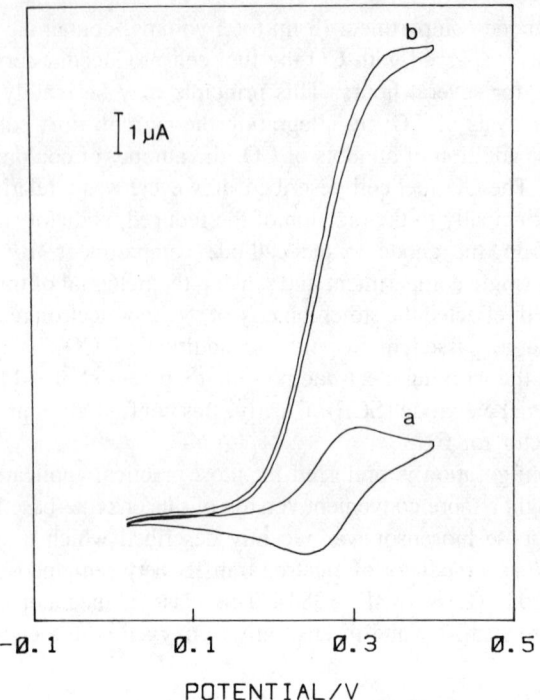

**Fig. 4.** Direct-current cyclic voltammogram of ferrocene monocarboxylic acid in argon-saturated Tris–HCl buffer containing (a) either no additions, CO or CO oxidoreductase and (b) CO and CO oxidoreductase.

and chemical properties suitable for enzyme-based sensors. The kinetics of the homogeneous reaction between ferrocene monocarboxylic acid and CO oxidoreductase were studied by using DC cyclic voltammetry. The effect on the reversible electrochemistry of ferrocene monocarboxylic acid of adding CO oxidoreductase in the presence of substrate is shown in Fig. 4. Enzyme was added

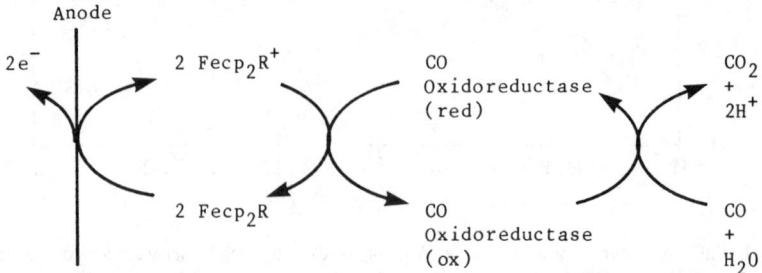

**Fig. 5.** Sequence indicated by enhanced anodic current observed in cyclic voltammetry experiments in the presence of ferricinium ion.

to final concentrations in the range 10 to 100 $\mu M$ and the enhanced anodic current obtained recorded as a function of scan rate. Although the cyclic voltammetry experiments showed no indication of direct electrochemistry of either CO oxidoreductase or its substrate, the enhanced anodic current obtained in the presence of the ferricinium ion ($Fecp_2R^+$) was indicative of the sequence shown in Fig. 5.

Using the theory developed by Nicholson and Shain (1964), quantitative kinetic data for the homogeneous reaction between a mediator and an enzyme may be calculated (Davis *et al.*, 1983; Cass *et al.*, 1984). A second-order homogeneous rate constant for the reaction between the ferricinium ion and the reduced CO oxidoreductase of $4.0 \times 10^5$ litre $mol^{-1} s^{-1}$ (pH 7, 20°C) was obtained. This rapid reaction indicated that a ferrocene derivative incorporated into an electrode would allow the construction of an effective enzyme electrode.

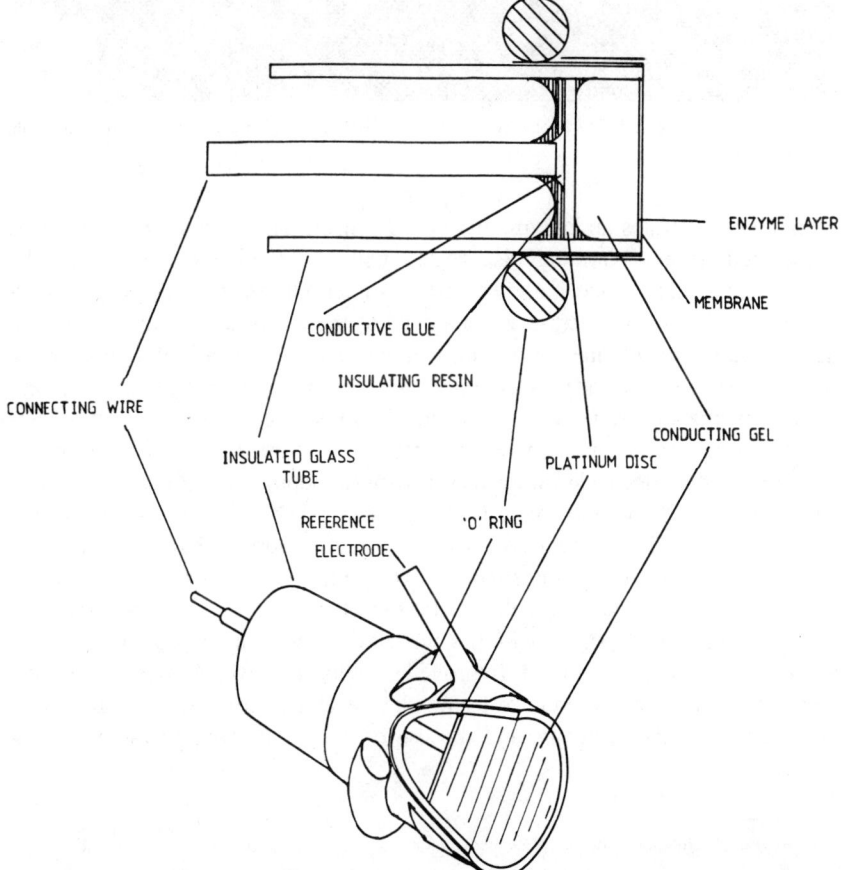

**Fig. 6.** An enzyme-based CO probe. From Turner *et al.* (1984).

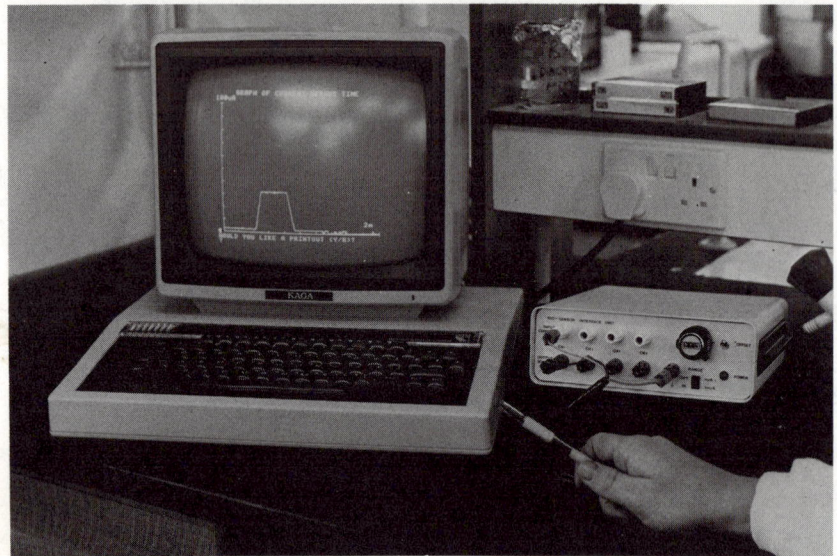

**Fig. 7.** An enzyme-based CO probe connected to a BBC microcomputer via a programmable interface.

Figure 6 illustrates an enzyme-based CO sensor. The working electrode was constructed from a platinum disc coated with a carbon paste containing the virtually insoluble ferrocene derivative $1,1'$-dimethylferrocene as mediator. Carbon monoxide oxidoreductase was retained at the electrode surface by using a gas-permeable membrane. A cylindrical piece of silver foil was used as a pseudo-reference electrode. A potential of $+150$ mV versus Ag/AgCl was applied to enzyme electrodes with a BBC microcomputer equipped with a programmable interface developed in our laboratory[1] (Figure 7). The amperometric responses of the electrodes were recorded by the computer. Probes responded very rapidly to CO both in solution and as a gas, reaching a steady-state current in less than 15 s. The current obtained was directly proportional to aqueous CO concentration up to approximately 60 $\mu M$ (Fig. 8), which is well above the $K_m$ of the enzyme in homogenous solution (apparent $K_m$ for CO $= 0.6$ $\mu M$). The probes were relatively stable with the current decreasing by approximately 12% per hour on continual operation. Improved stability, however, is a priority for the development of a commercially useful device and alternative immobilisation techniques, stabilising agents and the use of intact microorganisms are being investigated.

---

[1]The programmable biosensor interface package was developed by A.P.F.T. in association with ARTEK.

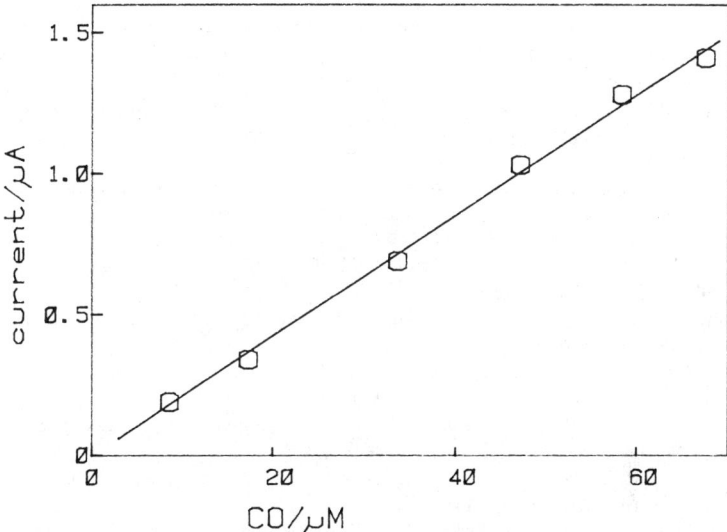

**Fig. 8.** Calibration curve for an enzyme-based CO probe. Carbon monoxide was added as aliquots of CO-saturated buffer and the steady-state output of the sensor were recorded.

The substantial currents produced by this type of amperometric biosensor coupled with its potential for mass production from simple materials suggest that this is a promising route to a cheap miniature CO sensor with high specificity. Its ready marriage with production techniques and established principles in the semi-conductor industry could prove expedient in the development of these devices.

## Acknowledgments

A.P.F.T. is a Senior Research Fellow of the British Diabetic Association. H.A.O.H. is a member of the Oxford Enzyme Group. The authors wish to thank Genetics International Inc. for financial assistance.

## References

Albery, W. J., Eddowes, M. J., Hill, H. A. O. and Hillman, A. R. (1981). Mechanism of the reduction and oxidation of cytochrome *c* at a modified gold electrode. *Journal of the American Chemical Society* **103**, 3904–3910.

Aston, W. J. and Turner, A. P. F. (1984). Biosensors and biofuel cells. *In* "Biotechnology and Genetic Engineering Reviews" Vol. 1 (Ed. G. E. Russell), pp. 89–120. Intercept Ltd., Newcastle upon Tyne.

Cass, A. E. G., Davis, G., Francis, D. G., Hill, H. A. O., Aston, W. J., Higgins, I. J., Plotkin, E. V., Scott, L. D. L. and Turner, A. P. F. (1984). Ferrocene-mediated enzyme electrode for amperometric determination of glucose. *Analytical Chemistry* **56**, 667–671.

Davis, G., Hill, H. A. O., Aston, W. J., Turner, A. P. F. and Higgins, I. J. (1983). Bioelectrochemical fuel cell and sensor based on a quinoprotein methanol dehydrogenase. *Enzyme and Microbial Technology* **5**, 383–388.

Diviès, C. (1975). Remarques sur l'oxidation de l'ethanol par une 'electrode microbienne' de *Acinetobacter xylinium. Annals of Microbiology* **126A**, 175–186.

Goodson, L. H. and Jacobs, W. B. (1974). Application of immobilised enzymes to detection and monitoring. *In* "Enzyme Engineering" Vol. 2 (Eds. E. K. Pye and L. B. Wingard, Jr.), pp. 393–400. Plenum Press, New York.

Guilbault, G. G. (1983). Determination of formaldehyde with an enzyme-coated piezoelectric crystal detector. *Analytical Chemistry* **55**, 1682–1684.

Hikuma, M., Kubo, T., Yasuda, T., Karube, I. and Suzuki, S. (1980). Ammonia electrode with immobilised nitrifying bacteria. *Analytical Chemistry* **52**, 1020–1024.

Hill, H. A. O., Walton, N. J. and Higgins, I. J. (1981). Electrochemical reduction of dioxygen using a terminal oxidase. *FEBS Letters* **126**, 282–284.

Karube, I., Okada, T. and Suzuki, S. (1982). A methane gas sensor based on methane oxidising bacteria. *Analytica Chimica Acta* **135**, 61–67.

Lyons, C. M., Williams, E. and Colby, J. (1983). Characterisation of novel strains of carboxydobacteria. *Society for General Microbiology Quarterly* **9**, (4), M7.

Meyer, O. and Schlegel, H. G. (1983). Biology of aerobic carbon monoxide-oxidising bacteria. *Annual Reviews in Microbiology* **37**, 277–310.

Nicholson, R. S. and Shain, I. (1964). Theory of stationary electrode polarography. *Analytical Chemistry* **36**, 706–723.

Okada, T., Karube, I. and Suzuki, S. (1981). Microbial sensor system which uses *Methylomonas* sp. for the determination of methane. *European Journal of Applied Microbiology and Biotechnology* **12**, 122–125.

Suzuki, S. and Karube, I. (1982). Microbial sensors for gas analysis. *In* "Enzyme Engineering" Vol. 6 (Eds. E. K. Pye and L. B. Wingard, Jr.), pp. 387–393. Plenum Press, New York.

Turner, A. P. F., Aston, W. J., Higgins, I. J., Davis, G. and Hill, H. A. O. (1982). Applied aspects of bioelectrochemistry: Fuel cells, sensors and bioorganic synthesis. *Biotechnology and Bioengineering Symposium* **12**, 401–412.

Turner, A. P. F., Ramsay, G. and Higgins, I. J. (1983). Applications of electron transfer between biological systems and electrodes. *Biochemical Society Transactions* **11**, 445–448.

Turner, A. P. F., Aston, W. J., Higgins, I. J., Bell, J. M., Colby, J., Davis, G. and Hill, H. A. O. (1984). Carbon monoxide: Acceptor, oxidoreductase from *Pseudomonas thermocarboxydovorans* strain C2 and its use in a carbon monoxide center. *Analytica Chimica Acta* **163**, 161–174

# Part V
# Methane

# 11

# Methane Oxidation by Microorganisms

HOWARD DALTON AND DAVID J. LEAK[1]

*Department of Biological Sciences, University of Warwick, Coventry, United Kingdom*

## Introduction

It has been estimated (Ehhalt, 1976) that around 500 million to 800 million tons of methane are produced annually and released into the atmosphere as an end product of anaerobic microbial degradation of organic material. This is matched by an equal amount produced from natural gas wells. In addition, there are very large proven resources under the Rocky Mountains in the United States as well as speculative reports of non-biogenic resources of methane inside the earth which could sustain our present levels of energy consumption for the next million years or so (Paterson, 1978). Clearly, methane represents an almost inexhaustible carbon and energy substrate for bacterial growth, a phenomenon which has been recognised since the early part of this century.

Research into microbial methane oxidation has moved very rapidly in the last decade or so due to three important findings. The first was by Whittenbury *et al.* (1970), who demonstrated that it was possible to isolate, in pure culture, a wide range of methane-utilizing bacteria (methanotrophs) from many different habitats. The second was by Ribbons and Michalover (1970), who were the first to demonstrate that extracts of *Methylococcus capsulatus* (Texas) could be prepared which retained methane-oxidizing activity and that such extracts required NADH, oxygen and methane for activity. The third was the discovery, in our laboratory (Colby *et al.*, 1977), that the enzyme responsible for the conversion of methane to methanol, methane monooxygenase (MMO), was also able to insert an atom of oxygen into a wide variety of organic compounds. These discoveries have led to a growing interest in methane oxidation in many groups throughout the world, partly due to the commercial potential afforded by the partial $O_2$-

---

[1]Present address: Centre for Biotechnology, Imperial College of Science and Technology, London SW7 2AZ, United Kingdom.

MICROBIAL GAS METABOLISM:
MECHANISTIC, METABOLIC
AND BIOTECHNOLOGICAL ASPECTS

dependent oxidation of organic compounds to products of increased value, an area of considerable patent interest (see Higgins *et al.*, 1980).

### The Methane-Oxidizing Bacteria

A pure culture of a methanotroph was first isolated by Söhngen in 1906 and named by him *Bacillus methanicus*. Reference to this organism disappeared shortly thereafter, although there were a number of cursory references to methanotrophs in subsequent years, until Dworkin and Foster (1956) reisolated the organism and named it *Pseudomonas methanica*. This organism was the subject of many of the pioneering studies on cometabolism by Foster's group in the late 1950s and has now been renamed as *Methylomonas methanica*. In subsequent years, before Whittenbury's group published their seminal work, three methanotrophs were isolated: *Methanomonas methanooxidans* (Brown *et al.*, 1964), *Pseudomonas methanitrificans* (Davis *et al.*, 1964) and *Methylococcus capsulatus* (Foster and Davis, 1966). All of these organisms were gram-negative, aerobic and capable of growth only on methane or methanol, being incapable of growth on substrates containing more than one carbon atom.

In 1970 Whittenbury *et al.* isolated over 100 strains of methanotrophs in pure culture. One of the consequences of the isolation of these strains was that a simple classification scheme was proposed which divided the methanotrophs into two types, I and II, based largely on internal membrane arrangements and cell shape. By some curious form of tautology a subsequent paper in the same issue of the journal (Davies and Whittenbury, 1970) reassigned the membrane types to different groups and this classification has been subsequently adopted in the literature. Essentially, organisms belonging to the groups *'Methylococcus'*, *'Methylomonas'* and *'Methylobacter'* possessed a type I membrane arrangement in which the membranes were aggregated into bundles throughout the cytoplasm, whereas membranes of the groups *'Methylosinus'* and *'Methylocystis'* were less ordered than the type I membranes, were generally peripherally arranged and enclosed a lumen of varying dimensions. A number of other features, mainly biochemical, which strengthen this original classification scheme have been added in subsequent years and are incorporated into Table 1.

After these isolation procedures for obligate methanotrophs were described, Patt *et al.* (1974) reported the isolation of a facultative methanotroph from Lake Mendota, Wisconsin, U.S.A., which was able to grow on multicarbon substrates as well as methane and methanol. They named the organism *Methylobacterium organophilum* XX. Unfortunately, it lost the ability to grow on methane after subculture on methanol or multicarbon substrates, possibly due to loss of a plasmid encoding the methane monooxygenase. This view was strengthened by the observation that a plasmid of high $M_r$ was present in cells grown on methane (Schilling, 1978, in Hanson, 1980), but could not be isolated from cells grown

**Table 1.** Tentative classification scheme for methane-oxidizing bacteria[a]

| Determinants | Type I | | Type II | |
|---|---|---|---|---|
| | Subgroup A | Subgroup B | Subgroup obligate[b] | Subgroup facultative[c] |
| Membrane arrangement | Bundles of vesicular disks | | Paired membranes around cell periphery | |
| Resting stages | Cysts (*Azotobacter*-like) | | Exospores or lipid cysts | |
| Major carbon assimilation pathway | RuMP (hexulosephosphate synthase +) | | Serine pathway (hydroxypyruvate reductase +; hexulosephosphate synthase −) | |
| TCA cycle | Incomplete (2-oxoglutarate dehydrogenase negative) | | Complete | |
| Nitrogenase | Some + | | + | |
| Predominant fatty acid C chain length | 16 | | 18 | |
| Presence of RuBP carboxylase | − | + | | − |
| DNA base ratio (%G+C) | 50–54 | 62.5 | 62.5+ (where tested) | |
| Isocitrate dehydrogenase | NAD- or NADP-dependent | NAD-dependent | NADP-dependent | |
| Cell shape | Rod and ? coccus | Coccus | Rod and vibrio | |
| Growth at 45 °C | Some + | + | − | |
| Presence of glutamate dehydrogenase when grown on ammonia | Present | Absent (uses alanine dehydrogenase) | Absent (uses GS/GOGAT) | |
| Examples | *Methylomonas methanica* and *Methylomonas albus* | *Methylococcus capsulatus* | *Methanomonas methanooxidans*, *Methylosinus trichosporium* (both obligate) and *Methylobacterium organophilum* (facultative) | |

[a]Not all strains classifiable into type I and type II have been shown to possess all the biochemical characteristics outlined in this scheme.

[b]Use methanol and formaldehyde as carbon and energy source, but not $C_2+$ compounds.

[c]Use variety of organic compounds, e.g. glucose as carbon and energy source.

on other substrates. Other workers have also reported the isolation of facultative methanotrophs (Patel *et al.*, 1978a; Lynch *et al.*, 1980). There is also a recent report (Zhao and Hanson, 1983) that a third type of methanotroph has been isolated which, although obligate in its requirements for methane or methanol, can be grown in the presence of 0.8% glucose. Under these circumstances, up to 84% of cellular carbon can be derived from the added glucose.

The existence of facultative methanotrophs was seriously questioned recently following a report by Lidstrom-O'Connor *et al.* (1983) on their own previously published isolate '*Methylobacterium ethanolicum*' H414. This isolate has now been shown to consist of a stable mixture of two methylotrophs; one is an obligate methanotroph (strain POC) and is similar to '*Methylocystis*' species as described by Whittenbury *et al.* (1970), whereas the other is a *Xanthobacter* species (strain H414) which fixes dinitrogen and grows on $H_2 + CO_2$, methanol and multicarbon substrates. The POC strain was also shown to contain three cryptic plasmids which were present only in this member of the consortium. Since the methanotroph would be present only in very low numbers on methanol or multicarbon substrates, this may explain why plasmids appeared in the *M. organophilum* XX strain when grown on methane but not when grown on other carbon compounds. The authenticity of other pure facultative methanotrophs must await independent verification.

The ability to use methane as a sole carbon and energy source is not restricted to the prokaryotes since there have been reports from Wolf and Hanson (1979, 1980) and Wolf (1981) that five methanotrophic yeast strains could be isolated by selective enrichment. The organisms proved difficult to grow in liquid volumes greater than 15 ml or when the cultures were forcibly aerated. These limitations, coupled with the generation times of 2 to 7 days, have not made them very attractive candidates for further study.

## Enzymology

*Catabolic Pathways*

Methane is sequentially oxidized by methanotrophs to methanol, formaldehyde, formate and finally carbon dioxide (Fig. 1) by a series of enzymes which are now fairly well characterized.

*Methane monooxygenase.* When methane oxidation was first demonstrated in extracts of *M. capsulatus* (Texas) by Ribbons and Michalover (1970), the enzyme was associated with particulate fractions of the cell by measurement of $CH_4$- and NADH-dependent oxygen consumption (Ribbons, 1975). Despite the fact that Higgins and Quayle (1970) had demonstrated that methanol was the first

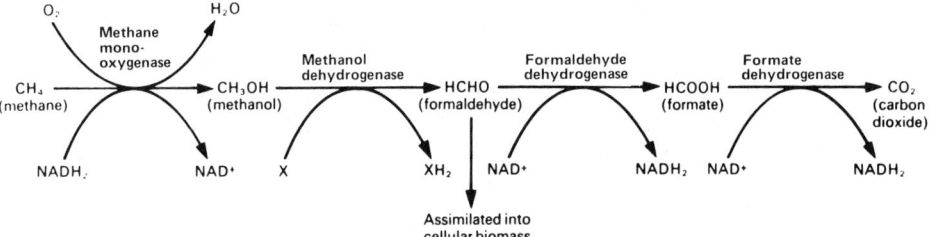

**Fig. 1.** Pathway of methane catabolism in methane-oxidizing bacteria; X is the quinone form of PQQ and $XH_2$ is the quinol form (see Fig. 3).

stable product of methane oxidation, the only product that could be detected in these particular preparations was formate. The high levels of methanol, formaldehyde and NADH oxidation in particles precluded attempts to define an unequivocal stoicheiometry for the system. Subsequently, Ferenci (1974) and Colby et al. (1975) reported that both carbon monoxide and bromomethane would serve as substrates for the particulate MMO from *M. methanica* and enabled the following stoicheiometry to be assigned for the reaction:

$$CH_4 + NADH + H^+ + O_2 \rightarrow CH_3OH + NAD^+ + H_2O$$

A particular preparation was also observed to catalyse methane oxidation in '*Methylosinus trichosporium*' OB3b by Tonge et al. (1975). Like the preparations from *M. capsulatus* (Texas) and *M. methanica,* the system from '*M. trichosporium*' OB3b was sensitive to a range of electron transport inhibitors and would use NADH as its electron donor. Unlike the other particulate preparations, the extract was reported to use ascorbate and methanol–methanol dehydrogenase (MDH) couple as alternative sources of electrons for the MMO, although ascorbate-driven activity could not be observed in more recent studies (Stirling and Dalton, 1979a; Scott et al., 1981b; Patel et al., 1982). The enzyme was resolved into three components (Tonge et al., 1977): a protein of $M_r$ 47,000 containing 1 atom of copper per mole, a protein of $M_r$ 9,000 and a cytochrome c ($M_r$ 13,000) containing 0.3–0.8 atom of copper per mole. The only effective donors for the purified system were ascorbate and the methanol–MDH couple; NADH was ineffective. It was proposed that electron transport intermediates involved in NADH utilization were separated from the MMO during purification but that the methanol–MDH couple (which does not produce NADH) could recycle electrons to the purified MMO. Stirling and Dalton (1979a) also investigated the '*M. trichosporium*' OB3b system and reported that the MMO was entirely soluble, would use only NADH or NADPH as electron donor and was insensitive to electron transport inhibitors, a variety of metal-chelating agents, cyanide and other inhibitors. In most respects, the preparation resembled the soluble system from *Methylococcus capsulatus* (Bath), which was originally isolated as a ''soluble" enzyme being non-sedimentable at 160,000 $g$ for 1 hr (Colby and Dalton, 1976). The latter enzyme was resolved into three components by DEAE–cellulose chromatography (Colby and Dalton, 1978) and these components have now been purified to near homogeneity (Colby and Dalton, 1979; Dalton, 1980; Woodland and Dalton, 1984). Component A has an $M_r$ of about 220,000 and is comprised of three subunits of $M_r$ 54,000, 42,000 and 17,000 which are present in stoicheiometric amounts. The protein contains non-haem iron (2 to 3 mole per mole), zinc (0.2 to 0.5 mole per mole) and no acid-labile sulphide. Based on electron paramagnetic resonance (EPR) studies, it appears that component A is the site of hydrocarbon substrate binding (Dalton, 1980; Woodland and Dalton, 1984).

Protein B is a low-molecular-weight ($M_r$ 20,000) acidic protein with a single polypeptide chain and is apparently devoid of metals or prosthetic groups. Protein C is also a single polypeptide of $M_r$ 39,000 to 44,000. It contains (per mole) 1 mole FAD and 1 mole each of non-haem iron and acid-labile sulphide. Protein C is the only protein in the complex that has any measurable catalytic activity in isolation: it will catalyse electron transfer from NADH or NADPH to a variety of acceptors such as cytochrome $c$, 2,6-dichlorophenolindephenol (DCPIP), potassium ferricyanide, oxygen (although it is a poor acceptor) and protein A. Based on studies of the reconstituted complex, it has been possible to propose a tentative scheme for electron transfer and substrate hydroxylation in this sytem (Fig. 2).

It is possible to obtain soluble MMO preparations from other organisms also. Scott *et al.* (1981a,b) have found that extracts of '*M. trichosporium*' OB3b do indeed possess a soluble MMO, but only when the cells are grown under meth-

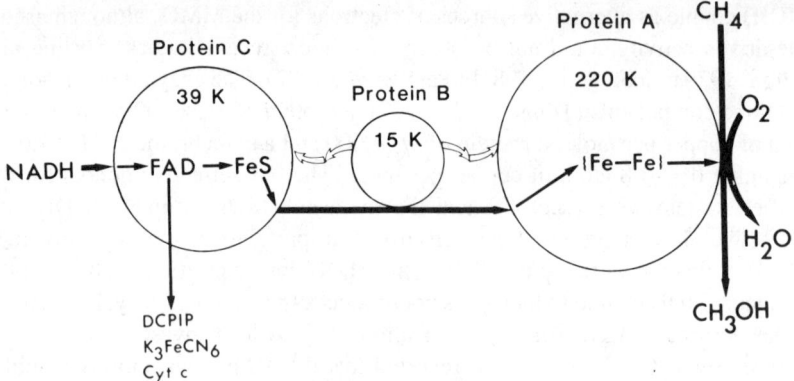

**Fig. 2.** Possible pathway of electron transfer from NADH to the A, B and C proteins of the methane monooxygenase complex. The role of protein B within the complex is not clear at present, but it does interact with protein A during electron transfer (compiled from Colby and Dalton, 1979; Woodland and Dalton, 1984; J. Lund, M. P. Woodland and H. Dalton, unpublished observations).

ane- or nitrate-limiting conditions; oxygen-limited conditions gave wholly particulate MMO. In each case, only reduced pyridine nucleotides would serve as electron donors for MMO; neither ascorbate or methanol would so serve. Two other reports of soluble MMO have recently appeared. One (Patel *et al.*, 1982) is in a facultative methanotroph, *Methylobacterium* sp. CRL26, where a crude separation on DEAE–cellulose gave three fractions which eluted in an identical manner to that reported for the *M. capsulatus* (Bath) system (Colby and Dalton, 1978). Maximum activity could be obtained only with all three fractions, although 50% activity was observed with A and C together, indicating contamination of A with protein B. The second (Allen *et al.*, 1983) is from a type II methanotroph SB-1, where the enzyme was resolved into two fractions by DEAE–Biogel chromatography. The presumed oxygenase component A was estimated to have an $M_r$ of 250,000 and to comprised of four subunits of $M_r$ 60,000 each, containing 0.5 g atom of iron per mole.

Although the prime catabolic role for the enzyme is to convert methane to methanol, it was observed several years ago that MMOs from *M. capsulatus* (Bath), '*M. trichosporium*' OB3b and *M. methanica* will catalyse a wide range of oxygenative reactions, including the oxidation of alkanes, alkenes, ethers, alicyclic, aromatic and heterocyclic compounds (Colby *et al.*, 1977; Stirling *et al.*, 1979). This ability is also manifest by whole cells (Stirling and Dalton, 1979b; Higgins *et al.*, 1979) and is covered in more detail below in the section on 'Exploitation of Methane-Oxidizing Bacteria'.

*Methanol dehydrogenase.* This enzyme catalyses the NAD-independent oxidation of methanol to formaldehyde and was initially characterized from methanol-grown *Pseudomonas* M27 by Anthony and Zatman (1964a,b, 1965). This organism will not grow on methane but the enzyme appears to be very similar to that isolated from the methanotroph *M. capsulatus* (Texas) (Patel and Hoare, 1971; Patel *et al.*, 1972, 1973; Wadzinksi and Ribbons, 1975). The methanol dehydrogenase from these sources had an $M_r$ of 120,000 and was comprised of two equal subunits of $M_r$ 60,000. The primary electron acceptor *in vitro* was phenazine methosulphate, and ammonia or methylamine was required as an activator at the optimum pH value of 9. A wide range of primary alcohols as well as formaldehyde will serve as substrates (Anthony and Zatman, 1967; Patel *et al.*, 1972; Sperl *et al.*, 1974; Patel and Felix, 1976) and the alternative name of primary alcohol dehydrogenase has been used to describe this enzyme. The enzyme from the methanotrophs *M. methanica* (Patel *et al.*, 1978b) and '*Methylosinus sporium*' (Patel and Felix, 1976) are single polypeptides of $M_r$ 60,000. Early studies with *Pseudomonas* M27 indicated that the prosthetic group of this enzyme was a pteridine (Anthony and Zatman, 1967), but subsequent independent characterization has shown this to be an orthoquinone compound, pyrroloquinoline quinone (PQQ, Fig. 3) (Salisbury *et al.*, 1979; Duine *et al.*, 1980).

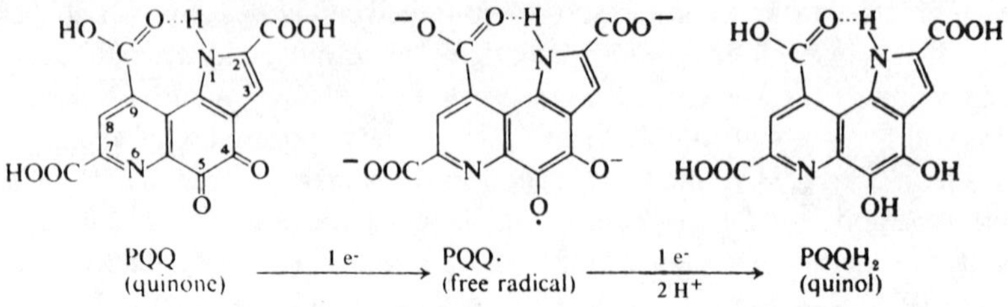

PQQ      1 e⁻      PQQ·      1 e⁻      PQQH₂
(quinone)    →    (free radical)    →    (quinol)

**Fig. 3.** Prosthetic group of methanol dehydrogenase; PQQ is pyrroloquinoline quinone and is the oxidized form. The free radical and quinol forms are the one-electron and two-electron reduction forms of the quinone, each of which can be extracted from methanol dehydrogenase (after Duine *et al.*, 1980).

PQQ has now been found in other dehydrogenases including alcohol, aldehyde, glucose and amine dehydrogenases, which have been collectively termed quinoproteins.

*In vivo*, methanol dehydrogenase is now thought to pass electrons directly to cytochrome *c* (Duine *et al.*, 1979; O'Keefe and Anthony, 1980) but this functional coupling occurs *in vitro* only if the enzyme is purified anaerobically and does not require an activator ($NH_3$); aerobically prepared MDH (or MDH oxidized by ferricyanide) does not reduce cytochrome *c*, but reduce Wurster's Blue in the presence of an activator (Duine *et al.*, 1979). Recently, Beardmore-Gray *et al.* (1983) showed that MDH from certain methanol-grown organisms interacts directly with the low-isoelectric-point form ($C_L$) of cytochrome *c* but not the high-isoelectric-point form ($C_H$) as electron acceptor. Furthermore, this interaction occurs under aerobic conditions in the absence of an activator and produces formaldehyde as the product. Formate can also be produced in this MDH-catalysed reaction since formaldehyde also serves as a substrate for the enzyme and is oxidized at the same rate as methanol (Beardmore-Gray *et al.*, 1983).

*Formaldehyde dehydrogenase.* Formaldehyde may be metabolized by at least four different enzymes in methanotrophs: two are involved in catabolic dissimilation to formate [$NAD(P)^+$-dependent and $NAD(P)^+$-independent dehydrogenases] and thence to $CO_2$, leading to a loss of carbon from the system, and two are involved in the assimilation of carbon via either the serine or hexulose monophosphate pathway, depending on the organism. An alternative role for the formaldehyde-assimilating enzyme in type I methanotrophs, possessing the hexulose monophosphate pathway, has been proposed in which some of the enzymes of this pathway, in conjunction with 6-phosphogluconate dehydrogenase, oxidize formaldehyde to $CO_2$ with the net generation of 2 moles of NADH (Strom *et al.*, 1974, Colby and Zatman, 1975) (Fig. 4). Definitive proof for the operation of this cycle *in vivo* is not available at present but the enzymic capability is present in *M. capsulatus* (Texas and Bath) and *M. methanica* (Strom *et al.*, 1974;

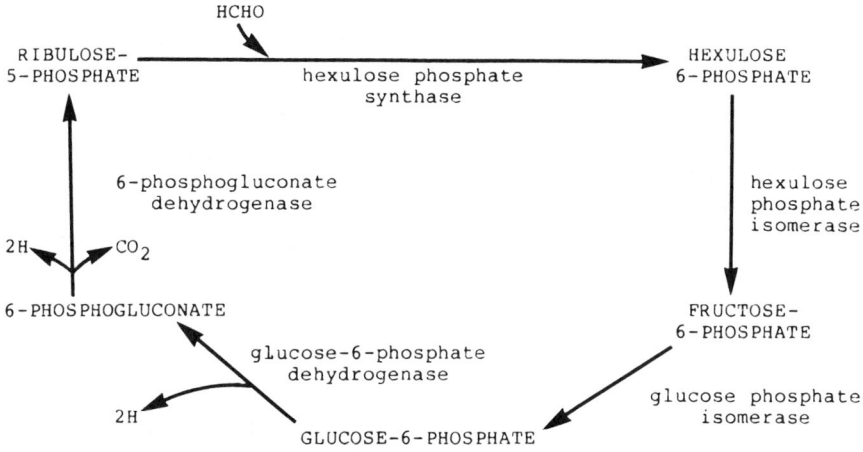

**Fig. 4.** Cyclic pathyway of formaldehyde oxidation.

Stirling, 1978). It has been pointed out that the metabolic significance of this dissimilatory ribulose monophosphate (RuMP) cycle in methanotrophs is questionable (Zatman, 1981), since the activities of the key enzymes, glucose-6-phosphate dehydrogenase (GPDH) and 6-phosphogluconate dehydrogenase (PGDH), are low and those of formaldehyde and formate dehydrogenases are high. In other RuMP-type non-methane-utilizers, which have low or non-existent formaldehyde and formate dehydrogenases, the dissimilatory cycle is probably significant since their $C_6$ dehydrogenases are present at high levels. Furthermore, Davey *et al.* (1972) reported that the GPDH and PGDH enzymes in type I methanotrophs were NADP$^+$-specific and may therefore serve as a source of NADPH for biosynthetic purposes rather than as a source of NADH. Both Stirling (1978) and Stanley and Dalton (1982) agree with Davey *et al.* (1972) that PGDH is NADP$^+$-specific but disagree on the specificity of the GPDH, which they find will use either NAD$^+$ or NADP$^+$ with almost equal facility. When assays were done at pH 8.0, Stirling (1978) observed rates for this enzyme that were comparable with the rates seen in non-methane-utilizing methylotrophs raising the possibility that this cycle could provide significant levels of reduced pyridine nucleotides. The NAD(P)$^+$-independent formaldehyde dehydrogenases use phenazine methosulphate (PMS) as electron acceptor and have beenfound in *M. capsulatus* (Texas) (Patel and Hoare, 1971) and '*M. sporium*' (Patel and Felix, 1976), the latter being purified as a PMS-dependent methanol dehydrogenase which also oxidized formaldehyde to formate. These enzymes also oxidized acetaldehyde. The enzymes from '*M. trischosporium*' (Patel *et al.*, 1980) and '*Methylomonas albus*' (BG8) (Leak and Dalton, 1983) also oxidized other aldehydes including straight-chain and aromatic aldehydes. NAD(P)$^+$-dependent enzymes were found in '*M. methanica*' (Johnson and Quayle, 1964; Harrington and Kallio, 1960), *M. capsulatus* (Bath) (Stirling and Dalton, 1978) and '*M. albus*' BG8 (Leak and Dalton, 1983) and may be present in '*M. trischosporium*' OB3b (Stirling and Dalton, 1979a). The enzyme from '*M. albus*' BG8 also required glutathione for activity and was specific for formaldehyde, whereas the enzyme from *M. capsulatus* (Bath) did not require glutathione and also oxidized glyoxal, glycoaldehyde and glyceraldehyde but not other aldehydes.

*Formate dehydrogenase.* In all methanotrophs tested, this enzyme is present as a soluble $NAD^+$-linked dehydrogenase which appears to be specific for formate and does not oxidize other acids. The activity of the enzyme in cell extracts is generally high, although there are some exceptions to this in type II organisms (see Zatman, 1981). It has been suggested that these organisms generate NADH from their complete tricarboxylic acid (TCA) cycle (type I organisms have an incomplete TCA cycle) to compensate for lack of formate dehydrogenase activity.

*Anabolic Pathways*

Carbon is assimilated at the level of formaldehyde via either a RuMP pathway or a serine pathway and can be correlated with the type of membrane system observed. Type I organisms use the RuMP pathway and type II organisms use the serine pathway. The pathway used by the methanotrophic yeasts has not been elucidated but is probably neither of the pathways used by bacteria since it is believed that carbon is assimilated as carbon dioxide and not as formaldehyde.

The biochemistry of the assimilation pathways has been extensively reviewed elsewhere (Quayle and Ferenci, 1978; Colby *et al.*, 1979; Quayle, 1980; Higgins *et al.*, 1981; Anthony, 1982) and therefore will not be considered in detail here. However, there are several features about each of the pathways which are worthy of comment.

*Ribulose monophosphate pathway.* There are a number of variables of the RuMP pathway in type I organisms, depending on whether fructose-6-phosphate is isomerized to glucose-6-phosphate [for the 2-keto-3-deoxy-6-phosphogluconate (KDPG) aldolase pathway] or phosphorylated to fructose 1,6-bisphosphate (for the FBP aldolase pathway). In the methanotrophs that have been studied in detail, *M. capsulatus* (Texas and Bath) and '*M. methanica*', it appears that the main route of carbon assimilation is the KDPG aldolase pathway, although the FBP aldolase route functions in '*M. methanica*' and thereby provides both glyceraldehyde-3-phosphate (G3P) for the rearrangement reactions and the $C_3$ components of the glycolytic sequence. In the *M. capsulatus* strains, FBP aldolase is either very low (Texas strain) or absent (Bath strain), which raises the question 'how do these organisms satisfy the demand for G3P in the rearrangement reactions and generate their glycolytic intermediates?' The answer to this dilemma was proposed by Stanley and Dalton (1982) when they investigated the role played by ribulosebisphosphate carboxylase (RuBPcase), previously observed to be present in *M. capsulatus* (Bath) (Taylor, 1977). They found that the combined action of phosphoribulokinase and RuBPcase produced 2 moles of 3-phosphoglycerate (PGA) from 1 mole of RuMP (Fig. 5). One mole of PGA was used for the synthesis of glyceraldehyde phosphate to replenish the arrangement reactions and the other mole of PGA was used for the synthesis of the glycolysis components. All other type I organisms which lacked the RuBPcase had significant levels of FBP aldolase and presumably used that route for the synthesis of $C_3$ intermediates.

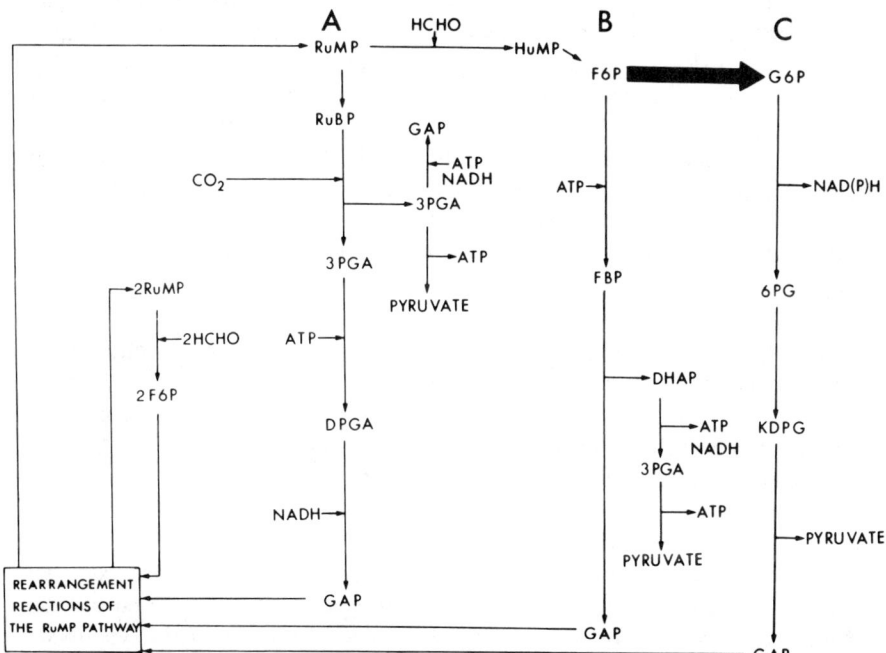

**Fig. 5.** Alternative routes of $C_6$ cleavage of the RuMP pathway in *M. capsulatus* (Bath). (A) RuBPcase route; (B) FBP aldolase route; (C) KDPG aldolase route. RuMP, ribulose monophosphate; HuMP, hexulose monophosphate; F6P, fructose-6-phosphate; G6P, glucose-6-phosphate; RuBP, ribulose bisphosphate; GAP, glyceraldehyde-3-phosphate; PGA, phosphoglyceric acid; FBP, fructose bisphosphate; 6PG, 6-phosphogluconic acid; DPGA, diphosphoglyceric acid; DHAP, dihydroxyacetone phosphate; KDPG, keto-deoxy-6-phosphogluconate. The bold arrow between F6P and G6P indicates that this reaction is much more active than the F6P to FBP enzyme.

It is interesting to note here that the primary structure of the large subunit of RuBPcase from three plant species bears a certain resemblance to that of class I FBP aldolase from rabbit muscle (Poulsen, 1981). Furthermore, they are enzymes of the same class (the lyases), the substrate of one is a stimulator–inhibitor of the other, they both form Schiff bases with pyridoxal-5'-phosphate, both are inactivated and modified by the same analogue, $N$-bromoacetyl ethanolamine phosphate, and the same types of amino acid chains are involved in catalysis by both enzymes. Although the class II FBP aldolases from bacteria have even more catalytic features in common with the RuBPcase, it is not possible to make any comparisons of the primary structures since they are not available for the bacterial aldolases. However, it is tempting to speculate on the evolutionary relationship between the aldolase and carboxylase in *M. capsulatus* based upon these similarities and the suggestions that the Calvin cycle may have arisen from glycolysis (McFadden, 1973) and that the RuMP cycle was a staging post in the development of this $CO_2$ fixation cycle (Quayle and Ferenci, 1978). If the RuMP cycle was a precursor of the Calvin cycle, and organisms, such as the autotrophs, which acquired this ability are evolutionarily more advanced, then *M. capsulatus* (Bath) is clearly a less primitive representative of the present-day methanotrophs. Unfortunately, *M. capsulatus* (Bath) does not grow autotrophically (Stanley and Dalton, 1982) because of its low FBP aldolase activity, which is necessary for a functional Calvin cycle. It is possible, therefore, that this organism represents an intermediate form between the primitive RuMP-type organisms and the chemolithotrophs such as *Nitrosomonas* [which incidentally also oxidize methane (Hyman and Wood, 1983) and have an incomplete TCA cycle].

*Serine pathway.* The key intermediates in this pathway of formaldehyde fixation are carboxylic acids and amino acids as opposed to the phosphorylated sugars in the RuMP pathway (Fig. 6). A number of enzymes are regarded as peculiar to this cycle—namely serine transhydroxymethylase, malyl-CoA lyase, hydroxypyruvate reductase and serine-glyoxylate aminotransferase. Very little work has been done on the occurrence of the complete pathway in methanotrophs and much evidence for its existence in the type II organisms has relied on the identification of one or two of the above key enzymes. This, of course, can give a false impression, particularly since several type I organisms which use the RuMP pathway have been shown to contain low levels of hydroxypyruvate reductase and malyl-CoA lyase, the exact role of which is still uncertain. There are basically two variants of the serine pathway; the icl$^+$ variant, in which isocitrate lyase is involved in the oxidation of acetyl CoA to glyoxylate, and the icl$^-$ variant, which lacks isocitrate lyase. Although Kortstee (1980, 1981) proposed that the icl$^-$ variants oxidized acetyl CoA via homoisocitrate, this work has not been substantiated by others (Bellion *et al.*, 1981) and still remains an enigma.

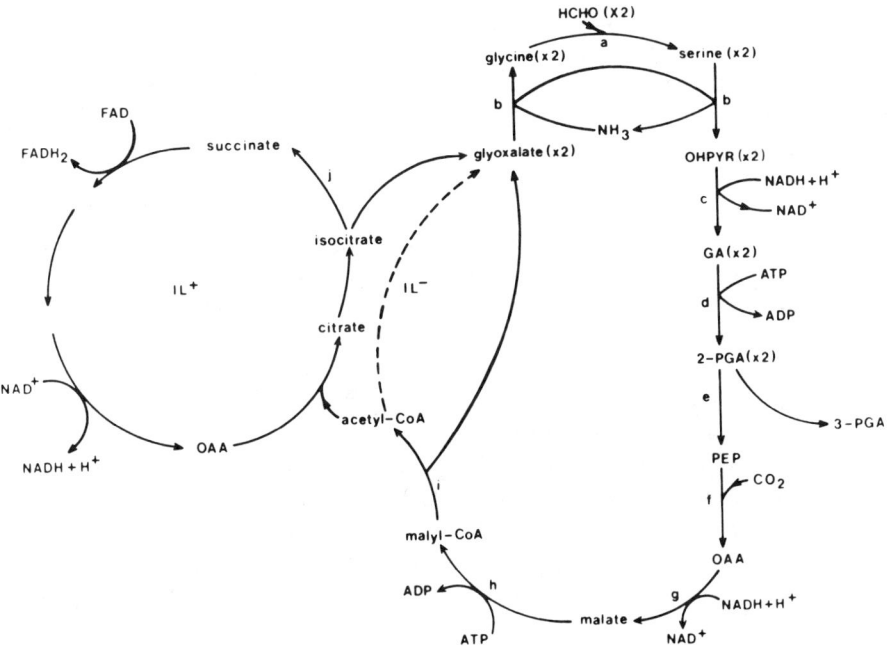

**Fig. 6.** The serine pathway: a, serine transhydroxymethylase; b, serine-glyoxylate aminotransferase; c, hydroxypyruvate reductase; d, glycerate kinase; e, phosphopyruvate hydratase; f, phosphoenolpyruvate carboxylase; g, malate dehydrogenase; h, malate thiokinase; i, malyl-CoA lyase; j, isocitrate lyase; OHPYR, hydroxypyruvate; GA, glyceric acid; PGA, phosphoglyceric acid; PEP, phosphoenol pyruvate; OAA, oxaloacetic acid; IL, isocitrate lyase; - - -, unknown reactions.

## Physiology

### Growth Energetics in Methanotrophs

Interest in the application of methane- or methanol-utilizing bacteria as a source of single-cell protein (SCP) has given rise to a plethora of experimental and theoretical studies of growth yields on $C_1$ substrates (see Anthony, 1982, p. 247 for summary). Coupled with the practical problems of low solubility and safety considerations, the lower theoretical and experimental yield than that achieved with methanol makes methane the less attractive substrate for the commercial development of SCP processes. However, the use of methane as a substrate is worth considering here for two reasons. Firstly, the commercial application of co-oxidation (see 'Exploitation of Methane-Oxidizing Bacteria') will require the optimization of growth yields combined with the optimization of expression of the MMO, and this dual consideration may differ from those applied to SCP production. Secondly, there has been considerable discrepancy in the literature concerning the growth yields obtainable with methane (see Leak, *et al.,* Chapter

12, this volume). In some cases these appear to be greater than theoretically possible, assuming that methane oxidation requires NADH as a cofactor (Table 2), and have been cited by Wolfe and Higgins (1979) in support of a scheme for electron recycling from methanol dehydrogenase. The problem is, therefore, fundamental to the elucidation of methane hydroxylation mechanisms *in vivo*, a subject which is discussed in the next section.

Considerations of type I and type II organisms with respect to the effect of different carbon assimilatory pathways on predicted cell yields is the same whether applied to growth on methane or methanol. The relative inefficiency of the type II pathway compared to the type I for the assimilation of formaldehyde results in a lower theoretical yield (Table 2), which has been borne out in practice in studies with methanol utilizers (Goldberg *et al.*, 1976; Rokem *et al.*, 1978). However, the assumption that growth on methanol and assimilation via the RuMP pathway represents the most attractive combination for the growth of methanotrophs may be erroneous, as measured yields from this combination have been surprisingly low (Linton and Vokes, 1978). This may be the result of poor coupling of NADH oxidation to ATP synthesis, a problem which would be much less significant when considering growth on methane (Anthony, 1982 pp. 266–268), transport phenomena or energy-consuming methanol oxidation by the

**Table 2.**   *Predicted yields on $C_1$ substrates[a]*

| Assimilation | $Y_s$ | $Y_o$ | CCE (%) |
|---|---|---|---|
| Methanol as substrate | | | |
| Serine pathway | 13.9–20.7 | 7.4–15.5 | 55–81 |
| RuMP(KDPG/TA) | 14.7–21.6 | 8.3–17.8 | 58–85 |
| RuMP (FBP/SBP) | 17.0–23.4 | 10.6–21.7 | 67–92 |
| Methane as substrate [hydroxylation requires NAD(P)H] | | | |
| Serine pathway | 9.2–10.4 | 2.8–3.3 | 36–41 |
| RuMP (KDPG/TA) | 9.7–12.0 | 3.0–4.0 | 38–47 |
| Methane as substrate (hydroxylation driven by Mdh) | | | |
| Serine pathway | 9.2–15.1 | 3.0–5.9 | 36–59 |
| RuMP pathway (KDPG) | 9.8–16.2 | 3.1–6.1 | 38–63 |

[a] $Y_s$ = grams dry weight (mole substrate)$^{-1}$, $Y_o$ = grams dry weight (g atom oxygen)$^{-1}$, CCE = carbon conversion efficiency, KDPG = ketodeoxyphosphogluconate (aldolase), TA = transaldolase, FBP = fructose bisphosphate (aldolase), SBP = sedoheptulose bisphosphatase, Mdh = methanol dehydrogenase. The range of predicted values corresponds to P/O ratios between 1 and 3 for NADH oxidation and 1 or 2 for other substrates.

MMO (Colby et al., 1977) which is known to be present under these conditions (Best and Higgins, 1981; Linton and Vokes, 1978). Growth on methane, which might be expected to be optimal for expression of the MMO, therefore becomes more attractive, especially if the high yields reported in some studies can be obtained routinely.

The disparity between theoretical and experimental figures for growth yields on methane probably arises from incorrect assumptions about the energy balance of the first two steps of methane dissimilation in vivo. Even assuming that the steps in the complete oxidation of formaldehyde can reduce $2NAD(P)^+$, current evidence for an absolute requirement for NAD(P)H for the initial hydroxylation of methane (Colby and Dalton, 1976) and the low energy yield from methanol oxidation (probably P/O = 1; Drozd and Wren, 1980; Dawson and Jones, 1981) suggest that the carbon conversion efficiency (CCE)—that is, the percentage of carbon substrate incorporated into biomass—would never be greater than 50% during growth on methane. For M. capsulatus (Bath), assimilating formaldehyde via the KDPG aldolase/transaldolase variant of the RuMP pathway, the theoretical figure is 38 to 47% for CCE, depending on the P/O ratio for respiration from different substrates (Table 2). Although recent studies with Methylococcus Sp NCIB 11083 closely match these figures, previous studies produced figures in excess of 60% for CCE for type I methanotrophs (Whittenbury et al., 1970; Harwood and Pirt, 1972). Although it is not known whether these differences were physiological or strain-dependent, it is evident that such high CCE values could only be achieved, theoretically, by assuming that the oxidation of methane through to formaldehyde was not consuming energy. The recent demonstration that the MMO in M. capsulatus (Bath) and 'M. trichosporium' OB3b can exist in two forms, dependent on growth conditions, may yield an explanation for these discrepancies. Preliminary findings have been presented in this volume (Leak et al., Chapter 12).

*Intracellular Location of the MMO*

The demonstration by Stirling and Dalton (1979a) that 'M. trichosporium' OB3b possessed a soluble MMO, which was substantially different from the particulate system originally described for this organism (Tonge et al., 1975, 1977), indicated that the difference between methane-oxidizing systems in M. capsulatus (Bath) (Colby and Dalton, 1978) and 'M. trichosporium' (Tonge et al., 1977) might have a physiological rather than a taxonomic basis. Scott et al. (1981a) subsequently examined this possibility with 'M. trichosporium' and confirmed that soluble or particulate enzyme systems can be obtained under defined growth conditions, 100% particulate activity being obtained under oxygen limitation, while 100% soluble activity was found under nitrate limitation in continuous culture. However, recent studies in our laboratory (Stanley et al., 1983) demonstrated that these conditions masked an underlying regulatory feature. During a study of the type II organism '*Methylocystis parvus*' OBBP in continuous culture it was observed that growth became limited at relatively low cell densities by some requirement other than carbon, oxygen or nitrogen. Supplementation of the growth medium with various trace elements established this as copper limitation. Although a role for copper in methane oxidation had previously been inferred from studies with other methanotrophs (Takeda *et al.*, 1976; Ohtomo *et al.*, 1977; Hubley *et al.*, 1975), growth limitation by copper had not previously been encountered with *M. capsulatus* (Bath) or '*M. trichosporium*' OB3b when grown under conditions identical to those used for '*M. parvus*' OBBP. Subsequent investigations demonstrated that both of these organisms avoid copper limitation by synthesis of a soluble MMO, which, at least in the case of *M. capsulatus* (Bath), is not a copper-containing enzyme (Woodland and Dalton, 1984). Thus, in a medium with excess copper, both *M. capsulatus* (Bath) and '*M. trichosporium*' OB3b made only particulate MMO, but a progressive decrease in copper concentration, or an increase in biomass concentration, resulted in the appearance of the soluble form of the MMO. To distinguish this phenomenon from growth limitation by copper, as observed with '*M. parvus*' OBBP, it has been referred to as copper stress (Stanley *et al.*, 1983). The independence of this switch in intracellular location of the MMO from other growth limitations was confirmed by demonstrating the transition between particulate and soluble forms with increasing biomass concentration during exponential growth of *M. capsulatus* (Bath). The biomass concentration at which this transition occurred could be altered by changing the concentration of copper in the growth medium (Stanley *et al.*, 1983).

Previous reports of copper limitation of growth (Takeda *et al.*, 1976) and a preliminary screen of representative type I and type II organisms indicate that the ability to avoid copper limitation may be restricted to a few strains of methanotroph. On the basis of this limited range of organisms, the inability to avoid copper limitation in the context described here (and presumably the inability to make a soluble MMO) is not a type-specific characteristic, examples being found in both type I and type II organisms. However, on the basis of the separate classification of *M. capsulatus* (Bath) into type X (Whittenbury and Dalton, 1981), a type I organism capable of making a soluble MMO has yet to be demonstrated.

*Characteristics of the Particulate MMO from M. capsulatus* (Bath)

Although a particulate MMO has been demonstrated in *M. capsulatus* (Texas) (Ribbons, 1975) with an inhibitor sensitivity profile similar to those described for *M. methanica* (Colby *et al.*, 1975) and '*M. trichosporium*' OB3b (Tonge *et al.*,

1975), the recent demonstration of a particulate system in *M. capsulatus* (Bath) allows comparison with the extensively characterized soluble MMO from the same organism.

Major differences were suggested by the polypeptide band profiles on sodium dodecyl sulphate–polyacrylamide gels (SDS–PAGE) of extracts from cells with soluble or particulate MMO (Fig. 7). During the transition from soluble to particulate MMO, three major bands corresponding to fraction A of the soluble MMO decreased in intensity with a concomitant increase in at least three bands of different $M_r$ in the particulate fraction.

The sensitivity of particulate MMO from *M. capsulatus* (Bath) to a wide range of inhibitors, notably to a number of electron transport inhibitors (Table 3), more closely resembled that of previously described particulate systems than of the soluble MMO from the same organism. Although sensitivity to copper-chelating agents such as thiourea suggests that copper is essential to this system, it is difficult to distinguish between direct inhibition of the MMO and inhibition of electron transport to the MMO in crude preparations. Determination of the role of copper will therefore require resolution of the components of the particulate system. Inhibition by thioglycollate, which is now routinely used as a stabilizing

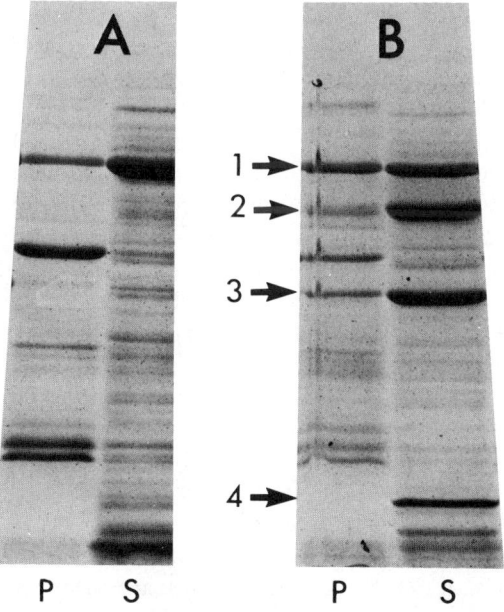

**Fig. 7.** Electrophoresis on 10–30% gradient SDS-polyacrylamide gel of particulate (P) and soluble (S) fractions of cell extracts from *M. capsulatus* (Bath). (A), 100% particulate MMO activity; (B) 100% soluble MMO activity; 1, methanol dehydrogenase; 2, 3 and 4, α, β, and γ subunits of fraction A of the soluble MMO (from Stanley *et al.*, 1983).

**Table 3.**  *Effect of inhibitors on particulate and soluble MMO
from Methylococcus capsulatus* (Bath)[a,b]

| Inhibitor (0.1 mM) | % Activity | |
| --- | --- | --- |
| | Soluble MMO | Particulate MMO |
| None | 100 | 100 |
| Potassium cyanide | 100 | 0 |
| 2-Mercaptoethanol | 100 | 0 |
| 2,2-Bipyridyl | 99 | 10 |
| Thiourea | 90 | 0 |
| Dithiothreitol | 100 | 25 |
| Imidazole | 81 | 42 |
| 8-Hydroxyquinoline | 29 | 0 |
| Thioglycollate | 100[c] | 40 |
| Ethyne (3% in air) | 0 | 0 |

[a]From Stanley *et al.* (1983).
[b]Samples were preincubated for 1 min with inhibitor before addition of
assay substrate (propylene).
[c]Required as stabilizing agent for the soluble MMO.

agent for the soluble MMO (Stanley and Dalton, 1982), provides a useful distinction between soluble and particulate enzymes.

*Requirement for two methane-oxidizing systems.* In considering the role of soluble and particulate methane-oxidizing systems, two features must be borne in mind. The first is that some methanotrophs appear incapable of avoiding copper stress, and growth of these organisms becomes copper-limited at a cell density dependent on the available copper concentration. Therefore, the ability to respond to copper stress by synthesis of a soluble MMO represents a selective advantage to the host organism when in competition for low copper resources. The second feature is that in those organisms capable of making both systems the particulate enzyme is made preferentially. Indeed, some particulate enzyme appears to be present in *M. capsulatus* (Bath) under conditions of copper stress, although *in vitro* activity is demonstrable only after addition of copper to the assay (S. D. Prior and H. Dalton, unpublished observations).

An explanation for this preference comes from recent studies in our laboratory, (Leak *et al.,* Chapter 12, this volume), which have demonstrated that the CCE of *M. capsulatus* (Bath) growing on methane in a high-copper medium (particulate MMO) may be up to 35% higher than in a low-copper medium (soluble MMO). The major changes involved in the transition from soluble to particulate enzyme systems suggest that this increase in efficiency is intimately associated with the process of methane oxidation or electron transport to the MMO *per se.*

Extensive studies of the soluble MMO from *M. capsulatus* (Bath) have demonstrated that NAD(P)H is the only effective electron donor for this enzyme. Theoretical calculations (Anthony, 1982) reveal that with this high NAD(P)H requirement for the initial hydroxylation of methane, an organism such as *M. capsulatus* (Bath) must completely oxidize 43 to 47% of its carbon substrate to produce the NAD(P)H necessary for carbon assimilation via the type I RuMP pathway (KDPG variant), whereas only 11 to 19% must be oxidized to satisfy the remaining ATP demand. As these calculations are based on the use of ammonia as a nitrogen source, the use of nitrate, requiring additional NADH for reduction to ammonia, would exacerbate this situation to the extent that the observed increase in growth efficiency during the transition from soluble to particulate MMO could not be achieved by an increased efficiency of ATP synthesis (Leak *et al.*, Chapter 12, this volume). An increase in growth efficiency of this magnitude must, therefore, arise either from a decreased demand for NADH for methane assimilation or from an increased efficiency of NADH generation during methane dissimilation. As it is assumed in the calculations that both formaldehyde and formate dehydrogenases reduce NAD(P) directly, the latter possibility could be achieved only by linking methanol dehydrogenase to NAD(P) reduction, either directly or by coupling an increased ATP yield from methanol oxidation to NAD(P) reduction via reversed electron transport.

A clue to the basis of this increased growth efficiency came from the observation that ethanol and higher primary alcohols incapable of reducing NAD(P) directly could act as electron donors for the MMO *in vivo*, both in organisms incapable of avoiding copper limitation (e.g. '*M. albus*' BG8, '*M. parvus*' OBBP) and in *M. capsulatus* (Bath) when grown in a high-copper medium (Leak and Dalton, 1983). However, this property was lost concomitantly with the transition from particulate to soluble MMO in *M. capsulatus* (Bath) when transferred from a high- to a low-copper medium (Stanley *et al.*, 1983). On the assumption that primary alcohols are oxidized by the methanol/primary alcohol dehydrogenase in methanotrophs, for which there is substantial evidence, this indicates that methanol, arising from the oxidation of methane, can act as an electron donor for further methane oxidation by the particulate, but not the soluble MMO. The details of the mechanism involved remain to be established. From studies with the particulate MMO from '*M. trichosporium*' OB3b, Tonge *et al.* (1977) have proposed that methanol dehydrogenase can recycle electrons to the MMO without the involvement of NADH. While such a scheme could adequately explain the improved efficiency of growth in *M. capsulatus* (Bath) with particulate MMO, irreproducibility (Scott *et al.*, 1981b) and the failure to demonstrate this in *M. capsulatus* (Bath) *in vitro* leave the situation unresolved. It should be noted that in studies of the ability of ethanol to act as electron donor for the MMO in a variety of organisms, the stoicheiometry of substrate oxidation to ethanol usage never exceeded 0.5 (Leak and Dalton, 1983), and it has been argued that this would be inconsistent with the scheme proposed by Tonge *et al.*

(1975). However, the alternative proposal, for reduction of NAD via reversed electron transport, also seems unlikely in view of the requirement for an MMO with major differences from the soluble enzyme as revealed by SDS–PAGE analysis of cell extracts. This topic is the subject of current investigations in our laboratory.

### Exploitation of Methane-Oxidizing Bacteria

Leadbetter and Foster (1959) were the first to observe that *Pseudomonas methanica* grown on methane was also able to oxidize higher alkanes to their homologous oxidation products. The oxidation products thus formed were not assimilated by the organism, so that the higher alkanes would not serve as growth substrates for the organism. This production of oxidized compounds from co-substrates presented to an organism growing on a different substrate was called 'co-oxidation' by Foster (1962). Since this pioneering work, there have been many examples of the transformation of non-growth substrates by microbes (see Dalton and Stirling, 1982) and the methane oxidizers have proved to be a good model system for study in this respect. The observation that MMO is the enzyme responsible for the insertion of oxygen into many different organic substrates as well as ammonia (Colby *et al.*, 1977; Dalton, 1977) (Table 4) has led to wide-

**Table 4.** *Some substrates and products of the soluble MMO from Methylococcus capsulatus (Bath)*

| | |
|---|---|
| Methane | Methanol |
| Ethane | Ethanol |
| Hexane | Hexan-1-ol, hexan-2-ol |
| Ethylene | Epoxyethane |
| Propylene | Epoxypropane |
| *trans*-But-2-ene | *trans*-2,3-Epoxybutane |
| | *trans*-2-Buten-1-ol |
| Chloromethane | Formaldehyde |
| Dichloromethane | Carbon monoxide |
| Trichloromethane | Carbon dioxide |
| Dimethyl ether | Methanol, formaldehyde |
| Diethyl ether | Ethanol, acetaldehyde |
| Cyclohexane | Cyclohexanol |
| Benzene | Phenol, hydroquinone |
| Toluene | Benzyl alcohol, *p*-cresol |
| Styrene | Styrene epoxide |
| Pyridine | Pyridine-*N*-oxide |
| Naphthalene | $\alpha$- and $\beta$-Naphthol |
| Carbon monoxide | Carbon dioxide |
| Ammonia | Hydroxylamine |

spread interest in their possible use as industrial catalysts in the production of certain oxygen-containing chemicals (Higgins *et al.*, 1980; May and Padgette, 1983). From an economic point of view, the use of whole cells (via co-oxidation) rather than active cell-free preparations is clearly advantageous due to the prohibitive cost of extracting the enzymes on a large scale and the need to regenerate the cofactor NADH in cell-free systems. In this respect, the transformation of alkenes to epoxides by using whole cells may be commercially significant, since it has been shown that methanotrophs will catalyse these conversions stoicheiometrically and excrete the product (Stirling and Dalton, 1979b; Higgins *et al.*, 1979; Hou *et al.*, 1979). The supply of reductant for this, and any other biotransformation by MMO, could prove to be critical for the commercial feasibility of such a process. In contrast to the reports from the Exxon group (Hou *et al.*, 1979) that cells grown in the presence of methanol do not catalyse this epoxidation reaction, both we (Dalton, 1980) and Best and Higgins (1981) have found that this reaction is indeed catalysed by methanol-grown cells and thus methanol can serve as a suitable reductant for the MMO reactions. It is possible, therefore, to envisage catalysis of alkenes to epoxides by methanotrophs grown in the presence of methanol in which the latter acts as a growth substrate and cofactor generator.

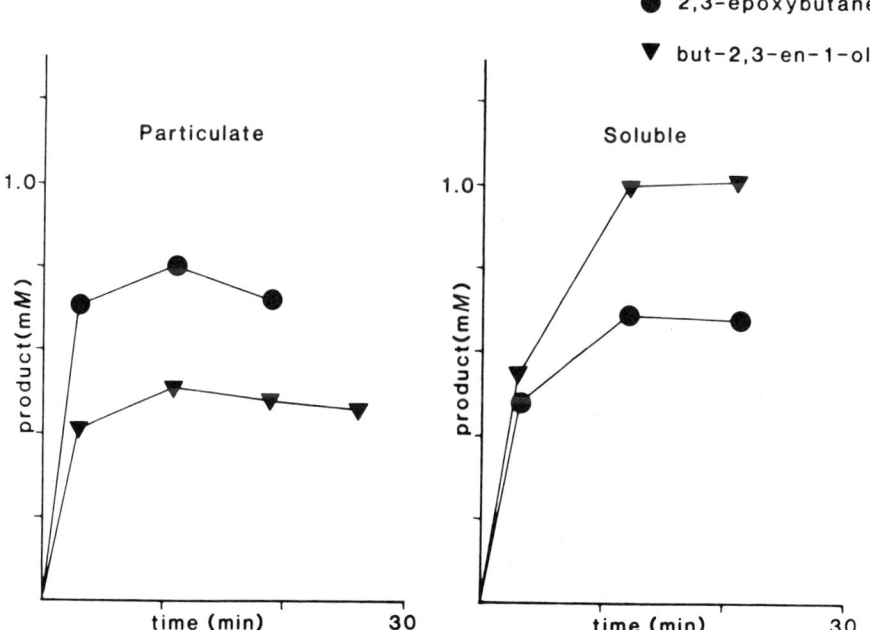

**Fig. 8.** Oxidation of *trans*-but-2-ene by cells of *M. capsulatus* (Bath) in which the MMO was present either in its particulate or soluble form.

From the results presented above, the presence of the particulate or soluble MMO is important when one considers how reducing power is to be supplied to the MMO, since it is possible that the particulate form of the enzyme may not require NADH as electron donor. Furthermore, we recently reinvestigated the oxidation of *trans*-but-2-ene by whole cells and cell extracts of *M. capsulatus* (Bath) and found that the ratio of different products obtained depends on whether the MMO enzyme is present in its soluble or particulate form. Cells in which MMO is present in its particulate form produced more epoxide than alcohol, whereas the opposite occurred when the enzyme was in its soluble form (Fig. 8). At present, we do not know why this should occur, but it is clear that the nature of the MMO can influence the specificity of product formation, and since the growth conditions influence which form of MMO is produced in the cell, then consideration must be given to the way in which the cells are grown before these organisms can be fully exploited.

## Acknowledgment

D. J. L. would like to thank Allelix Inc. for financial support.

## References

Allen, L. N., Olstein, A. D., Haber, C. L. and Hanson, R. S. (1983). Genetic and biochemical studies of representative Type II methylotrophic bacteria. *In* "Microbial Growth on C₁-Compounds" (Eds. R. L. Crawford and R. S. Hanson), pp. 236–243. American Society for Microbiology, Washington, D.C.

Anthony, C. (1982). "The Biochemistry of Methylotrophs." Academic Press, London.

Anthony, C. and Zatman, L. J. (1964a). The microbial oxidation of methanol. I. Isolation and properties of *Pseudomonas* M27. *Biochemical Journal* **92**, 609–613.

Anthony, C. and Zatman, L. J. (1964b). The microbial oxidation of methanol. II. The methanol-oxidizing enzyme of *Pseudomonas* M27. *Biochemical Journal* **92**, 614–621.

Anthony, C. and Zatman, L. J. (1965). The microbial oxidation of methanol. The alcohol dehydrogenase of *Pseudomonas* M27. *Biochemical Journal* **96**, 808–812.

Anthony, C. and Zatman, L. J. (1967). The microbial oxidation of methanol. The prosthetic group of the alcohol dehydrogenase of *Pseudomonas* M27. *Biochemical Journal* **104**, 960–969.

Beardmore-Gray, M., O'Keefe, D. T. and Anthony, C. (1983). The methanol:cytochrome c oxidoreductase activity of methylotrophs. *Journal of General Microbiology* **129**, 923–933.

Bellion, E., Bolbot, J. A. and Lash, T. D. (1981). Generation of glyoxylate in methylotrophic bacteria. *Current Microbiology* **6**, 367–372.

Best, D. J. and Higgins, H. I. (1981). Methane-oxidising activity and membrane morphology in a methanol-grown obligate methanotroph, *Methylosinus trichosporium* OB3b. *Journal of General Microbiology* **125**, 73–84.

Brown, L. R., Strawinski, R. J. and McCleskey, C. S. (1964). The isolation and characterization of *Methanomonas methanooxidans* Brown and Strawinski. *Canadian Journal of Microbiology* **10**, 791–800.

Colby, J. and Dalton, H. (1976). Some properties of a soluble methane monooxygenase from *Methylococcus capsulatus* strain Bath. *Biochemical Journal* **157**, 495–497.

Colby, J. and Dalton, H. (1978). Resolution of the methane monooxygenase of *Methylococcus capsulatus* (Bath) into three components. Purification and properties of component C, a flavoprotein. *Biochemical Journal* **171**, 461–468.

Colby, J. and Dalton, H. (1979). Characterization of the second prosthetic group of the flavoenzyme NADH–acceptor reductase (component C) of the methane monooxygenase from *Methylococcus capsulatus* (Bath). *Biochemical Journal* **177**, 903–908.

Colby, J. and Zatman, L. J. (1975). Tricarboxylic acid cycle and related enzymes in restricted facultative methylotrophs. *Biochemical Journal* **148**, 505–511.

Colby, J., Dalton, H. and Whittenbury, R. (1975). An improved assay for bacterial methane monooxygenase: Some properties of the enzyme from *Methylomonas methanica*. *Biochemical Journal* **151**, 459–462.

Colby, J., Stirling, D. I. and Dalton, H. (1977). The soluble methane monooxygenase of *Methylococcus capsulatus* (Bath). Its ability to oxygenate *n*-alkanes, *n*-alkenes, ether and alicyclic, aromatic and heterocyclic compounds. *Biochemical Journal* **165**, 394–402.

Colby, J., Dalton, H. and Whittenbury, R. (1979). Biological and biochemical aspects of microbial growth on C₁ compounds. *Annual Review of Microbiology* **33**, 481–517.

Dalton, H. (1977). Ammonia oxidation by the methane oxidizing bacterium *Methylococcus capsulatus* strain Bath. *Archives of Microbiology* **114**, 273–279.

Dalton, H. (1980). Oxidation of hydrocarbons by methane monooxygenase from a variety of microbes. *Advances in Applied Microbiology* **26**, 71–87.

Dalton, H. and Stirling, D. I. (1982). Co-metabolism. *Philosophical Transactions of the Royal Society of London Series B* **297**, 481–496.

Davey, J. F., Whittenbury, R. and Wilkinson, J. F. (1972). The distribution in the methylobacteria of some key enzymes concerned with intermediary metabolism. *Archiv für Mikrobiologie* **87**, 359–366.

Davies, S. L. and Whittenbury, R. (1970). Fine structure of methane and other hydrocarbon utilizing bacteria. *Journal of General Microbiology* **61**, 227–232.

Davis, J. B., Coty, V. F. and Stanley, J. P. (1964). Atmospheric nitrogen fixation by methane oxidizing bacteria. *Journal of Bacteriology* **88**, 468–472.

Dawson, M. J. and Jones, C. W. (1981). Energy conservation in the terminal region of the respiratory chain of the methylotrophic bacterium *Methylophilus methylotrophus*. *European Journal of Biochemistry* **118**, 113–118.

Drozd, J. W. and Wren, S. J. (1980). Growth energetics in the production of bacterial single cell protein from methanol. *Biotechnology and Bioengineering* **22**, 352–362.

Duine, J. A., Frank, J. and DeRuiter, L. G. (1979). Isolation of a methanol dehydrogenase with a functional coupling to cytochrome *c*. *Journal of General Microbiology* **115**, 523–526.

Duine, J. A., Frank, J. and DeRuiter, L. G. (1980). The prosthetic group of methanol dehydrogenase. Purification and some of its properties. *European Journal of Biochemistry* **108**, 187–192.

Dworkin, M. and Foster, J. W. (1956). Studies on *Pseudomonas methanica* (Sohngen) nov. comb. *Journal of Bacteriology* **72**, 646–659.

Ehhalt, D. H. (1976). The atmospheric cycle of methane. *In* "Microbial Production and Utilization of Gases (H₂, CH₄, CO)" (Eds. H. G. Schlegel, G. Gottschalk and N. Pfennig), pp. 13–36. Akademie der Wissenschaften, Göttingen.

Ferenci, T. (1974). Carbon monoxide-stimulated respiration in methane-utilizing bacteria. *FEBS Letters* **41**, 94–98.

Foster, J. W. (1962). Hydrocarbons as substrates for micro-organisms. *Antonie van Leeuwenhoek*, **28**, 241–274.

Foster, J. W. and Davis, R. H. (1966). A methane-dependent coccus, with notes on classification and nomenclature of obligate methane-utilizing bacteria. *Journal of Bacteriology* **91**, 1924–1931.

Goldberg, I., Rock, J. S., Ben-Bassat, A. and Mateles, R. I. (1976). Bacterial yields on methanol, methylamine, formaldehyde and formate. *Biotechnology and Bioengineering* **18**, 1657–1668.

Hanson, R. S. (1980). Ecology and diversity of methylotrophic organisms. *Advances in Applied Microbiology* **26**, 3–39.

Harrington, A. A. and Kallio, R. E. (1960). Oxidation of methane and formaldehyde by *Pseudomonas methanica*. *Canadian Journal of Microbiology* **6**, 1–7.

Harwood, J. H. and Pirt, S. J. (1972). Quantitative aspects of growth of the methane oxidising bacterium *Methlococcus capsulatus* on methane in shake flasks and continuous chemostat culture. *Journal of Applied Bacteriology* **35**, 597–607.

Higgins, I. J. and Quayle, J. R. (1970). Oxygenation of methane by methane-grown *Pseudomonas methanica* and *Methanomonas methanooxidans*. *Biochemical Journal* **118**, 201–208.

Higgins, I. J., Hammond, R. C., Sariaslani, F. S., Best, D., Davies, M. M., Tryhorn, S. E. and Taylor, F. (1979). Biotransformation of hydrocarbons and related compounds by whole organism suspension of methane-grown *Methylosinus trichosporium* OB3b. *Biochemical and Biophysical Research Communications* **89**, 671–677.

Higgins, I. J., Best, D. J. and Hammond, R. C. (1980). New findings in methane-utilizing bacteria highlight their importance in the biosphere and their commercial potential. *Nature (London)* **286**, 561–564.

Higgins, I. J., Best, D. J., Hammond, R. C. and Scott, D. (1981). Methane-oxidizing microorganisms. *Microbiological Reviews* **45**, 556–590.

Hou, C. T., Patel, R. N., Laskin, A. and Barnabe, N. (1979). Microbial oxidation of gaseous hydrocarbons: Epoxidation of $C_2$ to $C_4$ *n*-alkenes by methylotrophic bacteria. *Applied and Environmental Microbiology* **38**, 127–134.

Hubley, J. H., Thomson, A. W. and Wilkinson, J. F. (1975). Specific inhibitors of methane oxidation in *Methylosinus trichosporium*. *Archives of Microbiology* **102**, 199–202.

Hyman, M. R. and Wood, P. M. (1983). Methane oxidation by *Nitrosomonas europaea*. *Biochemical Journal* **212**, 31–37.

Johnson, P. A. and Quayle, J. R. (1964). Microbial growth on $C_1$ compounds. Oxidation of methanol, formaldehyde and formate by methane-grown *Pseudomonas* AM1. *Biochemical Journal* **93**, 281–290.

Kortstee, G. J. J. (1980). The homoisocitrate–glyoxylate cycle in pink, facultative methylotrophs. *FEMS Microbiology Letters* **8**, 59–65.

Kortstee, G. J. J. (1981). The second part of the ICL⁻ pathway. *In* "Microbial Growth on $C_1$ Compounds" (Ed. H. Dalton), pp. 211–219. Heyden, London.

Leadbetter, E. R. and Foster, J. W. (1959). Oxidation products formed from gaseous alkanes by the bacterium *Pseudomonas methanica*. *Archives of Biochemistry and Biophysics* **82**, 491–492.

Leak, D. J. and Dalton, H. (1983). *In vivo* studies of primary alcohols, aldehydes and carboxylic acids as electron donors for the methane monooxygenase in a variety of methanotrophs. *Journal of General Microbiology* **129**, 3487–3497.

Lidstrom-O'Connor, M. E., Fulton, G. L. and Wopat, A. E. (1983). '*Methylobacterum ethanolicum*': A syntrophic association of two methylotrophic bacteria. *Journal of General Microbiology* **129**, 3139–3148.

Linton, J. D. and Vokes, J. (1978). Growth of the methane-utilizing bacterium *Methylococcus* NCIB 11083 in mineral salts medium with methanol as the sole source of carbon. *FEMS Microbiology Letters* **4**, 125–128.

Lynch, M. J., Wopat, A. E. and O'Connor, M. L. (1980). Characterization of two new facultative methanotrophs. *Applied and Environmental Microbiology* **40**, 400–407.

McFadden, B. A. (1973). Autotrophic $CO_2$ assimilation and the evolution of ribulose diphosphate carboxylase. *Bacteriological Reviews* **37**, 289–319.

May, S. W. and Padgette, S. R. (1983). Oxidoreductase enzymes in biotechnology: Current status and future potential. *Biotechnology* **1**, 677–686.

Ohtomo, T., Iizuka, H. and Takeda, K. (1977). Effect of copper sulfate on the morphologic transition of a strain of *Methanomonas margaritae*. *European Journal of Applied Microbiology* **4**, 267–272.

O'Keefe, D. T. and Anthony, C. (1980). The interaction between methanol dehydrogenase and the autoreducible cytochromes *c* of the facultative methylotroph *Pseudomonas* AM1. *Biochemical Journal* **190**, 481–484.

Patel, R. N. and Felix, A. (1976). Microbial oxidation of methane and methanol: Crystallization and properties of methanol dehydrogenase from *Methylosinus sporium*. *Journal of Bacteriology* **128**, 413–424.

Patel, R. N. and Hoare, D. S. (1971). Physiological studies of methane and methanol oxidizing bacteria: Oxidation of C$_1$ compounds by *Methylococcus capsulatus*. *Journal of Bacteriology* **107**, 187–192.

Patel, R. N., Bose, H. R., Mandy, W. J. and Hoare, D. S. (1972). Physiological studies of methane and methanol-oxidizing bacteria: Comparison of a primary alcohol dehydrogenase from *Methylococcus capsulatus* (Texas strain) and *Pseudomonas* species M.27. *Journal of Bacteriology* **100**, 570–577.

Patel, R. N., Mandy, W. J. and Hoare, D. S. (1973). Physiological studies of methane and methanol-oxidizing bacteria: Immunochemical comparison of a primary alcohol dehydrogenase from *Methylococcus capsulatus* and *Pseudomonas* sp. M.27. *Journal of Bacteriology* **113**, 937–945.

Patel, R. N., Hou, C. T. and Felix, A. (1978a). Microbial oxidation of methane and methanol: Isolation of methane-utilizing bacteria and characterization of a facultative methane-utilizing isolate. *Journal of Bacteriology* **136**, 352–358.

Patel, R. N., Hou, C. T. and Felix, A. (1978b). Microbial oxidation of methane and methanol: Crystallization of methanol dehydrogenase and properties of holo- and apo-methanol dehydrogenase from *Methylomonas methanica*. *Journal of Bacteriology* **133**, 641–649.

Patel, R. N., Hou, C. T., Derelanko, P. and Felix, A. (1980). Purification and properties of a heme-containing aldehyde dehydrogenase from *Methylosinus trichosporium*. *Archives of Biochemistry and Biophysics* **203**, 654–662.

Patel, R. N., Hou, C. T., Laskin, A. I. and Felix, A. (1982). Microbial oxidation of hydrocarbons: Properties of a soluble methane monooxygenase from a facultative methane-utilizing organism, *Methylobacterium* sp. strain CRL-26. *Applied and Environmental Microbiology* **44**, 1130–1137.

Paterson, D. (1978). Methane from the bowels of the earth. *New Scientist* **1111**, 896–898.

Patt, T. E., Cole, G. C., Bland, J. and Hanson R. S. (1974). Isolation and characterization of bacteria that grow on methane and organic compounds as sole sources of carbon and energy. *Journal of Bacteriology* **120**, 955–964.

Poulsen, C. (1981). Comments on the structure and function of the large subunit of the enzyme ribulose bisphosphate carboxylase–oxygenase. *Carlsberg Research Communications* **46**, 259–278.

Quayle, J. R. (1980). Microbial assimilation of C$_1$ compounds. *Biochemical Society Transactions* **8**, 1–10.

Quayle, J. R. and Ferenci, T. (1978). Evolutionary aspects of autotrophy. *Microbiological Reviews* **42**, 251–273.

Ribbons, D. W. (1975). Oxidation of C$_1$ compounds by particulate fractions from *Methylococcus capsulatus:* Distribution and properties of methane-dependent reduced nicotinamide adenine dinucleotide oxidase (methane hydroxylase). *Journal of Bacteriology* **122**, 1351–1363.

Ribbons, D. W. and Michelover, J. L. (1970). Methane oxidation by cell-free extracts of *Methylococcus capsulatus*. *FEBS Letters* **11**, 41–44.

Rokem, J. S., Goldberg, I. and Mateles, R. I. (1978). Maintenance requirements for bacteria growing on C$_1$-compounds. *Biotechnology and Bioengineering* **20**, 1557–1564.

Salisbury, S. A., Forrest, H. S., Cruse, W. B. T. and Kennard, O. (1979). A novel coenzyme from bacterial dehydrogenases. *Nature (London)* **280**, 843–844.

Scott, D., Best, D. J. and Higgins, I. J. (1981a). Intracytoplasmic membranes in oxygen-limited chemostat cultures of *Methylosinus trichosporium* Ob3b: Biocatalytic implications of physiologically balanced growth. *Biotechnology Letters* **3**, 641–644.

Scott, D., Brannan, J. and Higgins, I. J. (1981b). The effect of growth conditions on intracytoplasmic membranes and methane mono-oxygenase activities in *Methylosinus trichosporium* OB3b. *Journal of General Microbiology* **125**, 63–72.

Söhngen, N. L. (1906). Uber bakterien, welch methan als kohlenstoffnahrung und energiequelle gebrauchen. *Zentralblatt für Bakreilologie, Mikrobiologie und Hygiene, Abreilung 1, Originale C* **15**, 513–517.

Sperl, G. T., Forrest, H. S. and Gibson, D. T. (1974) Substrate specificity of the purified primary alcohol dehydrogenase from methanol-oxidizing bacteria. *Journal of Bacteriology* **118**, 541–550.

Stanley, S. H. and Dalton, H. (1982). Role of ribulose-1,5-bisphosphate carboxylase/oxygenase in *Methylococcus capsulatus* (Bath). *Journal of General Microbiology* **128**, 2927–2935.

Stanley, S. H., Prior, S. D., Leak, D. J. and Dalton, H. (1983). Copper stress underlies the fundamental change in intracellular location of methane mono-oxygenase in methane-oxidizing organisms: Studies in batch and continuous cultures. *Biotechnology Letters* **5**, 487–492.

Stirling, D. I. (1978). Oxidation of carbon compounds by *Methylococcus capsulatus*. Ph.D. Thesis, University of Warwick.

Stirling, D. I. and Dalton, H. (1978). Purification and properties of an NAD(P)⁺-linked formaldehyde dehydrogenase from *Methylococcus capsulatus* (Bath). *Journal of General Microbiology* **107**, 19–29.

Stirling, D. I. and Dalton, H. (1979a). Properties of the methane monooxygenase from extracts of *Methylosinus trichosporium* OB3b and evidence for its similarity to the enzyme from *Methylococcus capsulatus* (Bath). *European Journal of Biochemistry* **96**, 205–212.

Stirling, D. I. and Dalton, H. (1979b). The fortuitous oxidation and cometabolism of various carbon compounds by whole-cell suspensions of *Methylococcus capsulatus* (Bath). *FEMS Microbiology Letters* **5**, 315–318.

Stirling, D. I., Colby, J. and Dalton, H. (1979). A comparison of the substrate and electron donor specificities of the methane monooxygenases from three strains of methane-oxidizing bacteria. *Biochemical Journal* **177**, 361–364.

Strom, T., Ferenci, T. and Quayle, J. R. (1974). The carbon assimilation pathways of *Methylococcus capsulatus*, *Pseudomonas methanica* and *Methylosinus trichosporium* (OB3b) during growth on methane. *Biochemical Journal* **144**, 465–476.

Takeda, K., Tezuka, C., Fukuoka, S. and Takahara, Y. (1976). Role of copper ions in methane oxidation by *Methanomonas margaritae*. *Journal of Fermentation Technology* **54**, 557–562.

Taylor, S. C. (1977). Evidence for the presence of ribulose-1,5-bisphosphate carboxylase and phosphoribulokinase in *Methylococcus capsulatus* (Bath). *FEMS Microbiology Letters* **2**, 305–307.

Tonge, G. M., Harrison, D. E. F., Knowles, C. J. and Higgins, I. J. (1975). Properties and partial purification of the methane-oxidizing enzyme system from *Methylosinus trichosporium*. *FEBS Letters* **58**, 293–299.

Tonge, G. M., Harrison, D. E. F. and Higgins, I. J. (1977). Purification and properties of the methane mono-oxygenase enzyme system from *Methylosinus trichosporium* OB3b. *Biochemical Journal* **161**, 333–344.

Wadzinski, A. M. and Ribbons, D. W. (1975). Oxidation of C₁ compounds by particulate fractions from *Methylococcus capsulatus:* Properties of methanol oxidase and methanol dehydrogenase. *Journal of Bacteriology* **122**, 1364–1374.

Whittenbury, R. and Dalton, H. (1981). The methylotrophic bacteria. *In* "The Prokaryotes" (Eds. M. P. Starr, H. Stolp, H. G. Truper, A. Balows and H. G. Schlegel), pp. 894–902. Springer-Verlag, Berlin and New York.

Whittenbury, R., Phillips, K. C. and Wilkinson, J. F. (1970). Enrichment, isolation and some properties of methane-utilising bacteria. *Journal of General Microbiology* **61**, 205–218.

Wolf, H. J. (1981). Biochemical characterization of methane-oxidizing yeast. *In* "Microbial Growth on $C_1$ Compounds" (Ed. H. Dalton), pp. 202–210. Heyden, London.

Wolf, H. J. and Hanson, R. S. (1979). Isolation and characterisation of methane utilising yeasts. *Journal of General Microbiology* **114**, 187–194.

Wolf, H. J. and Hanson, R. S. (1980). Identification of methane-utilising yeasts. *FEMS Microbiology Letters* **7**, 177–179.

Wolfe, R. S. and Higgins, I. J. (1979). Microbial biochemistry of methane—a study in contrasts. *International Review of Biochemistry* **21**, 267–353.

Woodland, M. P. and Dalton, H. (1984). Purification and properties of component A of the methane mono-oxygenase from *Methylococcus capsulatus* (Bath). *Journal of Biological Chemistry* **259**, 53–59.

Zatman, L. J. (1981). A search for patterns in methylotrophic pathways. *In* "Microbial Growth on $C_1$ Compounds" (Ed. H. Dalton), pp. 42–54. Heyden, London.

Zhao, S. J. and Hanson, R. S. (1983). *Methylomonas* sp. 761M: An unusual Type I methanotroph. *In* "Microbial Growth on $C_1$-Compounds" (Eds. R. L. Crawford and R. S. Hanson), pp. 262–268. American Society for Microbiology, Washington, D. C.

# Discussion

*J. Colby:* Do cells having particulate MMO have component C in the soluble extract?

*H. Dalton:* Apparently not. When we have added back the other soluble components (both separately and in combination) to the extract, which has solely particulate activity, we have not observed any soluble activity. It is possible, however, to manipulate conditions so that both particulate and soluble activities coexist in the cell by choosing a certain copper regime in the culture. Under these conditions there will be soluble component C as well as particulate MMO activity.

*C. Anthony:* Do you think that the two methane monooxygenases are completely different systems?

*Dalton:* The evidence we have to date does suggest this although we cannot be certain. The migration of the major polypeptides in SDS gels, the inhibitor profiles and substrate specificity of the two forms appear to be quite different, but until we can release the particulate form from the membrane and study its physico-chemical properties we are not in a position to confirm that they are completely different systems.

*Anthony:* How do you reconcile your results on the effects of growth conditions on the oxygenases with those published by Higgins' group?

*Dalton:* This is not an easy question to answer since there are many differences in the growth conditions used between the two groups. One thing that might be important here is biomass concentration. Scott *et al.* [*Biotechnology Letters* **3**, 641–644 (1981)] observed particulate MMO activity in chemostat cultures under oxygen limitation at low cell densities ($0.2$ g $l^{-1}$ and soluble activity under nitrate limitation at high cell densities ($2$-$5$ g $l^{-1}$). We would interpret this as being due to copper availability. In the former case, the cells were grown under copper-sufficient conditions and in the latter case were probably copper-stressed. Why they should observe the coexistence of soluble and particulate enzymes under methane limitation at low cell density ($0.2$ g $l^{-1}$) is unclear at present.

*A. P. F. Turner:* Do you see any future for the commercial exploitation of cell-free MMO for biotransformations?

*Dalton:* It is quite clear that no bulk chemical could ever be made using the cell-free system in its present form. The poor stability of the enzyme *in vitro*, the high cost of extraction of the enzyme and supply of reducing power would make economic nonsense. As far as high-value products are concerned, I think the same constraints would apply unless the product was of very high value indeed. However, if the problems of stability and provision of reducing power could be overcome then there might well be the possibility of using the MMO for the production of relatively high value added products. There is, however, a potentially interesting possibility in the use of the enzyme in biosensors for the detection of certain organic molecules, but here one must be certain that there are no contaminating molecules around that could serve as alternative MMO substrates.

*Turner:* Have you any explanation for the reports of ascorbate-linked MMO activity?

*Dalton:* No I don't. We have not observed this activity in any of the active MMO preparations that we have made from the methanotroph in our laboratory. Maybe this question should be asked of those who have reported it.

*I. R. Booth:* Can the soluble form of MMO be provoked to reassociate with the membrane on addition of copper to the culture?

*Dalton:* This presupposes that the two forms of the enzyme are structurally similar and we have no evidence for that. All I can say is that we can certainly observe the particulate form of the enzyme after 20 min following copper addition to a steady-state culture containing entirely soluble MMO. The soluble activity drops off dramatically during this period, although the proteins, as measured by SDS gel electrophoresis, disappear more slowly.

*Booth:* Could the role of copper in methane oxidizers be analogous to that of calcium in *Azotobacter*, as a signal for membrane proliferation?

*Dalton:* Yes, I guess it could, although I'm not sure that we really understand how calcium effects this change in *Azotobacter*.

*O. Meyer:* Do you have any information on the $K_m$ and $V_{max}$ values for methane and carbon monoxide of methane-oxidizing bacteria, and do these figures support the assumption that methane oxidizers can remove methane and/or carbon monoxide in their natural habitat?

*Dalton:* As far as methane is concerned there have been quite a number of measurements of these values. The $K_m$ values have ranged from about 5 to 50 $\mu M$ and $V_{max}$ values from 100 to 550 nmol min$^{-1}$ mg cells. For carbon monoxide fewer measurements have been made, mostly from Tom Ferenci and ourselves. Here the $K_m$ value was 2.7 $\mu M$ and $V_{max}$ between 80 and 520 nmol min$^{-1}$ mg cells. However, I think the recent re-evaluation of these $K_m$ values by Joergensen and Degn [*FEMS Microbiology Letters* **20,** 331–335 (1983)] has cast some doubt upon the accuracy of the previously measured $K_m$ values for methane at least. What they have found, using membrane inlet mass spectrometry, is that the $K_m$ values for methane oxidizers can be as low as 0.8 $\mu M$. Implicit in this observation is the possibility that the $K_m$ measurements for carbon monoxide may also be orders of magnitude lower. Now if one is to look at the measured values for CO and CH$_4$ in the environment you have to realize that these values can vary considerably from one environment to another and that it may be misleading to use a static value when we are really looking at a dynamic process. For example, the concentration of methane can be about 1 m$M$ in a lake sediment but zero near the surface of the lake. Since one can readily isolate methane oxidizers from stratified lakes where they effectively remove all the methane, we have little doubt that they are the agents responsible for methane consumption in the environment. There is also a very good correlation between their presence in the environment and the incidence of methane from natural gas leaks. As far as their ability to remove CO from the environment is concerned, one cannot be definitive about this but certainly the evidence to date does suggest that they are very good candidates. It appears that their $K_m$ for carbon monoxide is at least 10-fold lower than that of the carboxydobacteria (and may be lower if the analysis was done by mass spectroscopy) and might approach the natural soil level of about 0.22 to 0.35 $\mu M$. Here again, however, I should stress that it is a dynamic process and that *in situ* measurements of CO concentraion might not be the most relevant feature here.

# 12

# Implications of the Nature of Methane Monooxygenase on Carbon Assimilation in Methanotrophs

DAVID J. LEAK,[1] STEPHEN H. STANLEY AND HOWARD DALTON

*Department of Biological Sciences, University of Warwick, Coventry, United Kingdom*

## Introduction

Published carbon conversion efficiences (CCE) for growth of methanotrophs on methane vary between about 19 and 70% (Table 1). Although theoretical considerations indicate that growth of type II organisms (for a survey, see Dalton and Leak, Chapter 11, this volume) assimilating carbon via the serine pathway, would be less efficient than growth of type I organisms, assimilating carbon via the ribulose monophosphate (RuMP) pathway, large differences are also evident from studies with different strains of the same species of the type I organism *Methylococcus capsulatus*. While estimates of cell yields on gaseous substrates are open to many sources of error, the discrepancies involved suggest that the explanation is not trivial, but may have some physiological basis.

Variations in growth conditions can have a profound effect on the nature of the methane monooxygenase (MMO) in some methanotrophs, and this might affect the energetics of methane dissimilation in these organisms. Following the contradictory evidence for a membrane-bound (Tonge *et al.*, 1975) and soluble (Stirling and Dalton, 1979) MMO in the type II methanotroph '*Methylosinus trichosporium*' OB3b, Scott *et al.* (1981) demonstrated that both forms can indeed be obtained in this organism, in which they defined growth conditions for the expression of either form. Stanley *et al.* (1983) subsequently found that *Methylococcus capsulatus* (Bath) also produces both soluble and membrane-bound (particulate) forms of the MMO. However, these authors demonstrated

---

[1]Present address: Centre for Biotechnology, Imperial College of Science and Technology, London SW7 2AZ, United Kingdom.

MICROBIAL GAS METABOLISM:
MECHANISTIC, METABOLIC
AND BIOTECHNOLOGICAL ASPECTS

**Table 1.** Published yields on methane[a]

| Organism | Type | $Y_{CH_4}$ | $O_2/CH_4$ | CCE (%) | Culture | References |
|---|---|---|---|---|---|---|
| Mixed | | 1.11 | 2.3 | 69.6 | Batch | Hamer et al. (1967) |
| Mixed | | 0.5–0.7 | | 31.3–43.9 | Batch (CH$_4$ LIM) | Vary and Johnson (1967) |
| Mixed | | 0.7 | | 43.9 | Batch (O$_2$ LIM) | |
| Methylomonas sp. | I | 1.01 | 1.44 | 63.3 | Batch | Silverman and Ooyama (1968) |
| Various | I and II | 1.0–1.1 | 1.0–1.1 | 62.6–68.9 | Batch | Whittenbury et al. (1970) |
| Mixed | | 0.90 | 1.7 | 56.4 | CC | Bewersdorff and Dostalek (1971) |
| Mixed | | 0.62 | 1.44 | 38.8 | CC | Sheehan and Johnson (1971) |
| M. cap (Texas) | I | 1.01 | 1.8 | 63.3 | CC (CH$_4$ LIM) | Harwood and Pirt (1972) |
| M. cap (Texas) | I | 0.31 | 0.3 | 19.4 | CC (O$_2$ LIM) | |
| Mixed | | 0.99 | 1.24 | 62 | CC (CH$_4$ LIM) | Wilkinson et al. (1974) |
| Mixed | | 0.80 | 1.71 | 50.1 | CC (O$_2$ LIM) | |
| M. cap (Bath) | I (X) | 0.54–0.66 | | 39–47 | CC (CH$_4$ LIM) | Stanley (1977) |
| Methylococcus sp. NCIB 11083 | I | 0.75–0.78 | 1.52 | 46–47.8 | CC (CH$_4$ LIM) | Linton and Vokes (1978) |
| Methylococcus sp. NCIB 11083 | I | 0.83 | | 51 | CC | |

[a] M. cap., Methylococcus capsulatus; CC, continuous culture; CCE, carbon conversion efficiency; LIM, limitation; $Y_{CH_4}$, grams dry weight cells per gram methane consumed.

that expression of one or other form was determined by the availability of copper in the growth medium. In a high-copper medium, the organism preferentially synthesized the particulate MMO, the soluble enzyme being expressed in response to copper 'stress'. Extension of these studies to include '*M. trichosporium*' OB3b revealed that the intracellular location of the MMO in this organism was also determined primarily by copper availability.

The soluble and particulate MMO systems differ in stability and sensitivity to inhibitors. Major differences are revealed by sodium dodecyl sulphate–polyacrylamide gel electrophoresis of the protein fractions from cells with soluble or particulate MMO; the disappearance of the α, β and γ subunits of fraction A of the soluble MMO during transition from the soluble to the particulate enzyme system is particularly notable. Increased or *de novo* synthesis of at least three polypeptides associated with the particulate fraction is also evident during this transition. At present, it is not known whether some of these are components of a novel particulate MMO, proteins associated with membrane formation or novel electron transfer proteins. The latter possibility was suggested by the observation that the oxidation of ethanol, presumably via the primary alcohol dehydrogenase, could drive the particulate but not the soluble MMO (Stanley *et al.*, 1983). Theoretical consideration (below) suggests that if this involved direct recycling of electrons from methanol dehydrogenase to MMO then the difference in energetic requirements of the soluble and particulate MMO could account for the differences in growth efficiencies previously observed.

## Theoretical Considerations

### Soluble MMO

Studies with the soluble MMO from *M. capsulatus* (Bath) have demonstrated that NAD(P)H is the only effective electron donor for this system (Colby and Dalton, 1976). Thus it can be seen (Fig. 1) that for every 2 moles of methane consumed, 1 mole has to be completely oxidised to carbon dioxide to provide enough NADH for the initial oxidation of the 2 moles of methane to methanol.

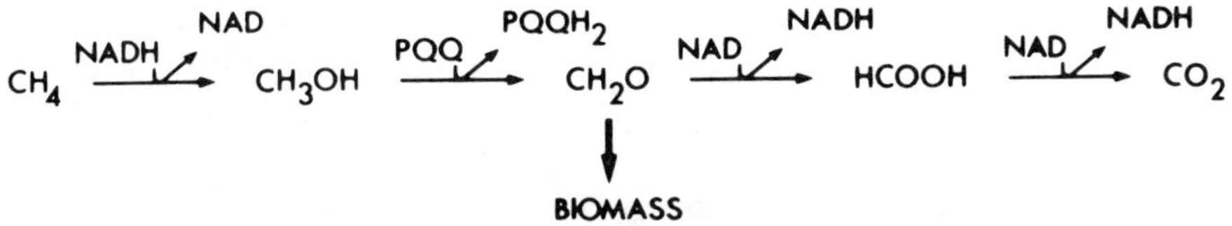

**Fig. 1.** Pathway of methane metabolism is *M. capsulatus* (Bath); PQQ and $PQQH_2$ refer to the oxidized and reduced forms of pyrroloquinoline quinone.

So the CCE (amount of carbon assimilated as a percentage of total carbon utilized) must be ≤50%, assuming $PQQH_2$ cannot reduce NAD.

From a consideration of the energy requirements of carbon assimilation and yield from dissimilation, Anthony (1982) suggested that the growth yield of methanotrophs may be limited more by the supply of NADH than by ATP, assuming growth with ammonia as the nitrogen source. Although this conclusion has been criticized by Harder *et al.* (1981), the use of nitrate as the nitrogen source would exacerbate this NADH limitation due to extra reductant required for nitrate assimilation. If it is assumed that the reduction of nitrate to ammonia requires 4NADH (Stouthamer, 1977), then the maximum CCE is reduced from 42 to 31.7% assuming a P/O ratio of 3 for oxidation of NADH and a P/O ratio of 1 for oxidation of methanol. The relative amounts of methane dissimilated for NADH and ATP production would be 67.4 and 0.9%, respectively, of the total consumed. Even if a P/O ratio as low as 1 was assumed for NADH oxidation, the relative amount of methane required for ATP synthesis would be only 2.2%.

*Particulate MMO*

From studies with the particulate MMO from *M. trichosporium* OB3b, the pathway in Fig. 2 has been suggested. A mechanism involving the direct recycling of electrons from methanol dehydrogenase to the MMO would make methane energetically equivalent to formaldehyde and obviously alleviate the problems of NADH limitation. Assuming Y = NAD and a P/O ratio of 3 for NADH oxidation, an organism with the RuMP pathway utilizing this mechanism of methane oxidation could have CCE ≤ 63% with ammonia as the nitrogen source. However, using nitrate as the nitrogen source, the efficiency would drop to ≤48%.

### Yields of *Methylococcus capsulatus* (Bath) on Methane

Preliminary experiments were carried out to define the continuous culture conditions necessary to obtain 100% soluble or particulate MMO activity. Maintaining constant conditions of gas flow, impeller speed and dilution rate, the effect of the concentration of $CuSO_4 \cdot 5H_2O$ in the growth medium was determined (Fig. 3).

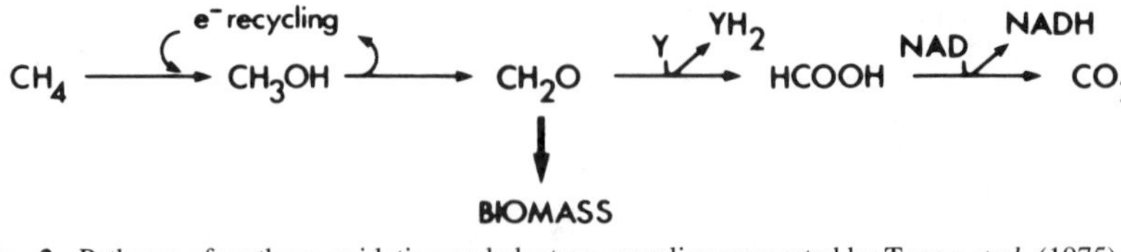

$$CH_4 \xrightarrow{\overbrace{\phantom{xxx}}^{e^- \text{recycling}}} CH_3OH \xrightarrow{\phantom{xx}} CH_2O \xrightarrow[Y]{YH_2} HCOOH \xrightarrow[NAD]{NADH} CO_2$$

BIOMASS

**Fig. 2.** Pathway of methane oxidation and electron recycling suggested by Tonge *et al.* (1975); Y is an electron acceptor for formaldehyde dehydrogenase.

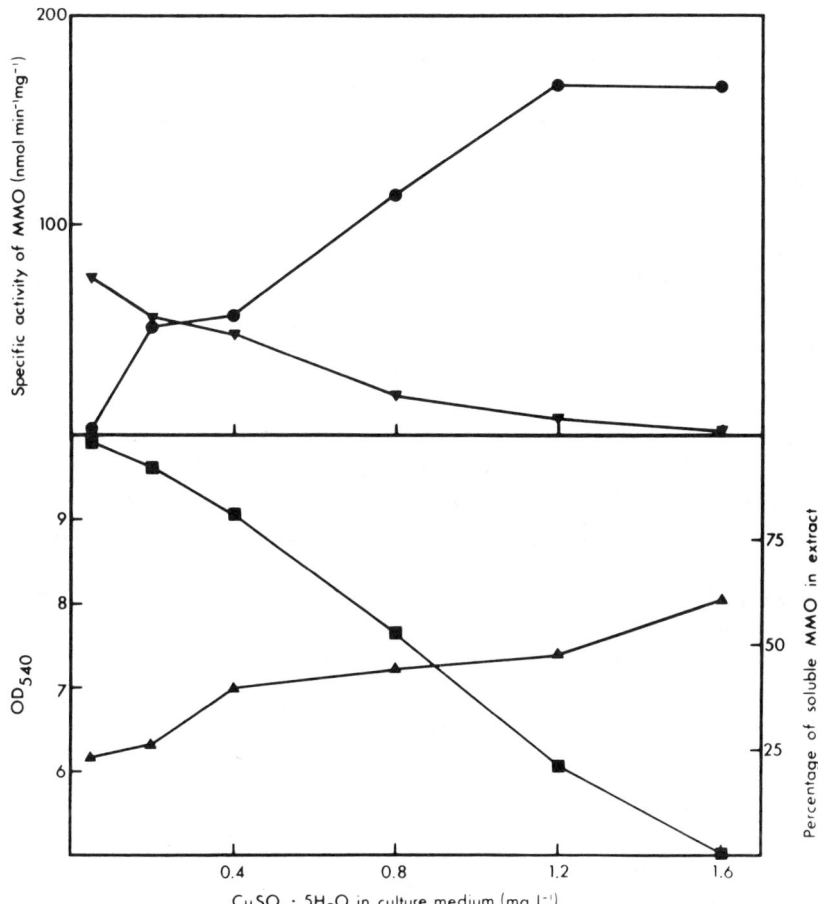

**Fig. 3.** Effect of copper concentration on the activity and intracellular location of the methane monooxygenase in *M. capsulatus* (Bath). *Methylococcus capsulatus* (Bath) was grown in an oxygen-limited continuous culture at a dilution rate of 0.1 $hr^{-1}$ on 0.2% (w/v) $KNO_3$, nitrate mineral salts medium (Stanley *et al.*, 1983). Particulate MMO activity (●) and soluble MMO activity (▼) found in cell extracts (upper panel). Steady state cell concentration measured as optical density at 540 n*M* (▲) and soluble MMO (■) as a percentage of total activity (lower panel).

Over the range 0.05 to 1.6 mg $1^{-1}$, the location of the MMO switched from almost 100% soluble to 100% particulate. During this transition, the cell density of the culture increased, indicating that growth with the particulate MMO was more efficient than with soluble MMO. It is therefore difficult to define accurately the concentration of copper necessary to produce soluble or particulate activity, as this is also cell density-dependent.

Table 2 shows the experimentally determined yields of *M. capsulatus* (Bath) on methane under various growth conditions. The experimental CCE for cells with a soluble MMO activity grown on nitrate agreed with the maximum theoretical values derived by Anthony (1982), assuming that assimilation of nitrate requires 4NADH. Cells grown on ammonia as the nitrogen source were 45% more efficient than cells grown on nitrate and the CCE was similar to that obtained for *Methylococcus* NCIB 11083. Both of these results are slightly higher than the maximum theoretical value of 42% calculated by Anthony (1982), assuming a P/O ratio of 3 for oxidation of NADH and 1 for oxidation of methanol.

Nitrate-grown cells with particulate MMO activity (1.5 mg added $CuSO_4 \cdot 5H_2O$ per litre of medium) assimilated 10 to 12% more carbon into cell material, representing an increase of 34 to 38% in CCE compared to cells with totally soluble MMO activity. From the estimate that, under these growth conditions, $\leqslant$ 2.2% of the carbon substrate has to be dissimilated for ATP production, it is evident that these higher values could not have been achieved by an increase in efficiency of ATP synthesis (that is, an increase in P/O ratio during NADH or methanol oxidation) *per se*. Therefore, the increase in growth efficiency in high-copper medium must arise from either an increased NADH yield from methane dissimilation or a decreased demand for NADH for assimilatory reactions. These could conceivably arise by (1) use of electron donors other than NADH for methane oxidation, (2) reverse electron transport from methanol dehydrogenase to reduce NAD or (3) a reduction in NADH demand for nitrate assimilation by, for example, reduction of nitrate by methanol dehydrogenase.

**Table 2.**    *Efficiency of biomass production from methane[a]*

| N source (%) | Limitation | Cu (mg litre$^{-1}$) | S/P (%) | ETOH | CCE (%) | $O_2/CH_4$ | $Y_{CH_4}{}^b$ |
|---|---|---|---|---|---|---|---|
| $KNO_3$ (0.2) | $O_2$ | 0.2 | S (98) | − | 31.3 | 1.47 | 0.50 |
| $KNO_3$ (0.2) | $O_2$ | 1.5 | P (100) | + | 43.3 | 1.45 | 0.69 |
| $KNO_3$ (0.2) | $CH_4$ | 0.05 | S (99) | − | 31.2 | 1.50 | 0.50 |
| $KNO_3$ (0.2) | $CH_4$ | 1.5 | P (99) | + | 41.9 | 1.41 | 0.67 |
| $NH_4Cl$ (0.1) | $O_2$ | 0.1 | S (98) | − | 45.5 | 1.57 | 0.73 |
| $NH_4Cl$ | $O_2$ | | | | | | |
| $NH_4Cl$ | $CH_4$ | To be investigated | | | | | |
| $NH_4Cl$ | $CH_4$ | | | | | | |

[a]Efficiencies calculated from analyses of $CH_4$ and $O_2$ usage and $CO_2$ production. Abbreviations: Cu, $CuSO_4 \cdot 5H_2O$; S/P, soluble or particulate MMO as percentage of total activity; ETOH, ethanol-driven MMO activity in whole cells; CCE, carbon conversion efficiency.
[b]$Y_{CH_4}$: yields computed assuming dry weight = 47% carbon [in fact, the measured cell yield was less than this, as approximately 7% of total organic carbon was found in the medium—mainly as a result of cell lysis, as observed by Drozd *et al.* (1978)].

The ability of NAD-independent methanol dehydrogenase to drive particulate MMO in whole cells has been demonstrated (Leak and Dalton, 1983), although whether this was a direct interaction via cytochrome $c$ (Tonge *et al.*, 1975) or involved reverse electron transport to NAD or some intermediate respiratory chain component remains to be established. The contribution of possibility 3 (above) is unclear at present, but future studies with ammonia as the nitrogen source may clarify this situation.

If the changes in yield are entirely due to the energy supply to the different forms of MMO, then a comparison of calculated and actual yields may provide some information on the energy supply to the particulate MMO. The CCE expected if the MMO is driven by electron recycling from methanol dehydrogenase via cytochrome $c$ (Tonge *et al.*, 1975) would be $\leq 48\%$ with nitrate as the nitrogen source. However, the experimental CCE was found to be 41.9 to 43.3%. Although this lower yield could be the result of a low P/O ratio for NADH oxidation (a situation which would not significantly affect yields with an NAD(P)H-requiring soluble MMO), an alternative possibility might be that the oxidation of methane via methanol to formaldehyde requires the input of one ATP (estimated yield 40.1%).

Although some aspects of energy supply to the particulate MMO remain unsolved, it is clear that growth conditions, particularly the concentration of copper in the medium, can have a considerable effect on the yield of methanotrophs growing on methane. Further studies with different carbon and nitrogen sources should establish whether these can be attributed to the different energy requirements of the soluble and particulate MMO.

## Acknowledgment

D. J. L. and S. H. S. are grateful to Allelix, Inc. for financial support.

## References

Anthony, C. (1982). "The Biochemistry of Methylotrophs." Academic Press, London.

Bewersdorff, M. and Dostalek, M. (1971). The use of methane for production of bacterial protein. *Biotechnology and Bioengineering* **13**, 49–62.

Colby, J. and Dalton, H. (1976). Some properties of a soluble methane mono-oxygenase from *Methylococcus capsulatus* strain Bath. *Biochemical Journal* **157**, 495–497.

Drozd, J. W., Linton, J. D., Downs, J. and Stephenson, R. J. (1978). An in situ assessment of the specific lysis rate in continuous cultures of *Methylococcus* sp. (NCIB 11083) grown on methane. *FEMS Microbiology Letters* **4**, 311–314.

Hamer, G., Heden, C. G. and Carenberg, C. O. (1967). Methane as a carbon substrate for the production of microbial cells. *Biotechnology and Bioengineering* **9**, 499–514.

Harder, W., van Dijken, J. P. and Roels, J. A. (1981). Utilization of energy in methylotrophs. *In* "Microbial Growth on $C_1$ Compounds" (Ed. H. Dalton), pp. 258–269. Heyden, London.

Harwood, J. H. and Pirt, S. J. (1972). Quantitative aspects of growth of the methane oxidising bacterium *Methylococcus capsulatus* on methane in shake flasks and continuous chemostat culture. *Journal of Applied Bacteriology* **35**, 597–607.

Leak, D. J. and Dalton, H. (1983). *In vivo* studies of primary alcohols, aldehydes and carboxylic acids as electron donors for the methane mono-oxygenase in a variety of methanotrophs. *Journal of General Microbiology* **129**, 3487–3497.

Linton, J. D. and Vokes, J. (1978). Growth of the methane-utilizing bacterium *Methylococcus* NCIB 11083 in mineral salts medium with methanol as the sole source of carbon. *FEMS Microbiology Letters* **4**, 125–128.

Scott, D., Brannan, J. and Higgins, I. J. (1981). The effect of growth conditions on intracytoplasmic membranes and methane mono-oxygenase activities in *Methylosinus trichosporium* OB3b. *Journal of General Microbiology* **125**, 63–72.

Sheehan, B. T. and Johnson, M. J. (1971). Production of bacterial cells from methane. *Applied Microbiology* **21**, 511–515.

Silverman, M. P. and Ooyama, V. I. (1968). Automatic apparatus for sampling and preparing gases for mass spectral analysis in studies of carbon isotope fractionation during methane metabolism. *Analytical Chemistry* **40**, 1833–1837.

Stanley, S. H. (1977). A study of the physiology and gas-limited growth of *Methylococcus capsulatus*. Ph.D. Thesis, University of Warwick, Coventry, U.K.

Stanley, S. H., Prior, S. D., Leak, D. J. and Dalton, H. (1983). Copper stress underlies the fundamental change in intracellular location of methane mono-oxygenase in methane-oxidizing organisms: Studies in batch and continuous cultures. *Biotechnology Letters* **5**, 487–492.

Stirling, D. I. and Dalton, H. (1979). Properties of the methane mono-oxygenase from extracts of *Methylosinus trichosporium* OB3b and evidence for its similarity to the enzyme from *Methylococcus capsulatus* (Bath). *European Journal of Biochemistry* **96**, 205–212.

Stouthamer, A. H. (1977). Theoretical calculation on the influence of the inorganic nitrogen source on parameters for aerobic growth of microorganisms. *Antonie van Leeuwenhock* **43**, 351–367.

Tonge, G. M., Harrison, D. E. F., Knowles, C. J. and Higgins, I. J. (1975). Properties and partial purification of the methane-oxidizing enzyme system from *Methylosinus trichosporium*. *FEMS Letters* **58**, 293–299.

Vary, P. S. and Johnson, M. J. (1967). Cells yields of bacteria grown on methane. *Applied Microbiology* **15**, 1473–1478.

Whittenbury, R., Phillips, K. C. and Wilkinson, J. F. (1970). Enrichment, isolation and some properties of methane-utilizing bacteria. *Journal of General Microbiology* **61**, 205–218.

Wilkinson, T. G., Topiwala, H. H. and Hamer, G. (1974). Interactions in a mixed bacterial population growing on methane in continuous culture. *Biotechnology and Bioengineering* **16**, 41–59.

# 13

# Electron Spin Resonance Properties of Component A of the Soluble Methane Monooxygenase from *Methylococcus capsulatus* (Bath)

MARC P. WOODLAND[*,1] AND RICHARD CAMMACK[†]

*Department of Biological Sciences, University of Warwick, Coventry, United Kingdom and †Department of Plant Sciences, King's College London, University of London, London, United Kingdom*

## Nature of the Enzyme

The soluble methane monooxygenase from *Methylococcus capsulatus* (Bath) is a multi-component enzyme system which catalyzes the *in vivo* oxidation of methane to methanol. In addition to this reaction, the enzyme has been shown to catalyze the *in vitro* oxidation of a wide variety of other substrates including various alkanes, alkenes, alicyclic, aromatic and heterocyclic hydrocarbons (Colby *et al.*, 1977).

The holoenzyme is believed to consist of three component proteins: component A ($M_r$ 210 000), thought to be the oxygenase; component B ($M_r$ 20 000), a small acidic protein, the role of which is not yet clear; and component C ($M_r$ 39 000), an iron–sulphur flavoprotein which acts as an NADH:acceptor reductase. The purification and properties of each of these three proteins has been described previously (Colby and Dalton, 1979; Dalton, 1980; Woodland and Dalton, 1984).

Component A was found to be an acidic, non-haem iron protein. Metal analysis gave values of 2.3 mol Fe and 0.35 mol Zn mol$^{-1}$. The protein is thought to consist of two copies of each of three subunits ($M_r$ 53,000, 39,000 and 17,000). The physicochemical properties of component A have been reported previously (Woodland and Dalton, 1984).

[1]Present address: Centre for Biotechnology, Imperial College of Science and Technology, London SW7 2AZ, United Kingdom.

MICROBIAL GAS METABOLISM:
MECHANISTIC, METABOLIC
AND BIOTECHNOLOGICAL ASPECTS

## Electron Spin Resonance Spectrum of Component A

As prepared, purified component A displayed ESR signals at $g = 4.3$, presumably due to rhombic iron, and a broad free radical at $g = 2.01$. Upon reduction with dithionite at room temperature for 2 min a third signal appeared, with principal $g$-values of $g_z = 1.78$, $g_y = 1.88$ and $g_x = 1.95$, while the $g = 4.3$ signal was slightly reduced in magnitude and the free radical was seemingly unaffected.

The temperature dependence of the $g = 1.95$ signal was examined. Above 17 K, the signal decreased in intensity and showed progressive line broadening, until at 35 K the signal was virtually undetectable. In this work, it was usual to record spectra around 10 K. Power saturation measurements showed the $g = 1.95$ signal to be almost unsaturated at this temperature (power for half-saturation = 119 mW).

In the initial studies of component A, it was reported that the addition of the substrate ethene to the reduced protein enhanced the peak-to-peak size of the $g = 1.95$ signal (Dalton, 1980). We have found that signal enhancement and an increase in the rhombic nature of the signal also occur in the presence of cyanomethane and methanol. This interaction with substrates suggests that protein A is the 'oxygenase' component of the methane monooxygenase.

## Electron Transfer Through the Enzyme

Anaerobic, visible spectrophotometric titrations of the protein components of the methane monooxygenase with NADH (M. P. Woodland, J. Lund and H. Dalton, unpublished data; Lund, 1983) clearly showed that electron transfer took place from NADH to protein C and then to protein A. To see if the reduction of component A in this manner generated a $g = 1.95$ signal, component A was incubated with catalytic quantities of components B and C in the presence of NADH at 4°C. As the role of component B appears to be part of the methane-oxidizing complex (although its role is not fully understood), this protein was also included in these experiments to ensure rapid electron transfer. Samples were removed at intervals and frozen as rapidly as possible. The ESR spectra obtained from these samples are given (Fig. 1).

The presence of the $g = 1.95$ signal in these spectra confirms results obtained from the anaerobic visible spectrophotometric titrations, namely that in the presence of components B and C electron transfer from NADH to component A can occur. On prolonged incubation the signal diminished; this is due to further reduction of component A to an ESR-silent state, the implications of which are discussed below. Control incubations lacking proteins B and C but containing NADH showed a complete absence of the $g = 1.95$ signal.

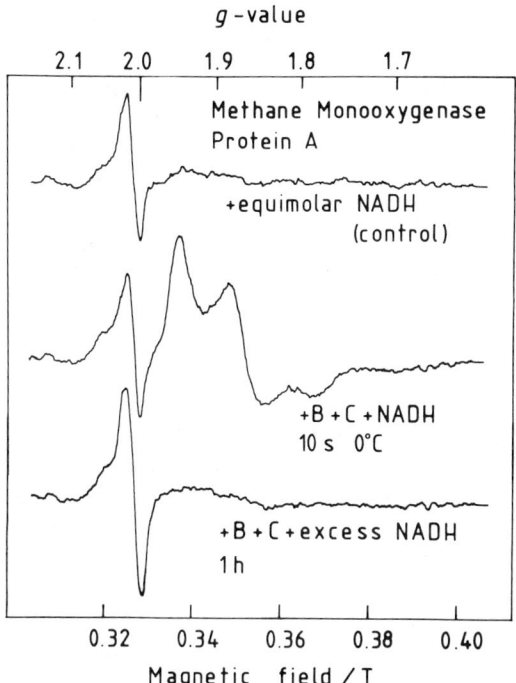

**Fig. 1.** Reduction of component A by NADH in the presence of catalytic amounts of components B and C. Component A (150 μM) in MOPS (20 mM, pH 7.0) was incubated with catalytic amounts of components B and C (15 μM) together with NADH (200 μM). Samples were transferred into quartz tubes at the times indicated, and rapidly frozen in a mixture of 2-methylbutane and methylcyclohexane (6:1, v:v) at −170°C. ESR spectra were recorded at a temperature of 10 K.

## Redox Properties of Component A

To further characterize the $g = 1.95$ signal, oxidation–reduction titrations were carried out on component A by the mediator-dye technique, using chemical reduction and ESR measurements (see Cammack et al., 1976). However, the rates of reduction and oxidation of the ESR-detectable centre were found to be slow. This was probably due to poor equilibration with the mediators used. Furthermore, it was not possible to allow sufficient equilibration time due to the instability of the protein. However, it was established that the appearance of the $g = 1.95$ signal required an applied redox potential (at pH 7.0) below $+100$ mV, with maximal intensity around 0 mV, and on further reduction below $-100$ mV the signal disappeared. The two-stage redox process was demonstrated to be reversible.

## Nature of the ESR-Active Centre

Previously it had been thought that component A might contain a [2Fe–2S] centre (Dalton, 1980). However, the spectra presented here and recently presented elsewhere (Woodland and Dalton, 1984) are not typical of those associated with classical iron–sulphur centres; on reduction, all known [2Fe–2S] and [4Fe–4S] centres give one g-value above 2. Furthermore, the most recent chemical analysis of the protein revealed <0.1 mol mol$^{-1}$ of acid-labile sulphur and no evidence was found to suggest the presence of an extrudable Fe–S centre (Woodland and Dalton, 1984).

A possible interpretation of the ESR signal and its redox behaviour is an antiferromagnetically coupled pair of high-spin iron atoms, by analogy with the two-iron ferredoxins (Gibson et al., 1966). In the oxidized state, both iron atoms are high-spin Fe(III) ($S = 5/2$) and the coupling between them produces a ground state with zero net spin. Upon reduction of one iron atom to high-spin Fe(II) ($S = 2$) the net spin of the coupled system is $S = 1/2$. It is this partially reduced state that gives rise to an ESR signal with a g-value less than 2.0. When both iron atoms are reduced to high-spin Fe(II), the coupling again gives zero net spin and the signal disappears. The temperature dependence of the signal is also explained by this model. The antiferromagnetic coupling leads to a set of spin states with $S = 1/2, 3/2, 5/2, 7/2$ and $9/2$ (see Gayda et al., 1976), of which the $S = 1/2$ state, which, we propose, gives the ESR signal, is lowest in energy. As the temperature is raised, however, the higher spin states become populated and the ESR signal intensity decreases.

This concept of an antiferromagnetically coupled iron centre is strengthened by consideration of two well-characterized proteins for which such a centre has been proposed. Both haemerythrin, a respiratory protein found in certain marine worms (Kurtz et al., 1978), and protein B2 from ribonucleotide reductase (Ehrenberg and Reichard, 1972) display spectroscopic similarities to component A. The electronic spectra of all three proteins are similar; in particular, protein B2 shows a variable peak at 410 nm ($\epsilon_m = 4.1 \times 10^3$) while component A shows a shoulder of variable intensity ($\epsilon_m = 2.6 \times 10^3$) at 406–410 nm (Woodland and Dalton, 1984). Perhaps most interestingly, semimethaemerythrin, the partially reduced form of methaemerythrin, gives an ESR signal which is very similar to that given by component A, with principal g-values of $g_z = 1.92$, $g_y = 1.82$ and $g_x = 1.57$. Furthermore, the lineshape and temperature dependence of the signal, the signal being lost above 34 K, are similar (Muhoberac et al., 1980).

Clearly, further studies are needed before the nature of the iron centre of component A is revealed; however, the information presented here would seem to suggest a form of antiferromagnetically coupled high-spin iron centre rather than a [2Fe–2S] centre as proposed initially.

## Acknowledgments

Financial support for this research was provided by an Extra-Mural Research Award from British Petroleum to M. P. Woodland and the Science and Engineering Research Council to R. Cammack. We thank Dr. D. S. Patil and Miss B. Dodemont for assistance with ESR and redox measurements and Dr. J. Lund for providing protein C.

## References

Cammack, R., Barber, M. J. and Bray, R. C. (1976). Oxidation–reduction potentials of molybdenum, flavin and iron–sulphur centres in milk xanthine oxidase. *Biochemical Journal* **157**, 469–478.

Colby, J. and Dalton, H. (1979). Characterisation of the second prosthetic group of the flavoenzyme NADH–acceptor reductase (component C) of the methane mono-oxygenase from *Methylococcus capsulatus* (Bath). *Biochemical Journal* **177**, 903–908.

Colby, J., Stirling, D. I. and Dalton, H. (1977). The soluble methane mono-oxygenase of *Methylococcus capsulatus* (Bath): Its ability to oxygenate *n*-alkanes, *n*-alkenes, ethers, and alicyclic, aromatic and heterocyclic compounds. *Biochemical Journal* **165**, 395–402.

Dalton, H. (1980). Oxidation of hydrocarbons by methane monooxygenases from a variety of microbes. *Advances in Applied Microbiology* **26**, 71–87.

Ehrenberg, A. and Reichard, P. (1972). Electron spin resonance of the iron-containing protein B2 from ribonucleotide reductase. *Journal of Biological Chemistry* **247**, 3485–3488.

Gayda, J. -P., Gibson, J. F., Cammack, R., Hall, D. O. and Mullinger, R. (1976). Spin lattice relaxation and exchange interaction in a 2-iron, 2-sulphur protein. *Biochimica et Biophysica Acta* **434**, 154–163.

Gibson, J. F., Hall, D. O., Thornley, J. H. M. and Whatley, F. R. (1966). The iron complex in spinach ferredoxin. *Proceedings of the National Academy of Sciences of the U.S.A.* **56**,987–990.

Kurtz, D. M., Shriver, D. F. and Klotz, I. M. (1978). Structural chemistry of hemerythrin. *Coordination Chemistry Reviews* **24**, 145–178.

Lund, J. (1983). Characterisation of component C of the soluble methane monooxygenase from *Methylococcus capsulatus* (Bath). Ph. D. Thesis, University of Warwick, Coventry, U.K.

Muhoberac, B. B., Wharton, D. C., Babcock, L. M., Harrington, P. C. and Wilkins, R. G. (1980). EPR spectroscopy of semi-methemerythrin. *Biochimica et Biophysica Acta* **626**, 337–345.

Woodland, M. P. and Dalton, H. (1984). Purification and characterisation of component A of the methane monooxygenase from *Methylococcus capsulatus* (Bath). *Journal of Biological Chemistry* **259**, 53–59.

# Part VI
# Nitrogen and Nitrogen Oxides

# 14

# Microbial Nitrous Oxide Metabolism and the Nitrous Oxide Electrode

S. J. FERGUSON

*Department of Biochemistry, University of Birmingham, Birmingham, United Kingdom*

## Introduction

Until recently, the study of synthesis and reduction of nitrous oxide has required use of gas chromatography and/or mass spectrometry as the methods of analysis for nitrous oxide. This article describes an electrode that can measure nitrous oxide concentrations continuously and rapidly. The introduction of a nitrous oxide electrode has come at a time when, as briefly outlined here, many bacterial species are being recognised as having a capability for synthesis and/or reduction of nitrous oxide.

## A Nitrous Oxide Electrode

A Clark-type electrode will respond to nitrous oxide as well as to oxygen, provided the cathode is made from silver instead of platinum and a polarising voltage of $-1.3$ V is used together with an alkaline electrolyte (Albery *et al.*, 1979; Alefounder and Ferguson, 1982). Such a silver cathode electrode is slightly more sensitive to gaseous nitrous oxide than to gaseous oxygen (Albery *et al.*, 1979) but, for aqueous solutions of these two gases, the electrode shows an approximately 10-fold greater sensitivity to oxygen (Alefounder and Ferguson, 1982). Nevertheless, this has not prevented study of the pattern of reduction of a mixture of oxygen and nitrous oxide by cells of *Paracoccus denitrificans*. Oxygen was clearly used in strict preference to nitrous oxide (Alefounder and Ferguson, 1982). Furthermore, most of the bacterial reactions that produce or consume nitrous oxide require anaerobic conditions, and so the problem of the sensitivity of the silver cathode to oxygen will not usually arise.

MICROBIAL GAS METABOLISM:
MECHANISTIC, METABOLIC
AND BIOTECHNOLOGICAL ASPECTS

## Organisms that Synthesise Nitrous Oxide

Denitrifying bacteria will, in general, only produce significant amounts of nitrous oxide if their nitrous oxide reductase is inhibited, although some strains of *Pseudomonas fluorescens* and *Pseudomonas chlororaphis* (Greenberg and Becker, 1977), as well as *Corynebacterium nephridii* (Hart *et al.*, 1965), produce nitrous oxide from nitrate because they lack nitrous oxide reductase activity.

As discussed by Bleakley and Tiedje (1982) and by Smith (1983), there is now evidence that a variety of organisms other than nitrifiers or denitrifiers can produce nitrous oxide. The mechanism and role of such synthesis are not clear. In the case of *Propionibacteria*, Kaspar (1982) has suggested that reduction of nitrite to nitrous oxide serves to remove the toxic nitrite rather than to provide energy for the cell. In contrast, Smith (1983) suggests that in *Escherichia coli* the rate of nitrous oxide production from nitrite is probably too slow to play a significant role in detoxification of nitrite, and that the action of nitrate reductase on nitrite is responsible for nitrous oxide production.

Of the nitrite reduced by *Klebsiella pneumoniae*, 95% is converted to ammonium ions, but there is evidence that 5% is reduced to nitrous oxide (Satoh *et al.*, 1983). One would expect different nitrite reductases to be involved in these two reactions, and this was confirmed by the finding that mutants unable to produce ammonium from nitrite were able to synthesise nitrous oxide (Satoh *et al.*, 1983).

The nitrifying bacteria are capable of producing nitrous oxide. A recent report suggests that, in soils, as much as 5% of the ammonium oxidised by *Nitrosomonas europaea* is converted to nitrous oxide (Lipschultz *et al.*, 1981). Production of nitrous oxide is most significant at oxygen concentrations below 2%. The mechanism for synthesis of nitrous oxide is unknown, but it is noteworthy that Miller and Wood (1983) recently described a nitrite reductase in *Nitrosomonas*. At low oxygen concentrations, this enzyme might permit the oxidation by cells of ammonia, using as oxidant the nitrite produced during previous exposure to high oxygen concentrations. Identification of the product of nitrite reduction as nitrous oxide is awaited.

It now seems that there are many potential sources of nitrous oxide in soils and other environments. A range of organisms might therefore be expected to be equipped to use nitrous oxide as a rather high-potential ($E^0$ $N_2O/N_2$ = $+1.35$ V) oxidant.

## Organisms that Reduce Nitrous Oxide

Nitrous oxide is recognised as an intermediate on the denitrifying pathway from nitrate to nitrogen, and accordingly denitrifying organisms will readily reduce added nitrous oxide, sometimes, as in the case of *P. denitrificans*, at rates higher than their nitrate reductase activity (Alefounder and Ferguson, 1982; Alefounder et al., 1983).

Wolinella (formerly Vibrio) succinogenes is not a denitrifier but an obligate anaerobe. It has been known since its discovery to gain energy for growth from coupling of the oxidation of either hydrogen or formate to the reduction of either fumarate to succinate or nitrate via nitrite to ammonia, but it has recently been demonstrated that W. succinogenes will grow anaerobically with nitrous oxide as sole added electron acceptor (Yoshinari, 1980). This organism, therefore, fails to be a denitrifier only because it lacks the type of dissimilatory nitrite reductase that converts nitrite to nitrous oxide.

The somewhat unexpected discovery of an energy-conserving pathway of electron flow to a nitrous oxide reductase in W. succinogenes raises the question of whether more organisms may possess an unsuspected capacity for reduction of nitrous oxide, which, because of its high solubility in water, may be relatively abundant in some aquatic environments. The availability of the nitrous oxide electrode should help investigation of this possibility.

Nitrogenase will reduce many small molecules that possess a triple bond, including nitrous oxide (Hardy and Knight, 1966). Presumably this means that nitrogen-fixing bacteria will reduce nitrous oxide, although the availability of dissolved nitrogen might prevent this reaction, depending on the relative affinity of the enzyme for the two substrates. The latter appears not to have been directly measured, but the nitrogen produced from reduction of nitrous oxide is not reduced through to ammonia (Hardy and Knight, 1966), suggesting that the enzyme might have a relatively high affinity for nitrous oxide. As nitrogen gas will normally be abundant, and in view of the large utilisation of ATP by nitrogenase, it is probable that the nitrous oxide reductase activity of nitrogenase has no physiological significance.

## Use of the Nitrous Oxide Electrode in an Alternative Assay for Nitrogenase?

Nitrogenase is usually assayed by measuring, by use of gas chromatography, the production of ethylene (ethene) from acetylene (ethyne). In principle, advantage could be taken of the reductase activity of nitrogenase to devise a new assay using the nitrous oxide electrode, which functions best under the anaerobic conditions that are required for nitrogenase activity. The advantage of the nitrous oxide electrode assay would be the simplicity of the apparatus required. Against this must be weighed the possibility that other enzymes, including unsuspected nitrous oxide reductases, might contribute to an observed consumption of nitrous oxide. However, the requirement of nitrogenase for ATP and a relatively power-ful reductant should help define the contribution of nitrogenase to observed reduction of nitrous oxide.

## Cellular Location of Nitrous Oxide Reductase

Degradation of the cell walls of both *P. denitrificans* and the denitrifying strain of *Rhodopseudomonas sphaeroides* results in loss of nitrous oxide reductase activity (Alefounder *et al.*, 1983; Urata *et al.*, 1982). These findings, together with the evidence that the protons required for reduction of nitrous oxide in *P. denitrificans* are taken from the periplasmic space (Boogerd *et al.*, 1981), indicate that the reductase is located in the periplasmic space. For *P. denitrificans*, such a location is consistent with the involvement of cytochrome *c*, which is located on the periplasmic surface of the cytoplasmic membrane, in catalysing electron flow from the electron transport chain to nitrous oxide (Alefounder and Ferguson, 1982; Ferguson, 1982). If *c*-type cytochromes should prove to have a widespread role as electron donors to the nitrous oxide reductase system, then a periplasmic location will probably prove general for nitrous oxide reductases because *c*-type cytochromes are almost certainly always found on the periplasmic surface of the cell membrane.

## Electron Transport Pathways to Nitrous Oxide

The pathway is best characterised in *P. denitrificans*. In this organism, electrons flow from the NADH or succinate dehydrogenases and through the ubiquinol–cytochrome *c* oxidoreductase system, from which cytochrome *c* transfers electrons either directly, or through as yet uncharacterised additional components, to the nitrous oxide reductase enzyme (Ferguson, 1982). Thus antimycin, by binding to the ubiquinol–cytochrome *c* oxidoreductase, inhibits reduction of nitrous oxide (Boogerd *et al.*, 1981; Alefounder and Ferguson, 1982) by all physiological substrates except methanol (Boogerd *et al.*, 1980). Methanol dehydrogenase, like the non-physiological substrate combination of ascorbate plus *N,N,N',N'*-tetramethyl-*p*-phenylenediamine, feeds electrons directly to cytochrome *c*. Thus, both methanol and ascorbate plus tetramethyl-*p*-phenylenediamine stimulate the reduction of nitrous oxide by *P. denitrificans* in an antimycin-insensitive manner (Boogerd *et al.*, 1980; Alefounder and Ferguson, 1982). Electron transport to nitrous oxide from NADH or succinate in *P. denitrificans* is clearly energy-conserving, as judged by both measurements of proton translocation (Boogerd *et al.*, 1981) and generation of a membrane potential (McCarthy *et al.*, 1981) with nitrous oxide as electron acceptor.

Rather less is known of the pathways of electron flow to nitrous oxide in other organisms. *Wolinella succinogenes* does not apparently possess the mitochondrial-like low molecular weight, but high potential, *c*-type cytochrome that is clearly involved in *P. denitrificans*. The route of electron flow from formate to nitrous oxide, which is energy-conserving (Yoshinari, 1980), remains to be elucidated. An antimycin-sensitive ubiquinol–cytochrome $c$ reductase is involved in light-driven cyclic electron flow in *Rps. sphaeroides*. It might, therefore, be expected that this reductase complex would also participate in the catalysis of electron flow from dehydrogenases via cytochrome $c$ to a nitrous oxide reductase in the denitrifying strain of *Rps. sphaeroides*. Should nitrous oxide reduction turn out to be insensitive to antimycin, it would imply that there must be an uncharacterised pathway for transferring electrons from ubiquinol to periplasmic reductases. This is actually a general problem in bacterial electron transport. Examples include electron flow to the periplasmic respiratory nitrate reductases of both *Rps. capsulata* (McEwan *et al.*, 1984) and *Rps. sphaeroides* (Sawada and Satoh, 1980) and the energy-conserving electron transport chain from formate to nitrite in *E. coli* (Motteram *et al.*, 1981).

## Molecular Nature of Nitrous Oxide Reductase

Anaerobic growth with added nitrate of both *Alcaligenes* sp. NCIB 11015 and *Pseudomonas perfectomarinus* on copper-deficient growth medium gave cells that were essentially unable to reduce nitrous oxide, and the latter rather than nitrogen was the end product of nitrate reduction (Iwasaki *et al.*, 1980; Matsubara *et al.*, 1982). These observations pointed to a role for copper in nitrous oxide reduction, an expectation that has been confirmed by the isolation from *Ps. perfectomarinus* of a homogeneous protein exhibiting nitrous oxide reductase activity with methyl viologen as immediate electron donor (Zumft and Matsubara, 1982). This protein was judged by sodium dodecyl sulphate–polyacrylamide gel electrophoresis and gel filtration to be a dimer of a polypeptide of molecular weight 62,000. Strikingly, 1 mole of enzyme contained 8 moles of copper. The physiological electron donor to this enzyme was not identified, although the fact that cells possessing nitrous oxide reductase activity have elevated levels of cytochrome $c$ is suggestive of such a role for this cytochrome.

A partially purified nitrous oxide reductase from *P. denitrificans* has been reported to have a molecular weight of 85,000 and lacks the chromophore observed in the *Ps. perfectomarinus* preparation (Kristjansson and Hollocher, 1981). The partially purified preparation from *P. denitrificans* apparently has a considerably higher specific activity than the enzyme from *Ps. perfectomarinus* (Hollocher, 1982) and therefore at present it cannot be decided whether one of the preparations is not a true respiratory nitrous oxide reductase or whether these two bacteria possess distinct types of nitrous oxide reductases. The purification of nitrous oxide reductases might in future be aided by use of the nitrous oxide electrode.

## Acknowledgment

Work from the author's laboratory was supported by the U.K. Science and Engineering Research Council.

## References

Albery, W. J., Brooks, W. N., Gibson, S. P., Heslop, M. W. and Hahn, C. E. W. (1979). An electrochemical method for the determination of $N_2O$. *Electrochimica Acta* **24**, 107–108.

Alefounder, P. R. and Ferguson, S. J. (1982). Electron transport-linked nitrous oxide synthesis and reduction by *Paracoccus denitrificans* monitored with an electrode. *Biochemical and Biophysical Research Communications* **104**, 1149–1155.

Alefounder, P. R., Greenfield, A. J., McCarthy, J. E. G. and Ferguson S. J. (1983). Selection and organisation of denitrifying electron-transfer pathways in *Paracoccus denitrificans*. *Biochimica et Biophysica Acta* **724**, 20–39.

Bleakley, B. H. and Tiedje, J. M. (1982). Nitrous oxide production by organisms other than nitrifiers or denitrifiers. *Applied and Environmental Microbiology* **44**, 1342–1348.

Boogerd, F. C., Van Verseveld, H. W. and Stouthamer, A. H. (1980). Electron transport to nitrous oxide in *Paracoccus denitrificans*. *FEBS Letters* **113**, 279–284.

Boogerd, F. C., Van Verseveld, H. W. and Stouthamer, A. H. (1981). Respiration-driven proton translocation with nitrite and nitrous oxide in *Paracoccus denitrificans*. *Biochimica et Biophysica Acta* **638**, 181–191.

Ferguson, S. J. (1982). Aspects of the organisation and control of bacterial electron transport. *Biochemical Society Transactions* **10**, 198–200.

Greenberg, E. P. and Becker, G. E. (1977). Nitrous oxide as end product of denitrification by strains of fluorescent pseudomonads. *Canadian Journal of Microbiology* **23**, 903–907.

Hardy, R. W. F. and Knight, E. (1966). Reduction of $N_2O$ by biological $N_2$-fixing systems. *Biochemical and Biophysical Research Communications* **23**, 409–414.

Hart, L. T., Larson, A. D. and McCleskey, C. S. (1965). Denitrification by *Corynebacterium nephridii*. *Journal of Bacteriology* **89**, 1104–1108.

Hollocher, T. C. (1982). The pathway of nitrogen and reductive enzymes of denitrification. *Antonie van Leeuwenhoek* **48**, 531–544.

Iwasaki, H., Saigo, T. and Matsubara, T. (1980). Copper as a controlling factor of anaerobic growth under $N_2O$ and biosynthesis of $N_2O$ reductase in denitrifying bacteria. *Plant and Cell Physiology* **21**, 1573–1584.

Kaspar, H. F. (1982). Nitrite reduction to nitrous oxide by propionibacteria: Detoxication mechanism. *Archives of Microbiology* **133**, 126–130.

Kristjansson, J. K. and Hollocher, T. C. (1981). Partial purification and characterisation of nitrous oxide reductase from *Paracoccus denitrificans*. *Current Microbiology* **6**, 247–251.

Lipschultz, F., Zafiriou, O. C., Wofsy, S. C., McElory, M. B., Valosi, F. W. and Watson, S. W. (1981). Production of NO and $N_2O$ by soil nitrifying bacteria. *Nature (London)* **294**, 641–643.

McCarthy, J. E. G., Ferguson, S. J. and Kell, D. B. (1981). Estimation with an ion-selective electrode of the membrane potential in cells of *Paracoccus denitrificans* from the uptake of the butyltriphenylphosphonium cation during aerobic and anaerobic respiration. *Biochemical Journal* **196**, 311–321.

McEwan, A. G., Jackson, J. B. and Ferguson, S. J. (1984). Rationalisation of properties of nitrate reductases in *Rhodopseudomonas capsulata*. *Archives of Microbiology* **137**, 344–349.

Matsubara, T., Frunzke, K. and Zumft, W. G. (1982). Modulation by copper of the products of nitrite respiration in *Pseudomonas perfectomarinus. Journal of Bacteriology* **149**, 816–823.

Miller, D. J. and Wood, P. M. (1983). The soluble cytochrome oxidase of *Nitrosomonas europaea. Journal of General Microbiology* **129**, 1645–1650.

Motteram, P. A. S., McCarthy, J. E. G., Ferguson, S. J., Jackson J. B. and Cole, J. A. (1981). Energy conservation during the formate-dependent reduction of nitrite by *Escherichia coli. FEMS Microbiology Letters* **12**, 317–320.

Satoh, T., Hom, S. S. M. and Shanmugam, K. T. (1983). Production of nitrous oxide from nitrite in *Klebsiella pneumoniae* mutants altered in nitrogen metabolism. *Journal of Bacteriology* **155**, 454–458.

Sawado, E. and Satoh, T. (1980). Periplasmic location of dissimilatory nitrate and nitrite reductases in a denitrifying phototrophic bacterium, *Rhodopseudomonas spheroides* forma sp. *denitrificans. Plant and Cell Physiology* **21**, 205–210.

Smith, M. S. (1983). Nitrous oxide production by *Escherichia coli* is correlated with nitrate reductase activity. *Applied and Environmental Microbiology* **45**, 1545–1547.

Urata, K., Shimada, K. and Satoh, T. (1982). Periplasmic location of nitrous oxide reductase in a photodenitrifier, *Rhodopseudomonas sphaeroides* forma sp. *denitrificans. Plant and Cell Physiology* **23**, 1121–1124.

Yoshinari, T. (1980). N₂O reduction by *Vibrio succinogenes. Applied and Environmental Microbiology* **39**, 81–84.

Zumft, W. G. and Matsubara, T. (1982). A novel kind of multi-copper protein as terminal oxidoreductase of nitrous oxide respiration in *Pseudomonas perfectomarinus. FEBS Letters* **148**, 107–112.

# 15

# The Basis for Preferential Electron Flow to Oxygen Rather Than Nitrogen Oxides in the Denitrifying Bacterium *Paracoccus denitrificans*

P. R. ALEFOUNDER, A. J. GREENFIELD, J. E. G. MCCARTHY AND
S. J. FERGUSON

*Department of Biochemistry, University of Birmingham, Birmingham, United Kingdom*

The availability of oxygen has long been known to inhibit dissimilatory nitrate reduction by bacteria. Less well documented are the effects of oxygen upon the activities of nitrite and nitrous oxide reductases. This article briefly summarises some recent findings concerning the possible mechanisms for, and advantages of, the preferential use of oxygen as electron acceptor by denitrifying bacteria. *Paracoccus denitrificans* grown under anaerobic conditions with nitrate as added electron acceptor has been the subject of recent studies in this area that are described here, but the findings with this bacterium will probably be applicable also to other denitrifying bacteria.

Recent work has shown that the inhibition of nitrate reduction when oxygen is available is not a consequence of a direct inhibitory effect of molecular oxygen (Alefounder *et al.*, 1981; Kucera *et al.*, 1981). This conclusion is based on the findings that: (1) the inhibitory effects of oxygen on nitrate reduction are partially relieved upon inhibition of electron flow to oxygen by either antimycin or hydroxylamine; (2) ferricyanide inhibits reduction of nitrate under anaerobic conditions but antimycin relieves its inhibitory effect; (3) nitrous oxide, and sometimes but not always nitrite, inhibits reduction of nitrate unless either antimycin or acetylene is added to inhibit reduction of nitrite or nitrous oxide (Alefounder *et al.*, 1981, 1983; Kucera *et al.*, 1981, 1983). An explanation for these findings is that nitrate reduction is controlled by the extent of oxidation of one or more components of the electron transfer chain (Alefounder *et al.*, 1983). It is envisaged that when such a component or components become relatively oxidised, for example in the presence of nitrous oxide, oxygen or ferricyanide, an

MICROBIAL GAS METABOLISM:
MECHANISTIC, METABOLIC
AND BIOTECHNOLOGICAL ASPECTS

inhibitory effect is exerted upon the reduction of nitrate (Alefounder *et al.*, 1983).

A matter requiring further study is the determination of the identity of the component(s) that has this putative role in controlling the reduction of nitrate. A candidate is ubiquinone because it is the last known common component before the respiratory chain to nitrate diverges from the electron transport pathways to oxygen, nitrous oxide or ferricyanide (Alefounder *et al.*, 1983; Ferguson, 1982). Unfortunately, direct measurement of the oxidation state of the ubiquinone/ ubiquinol couple in membranes is difficult because the difference in absorbance between the two molecules is obscured by the high background absorbance and light scattering by the cells. Indirect evidence that indeed onset of nitrate reduction is associated with an increased ratio of ubiquinol to ubiquinone has come from measurements with a fluorescent probe and observations on the extent of oxidation of the *b*-type cytochromes that are on the antimycin-sensitive pathway of electron flow (Alefounder *et al.*, 1983). By analogy with the mitochondrial respiratory chain, it is suggested that an increased extent of oxidation of these cytochromes correlates both with inhibition of nitrate reduction and a decreased ratio of ubiquinol to ubiquinone (Alefounder *et al.*, 1983).

Irrespective of the identity of the component(s) that controls nitrate reduction, a further matter to be understood is whether control is exerted on either the flow of electrons to nitrate reductase or the movement of nitrate to its presumed site of reduction at the cytoplasmic surface of the plasma membrane (Alefounder and Ferguson, 1980; Alefounder *et al.*, 1983). Two lines of evidence point to the latter mechanism. First, inside-out membrane vesicles, in which nitrate would have direct access to its reductase, can reduce oxygen and nitrate simultaneously and at similar rates. This applies also to vesicles prepared from a cytochrome *c*-deficient mutant of *P. denitrificans*. Hence the loss of control on nitrate reduction in vesicles cannot readily be attributed to a loss of cytochrome *c*, with concomitant inhibition of oxidase activity, during preparation of vesicles (Alefounder *et al.*, 1983). Second, the nitrate reductase in cells of *P. denitrificans* can only reduce chlorate, a substrate analogue, after a permeability barrier has been disrupted. Significantly, the titre of detergent that permits the appearance of chlorate reduction aso removes the control by oxygen on nitrate reduction (Alefounder and Ferguson, 1980).

There is evidence that the reductases for nitrite and nitrous oxide are located in the periplasmic space of *P. denitrificans* (Meijer *et al.*, 1979; Alefounder and Ferguson, 1980; Boogerd *et al.*,1981; Alefounder *et al.*, 1983). Thus, the observed inhibition of these enzymes by oxygen (e.g. Alefounder and Ferguson, 1982; Alefounder *et al.*, 1983) cannot be attributed to restriction of substrate access to the reductases. Furthermore, nitrous oxide probably freely permeates the cytoplasmic membrane.

Nitrite reductase activity of *P. denitrificans* cells is completely inhibited by oxygen when physiological reductants are the substrates. However, when a greater rate of supply of electrons is available from the substrate ascorbate plus *N,N,N',N'*-tetramethyl-*p*-phenylenediamine, nitrite reduction is possible in the presence of oxygen (Alefounder *et al.*, 1983). This observation, together with the information that the nitrite reductase has a lower $K_m$ value for nitrite than for oxygen, which is a non-physiological alternative substrate (Alefounder *et al.*, 1983), suggests that in the presence of oxygen the nitrite reductase is starved of electrons from physiological substrates by the two available oxidase enzymes (Ferguson, 1982; Alefounder *et al.*, 1983).

Under anaerobic conditions, the nitrous oxide reductase activity of *P. denitrificans* cells can be higher than the oxidase activity (Alefounder and Ferguson, 1982). Therefore, the observation that nitrous oxide reduction is completely inhibited by oxygen is unlikely to be explained by oxidase pathways dominating a competitive mechanism for available electrons. Rather, it seems more probable that the nitrous oxide enzyme is inactivated by molecular oxygen.

The present evidence is that three distinct mechanisms underlie the inhibitory effects of oxygen on the denitrifying reductase reactions. It seems, therefore, to be advantageous to the cell to use oxygen rather than the nitrogen oxidases as terminal electron acceptors. The reason for this preference is that the ATP yield for each pair of electrons flowing (P:2e ratio) to oxygen via cytochrome $aa_3$ is certainly, and via cytochrome *o* is probably, greater than the P:2e ratio when nitrate, nitrite or nitrous oxide is the electron acceptor. The organisation of the electron transfer chain, as well as the relative redox potentials of the acceptor couples, is the basis for this behaviour (Boogerd *et al.*, 1981; Ferguson, 1982). A second consideration might be that nitrate and nitrite are potentially valuable sources of cell nitrogen and therefore should be reduced by dissimilatory pathways only when unavoidable (Alefounder *et al.*, 1983).

There are some recent reports that denitrification by rhizobia (O'Hara *et al.*, 1983) and *Hyphomicrobium* sp. (Meiberg *et al.*, 1980; Hamer and Mechsner, 1983) can occur or even be stimulated in the presence of oxygen. Methanol was the substrate used in some studies with *Hyphomicrobium* sp.; this poses a problem for the cells because, by analogy with *P. denitrificans*, methanol can only be oxidised by nitrite or nitrous oxide and in each case the P:2e ratio is predicted to be zero (Alefounder and Ferguson, 1981). Thus, the availability of oxygen might enhance denitrification by generating formaldehyde via aerobic oxidation of methanol. As described in this article, dissimilatory nitrate and nitrite reduction is thought to be possible in the presence of oxygen provided the electron donors to the two reductases do not become too oxidised. A similar explanation may underlie the behaviour of rhizobia. The occurrence of nitrous oxide reduction in *Hyphomicrobium* sp. and rhizobia in the presence of oxygen is more difficult to account for in the context of the nature of the inhibition by oxygen of nitrous oxide reductase activity in *P. denitrificans* (Alefounder and Ferguson, 1982). A possibility to be investigated is that the nitrous oxide reductases of these two organisms are less sensitive to oxygen than their counterparts in *P. denitrificans*. A similar set of considerations might also apply to the recently described sulphide-oxidising bacterium *Thiosphaera pantotropha,* which is reported to denitrify under aerobic conditions (L. A. Robertson, personal communication).

## Acknowledgment

Work from the authors' laboratory was supported by the U.K. Science and Engineering Research Council.

## References

Alefounder, P. R. and Ferguson, S. J. (1980). The location of dissimilatory nitrite reductase and the control of dissimilatory nitrate reductase by oxygen in *Paracoccus denitrificans*. *Biochemical Journal* **192**, 231–240.

Alefounder, P. R. and Ferguson, S. J. (1981). A periplasmic location for methanol dehydrogenase from *Paracoccus denitrificans:* Implications for proton pumping by cytochrome $aa_3$. *Biochemical and Biophysical Research Communications* **98**, 778–774.

Alefounder, P. R. and Ferguson, S. J. (1982). Electron transport-linked nitrous oxide synthesis and reduction by *Paracoccus denitrificans* monitored with an electrode. *Biochemical and Biophysical Research Communications* **104**, 1149–1155.

Alefounder, P. R., McCarthy, J. E. G. and Ferguson, S. J. (1981). The basis of the control of nitrate reduction by oxygen in *Paracoccus denitrificans*. *FEMS Microbiology Letters* **12**, 321–326.

Alefounder, P. R., Greenfield, A. J., McCarthy, J. E. G. and Ferguson, S. J. (1983). Selection and organisation of dentrifying electron-transfer pathways in *Paracoccus denitrificans*. *Biochimica et Biophysica Acta* **724**, 20–39.

Boogerd, F. C., van Verseveld, H. W. and Stouthamer, A. H. (1981). Respiration driven proton translocation with nitrite and nitrous oxide in *Paracoccus denitrificans*. *Biochimica et Biophysica Acta* **638**, 181–191.

Ferguson, S. J. (1982). Aspects of the control and organisation of bacterial electron transport. *Biochemical Society Transactions* **10**, 198–200.

Hamer, G. and Mechsner, K. (1983). Denitrification by methane-utilizing mixed bacterial cultures. *EAWAG News* **14/15**, 12–16.

Kucera, I., Karlovsky, P. and Dadak, V. (1981). Control of nitrate respiration in *Paracoccus denitrificans* by oxygen. *FEMS Microbiology Letters* **12**, 391–394.

Kucera, I., Dadak, V. and Dobry, R. (1983). The distribution of redox equivalents in the anaerobic respiratory chain of *Paracoccus denitrificans*. *European Journal of Biochemistry* **130**, 359–364.

Meiberg, J. B. M., Bruinenberg, P. M. and Harder, W. (1980). Effect of dissolved oxygen tension on the metabolism of methylated amines in *Hyphomicrobium* X in the absence and presence of nitrate: Evidence for 'aerobic' dentrification. *Journal of General Microbiology* **120**, 453–463.

Meijer, E. M., van der Zwann, J. W. and Stouthamer, A. H. (1979). Location of the proton-consuming site in nitrite reduction and stoichiometries for proton pumping in anaerobically grown *Paracoccus denitrificans*. *FEMS Microbiology Letters* **5**, 369–372.

O'Hara, G. W., Daniel, R. M. and Steele, R. W. (1983). Effect of oxygen on the synthesis, activity and breakdown of the rhizobium denitrification system. *Journal of General Microbiology* **129**, 2405–2412.

# 16

# Reactions of Some Nitrogen Oxyanions and Nitric Oxide with Cytochrome Oxidase *d* from Oxygen-Limited *Escherichia coli* K12

JULIA A. M. HUBBARD,*·† MARTIN N. HUGHES* AND ROBERT K. POOLE†

*Departments of *Chemistry and †Microbiology, Queen Elizabeth College, University of London, London, United Kingdom*

## Introduction

*Escherichia coli*, like many other bacteria, produces different respiratory chain components depending upon the conditions of growth. During oxygen-sufficient growth, *E. coli* produces cytochrome *o* as its main terminal oxidase. However, cytochrome *d* is produced when oxygen supply is restricted and under a variety of apparently unrelated growth conditions (Poole, 1983). This is probably due in part to the greater affinity for oxygen of cytochrome *d* compared to cytochrome *o* (Rice and Hempfling, 1978).

An oxidase having some spectral properties similar to *E. coli* cytochrome *d* is found in the denitrifying bacterium *Pseudomonas aeruginosa* and other bacteria. The principal role *in vivo* of this cytochrome $cd_1$, however, is nitrite reductase activity (see Poole, 1983).

Recently, it has been reported that *E. coli* is able to perform some of the reactions of denitrification, and nitrous oxide production from added nitrate or nitrite has been observed. The enzymes involved have not been identified, but $N_2O$ production is increased during the stationary phase of growth, when cytochrome *d* is also produced (Bleakley and Tiedje, 1982).

The experiments reported here illustrate the reactions of cytochrome *d* with various nitrogen-containing compounds and suggest in each case the formation of a nitrosyl–cytochrome *d* complex, that is, an iron-bound $NO^-$ group.

MICROBIAL GAS METABOLISM:
MECHANISTIC, METABOLIC
AND BIOTECHNOLOGICAL ASPECTS

## Experimental

*Effects of the Ligands on Spectral Properties of*
*Cytochrome* d

Reduced cytochrome *d* in oxygen-limited *E. coli* reacts with nitrate (Hubbard *et al.*, 1983), nitrite, trioxodinitrate and nitric oxide, with the formation of a spectrally similar form in each case. In the reduced-minus-oxidised difference spectrum, this is seen as the loss of the reduced form (absorbance at 630 nm) with an increase in absorbance above 640 nm.

The absolute spectra (Fig. 1) show that addition of nitrate to reduced cytochrome *d* results in the formation of a broad band centered around 641 to 645 nm, which makes a simple interpretation of the reduced + nitrate-minus-oxidized difference spectrum difficult. Careful observation of the final difference spectrum shows that a shift of the absorbance maximum from 630 to 636 nm has occurred. This, and the increase in absorbance above 640 nm observed in the difference spectrum, may be due to loss of the reduced form of cytochrome *d* and production of the 645-nm band. It is probable that this complex is a cytochrome

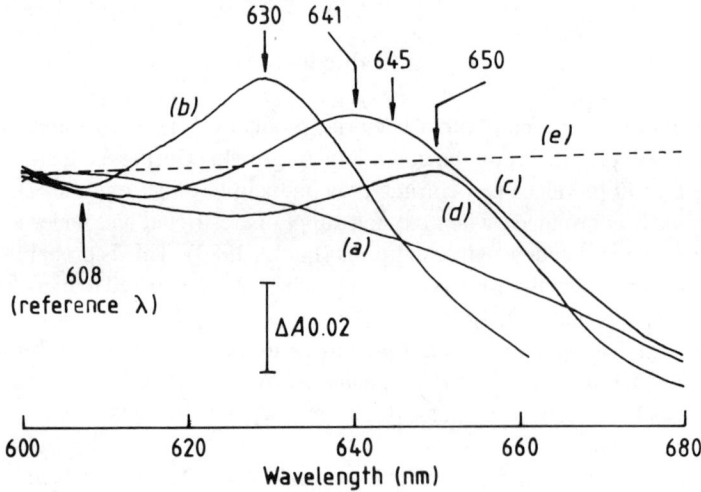

**Fig. 1.** Absolute spectra of forms of cytochrome *d*. Spectra of the reduced (b), reduced plus nitrate (c) and oxygenated (d) forms of cytochrome *d* were obtained with milk as a suitably opaque and scattering reference sample, which had no absorbance maxima in the region 500 to 700 nm as shown by spectrum (a), which is the difference spectrum of the milk sample, using water as a reference sample. The dashed line (e) is a milk-minus-milk baseline. The peaks at 630, 641–645 and 650 nm are due to the reduced (prepared by addition of 3.3 m*M* succinate, pH 7.2), reduced plus nitrate (160 μ*M* NaNO₃), and oxygenated (formed by reaction of membrane particles with a few microlitres 100 vol H₂O₂) forms of cytochrome *d*, respectively. The protein concentration was 8.7 mg ml⁻¹, path length 10 mm and spectral band width 4 nm.

*d*–nitrosyl complex, as previously suggested (Meyer, 1973). The production of this complex from nitrate and nitrite could occur by reduction of these compounds to $NO^-$ which is then bound to the metal. The cytochrome *d*–nitrosyl complex observed here may be an intermediate in the production of $N_2O$ by these membrane particles. The presence of the 645-nm band would confuse attempts to measure rates of loss of the reduced form of cytochrome *d* at 630 nm, except for the first few minutes, due to the overlap of the two bands.

## Kinetics of Ligand Binding to Cytochrome d

The same qualitative changes in cytochrome *d* resulted from the addition of either nitrate, nitrite, trioxodinitrate or nitric oxide. However, the time for the reaction to occur was dependent upon the nitrogen compound used. Nitric oxide

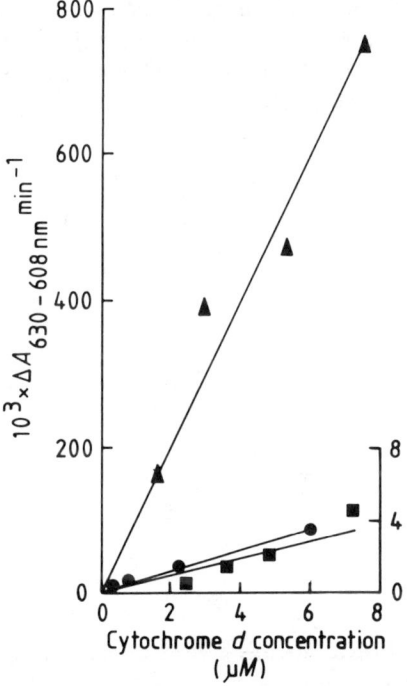

**Fig. 2.** Kinetics of the reaction of cytochrome *d* with nitrogen oxyanions at various concentrations. Cytochrome *d* concentration was kept constant (2.3 μ*M*). The initial rate of loss of reduced cytochrome *d* (which was also the maximum rate) is shown for a range of concentrations of added nitrite (●) and trioxodinitrate (▲). The initial rate of reaction with nitrate (■) was slow, and the low signal-to-noise ratio precluded accurate measurements. Therefore the rate shown, which is the maximum rate, was not obtained until several minutes after the reaction had started. The protein concentration was 5.4 mg ml$^{-1}$. Path length was 10 mm and spectral band width 4 nm.

caused an "immediate" reaction (within the time required to scan the spectrum, about 2 min). The kinetics of the slower reactions with $NO_3^-$, $NO_2^-$ and $N_2O_3^{2-}$ were studied in more detail in the following way. The rate of loss of the reduced form of cytochrome $d$ was measured by continuously monitoring the difference in absorbance between 630 and 608 nm, using a dual-wavelength spectrophotometer, after the nitrogen compound was added to succinate-reduced membrane particles. Temperature (19 to 20°C) was regulated by a circulating-water jacket and was monitored by a thermocouple inside the cuvette.

Rates were measured by using (1) a fixed concentration of cytochrome $d$ (2.3 $\mu M$) and a wide range of concentrations of sodium nitrate, nitrite or trioxodini-trate (Fig. 3), or (2) a fixed concentration of nitrate (2.0 m$M$), nitrite (0.1 m$M$) or trioxodinitrate (3.33 m$M$) and a range of concentrations of cytochrome $d$ (Fig. 2).

Figure 3 shows that, for nitrite and trioxodinitrate, the rate of loss of the reduced form of cytochrome $d$ was dependent on the concentration of the nitrogen compound used. However, the results are different for the reaction with nitrate, as increases in $NO_3^-$ above about 0.8 m$M$ had no further effect on the rate, which levelled off. This may reflect the saturation of a nitrate reductase,

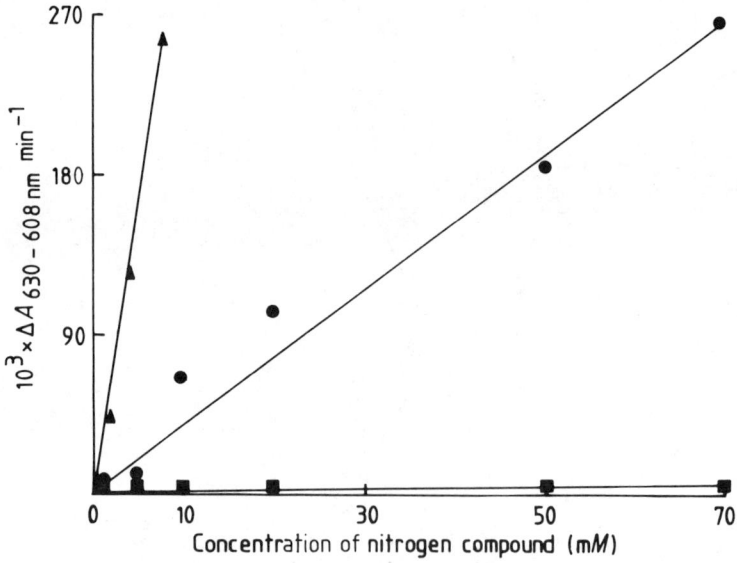

**Fig. 3.** Kinetics of the reaction of cytochrome $d$ with nitrogen oxyanions: effect of cytochrome concentration. Several concentrations of cytochrome $d$ were used with either nitrate (■, 2.0m$M$), nitrite (●, 0.1 m$M$) or trioxodinitrate (▲, 3.33 m$M$). Cytochrome $d$ concentration was calculated by using an extinction coefficient of 8.5 m$M^{-1}$ cm $^{-1}$ with $\Delta A$ 615–630 nm (Haddock and Schairer, 1973). The protein concentration was 1.15 to 12.8 mg ml$^{-1}$. Path length was 10 mm and spectral band width 4 nm.

which catalyses the reduction of nitrate to nitrite. The nitrite is then reduced further to give the nitrosyl complex.

Nitrate reductase activity [0.11 μmol benzyl viologen oxidized (mg protein)$^{-1}$ s$^{-1}$ (H. D. Williams, unpublished work)] has been measured in these membranes and in membranes from cells grown anaerobically with nitrate [17.1 μmol benzyl viologen oxidized (mg protein)$^{-1}$ s$^{-1}$], using the method of Jones and Garland (1977).

## Discussion

Trioxodinitrate has been proposed as an intermediate in denitrification (Averill and Tiedje, 1982), partly to accommodate the formation of an N–N bond, which is necessary for the production of $N_2O$.

$$Fe(NO_2^-) \; \rightleftharpoons \; Fe(NO^+) \overset{NO_2^-}{\longrightarrow} Fe(N_2O_3) \overset{2\epsilon}{\rightarrow} Fe(N_2O_3^{2-}) \overset{2\epsilon}{\rightarrow} Fe(N_2O)$$

If such a scheme is correct, and provided added trioxodinitrate has access to the site, then the reduction of trioxodinitrate should give $N_2O$. However, this does not appear to take place, and the reaction of trioxodinitrate with reduced cytochrome $d$ also appears to give the nitrosyl complex. This complex could be formed either by direct reaction with the trioxodinitrate or by reaction of HNO formed by decomposition of $HN_2O_3^-$ in solution.

$$HN_2O_3^- \rightleftharpoons HNO + NO_2^-$$

Comparison of the rate of self-decomposition of trioxodinitrate (Hughes and Wimbledon, 1976) with the rate of reaction of trioxodinitrate with reduced cytochrome $d$ indicates that it is unlikely that membrane particles catalyse the decomposition of trioxodinitrate, as the rate of self-decomposition is 10- to 50-fold greater than the rate of loss of the reduced cytochrome. The nitrosyl complex is therefore probably formed by the products of self-decomposition of trioxodinitrate in solution. These results do not support the proposal that trioxodinitrate is an intermediate in denitrification in *E. coli* (see also Garber *et al.*, 1983).

The findings allow an explanation of the spectrum of the cytochrome $d$ sometimes observed following anaerobic growth on nitrate. The spectrum of such cells is unusual in having the absorbance maximum of the reduced cytochrome $d$ shifted 10 nm to the red, compared to cells grown anaerobically with fumarate (Haddock *et al.*, 1976). A possible explanation of this shift, and the weak trough also seen in the spectrum of cytochrome $d$ of the nitrate-grown cells, is that a cytochrome $d$–nitrosyl complex is present. This complex may be the result of the reaction with cytochrome $d$ of reduced nitrogen compounds, formed by the action of nitrate reductase.

236     JULIA A. M. HUBBARD *ET AL.*

## Acknowledgments

We thank the Royal Society and the Science and Engineering Research Council for financial support and Huw Williams for the nitrate reductase assays.

## References

Averill, B. A. and Tiedje, J. M. (1982). The chemical mechanism of microbial denitrification. *FEBS Letters* **138**, 8–12.

Bleakley, B. H. and Tiedje, J. M. (1982). Nitrous oxide production by organisms other than nitrifiers or denitrifiers. *Applied and Environmental Microbiology* **44**, 1342–1348.

Garber, E. A. E., Wherli, S. and Hollocher, T. C. (1983). $^{15}$N-Tracer and NMR studies on the pathway of denitrification. Evidence against trioxodinitrate but for nitroxyl as an intermediate. *Journal of Biological Chemistry* **258**, 3587–3591.

Haddock, B. A. and Schairer, H. U. (1973). Electron-transport chains of *Escherichia coli*. Reconstitution of respiration in a 5-aminolaevulinic acid-requiring mutant. *European Journal of Biochemistry* **35**, 34–45.

Haddock, B. A., Downie, J. A. and Garland, P. B. (1976). Kinetic characterization of the membrane-bound cytochromes of *Escherichia coli* grown under a variety of conditions by using a stopped-flow dual-wavelength spectrophotometer. *Biochemical Journal* **154**, 285–294.

Hubbard, J. A. M., Hughes, M. N. and Poole, R. K. (1983). Nitrate, but not silver, ions induce spectral changes in *Escherichia coli* cytochrome *d*. *FEBS Letters* **164**, 241–243.

Hughes, M. N. and Wimbledon, P. E. (1976). The chemistry of trioxodinitrates. Part 1. Decomposition of sodium trioxodinitrate (Angeli's salt) in aqueous solution. *Journal of the Chemical Society, Dalton Transactions*, 703–707.

Jones, R. W. and Garland, P. B. (1977). Sites and specificity of the reaction of bipyridylium compounds with anaerobic respiratory enzymes of *Escherichia coli*. Effects of permeability barriers imposed by the cytoplasmic membrane. *Biochemical Journal* **164**, 199–211.

Meyer, D. J. (1973). Interaction of cytochrome oxidases $aa_3$ and *d* with nitrite. *Nature (London), New Biology* **245**, 276–277.

Poole, R. K. (1983). Bacterial cytochrome oxidases. A structurally and functionally diverse group of electron-transfer proteins. *Biochimica et Biophysica Acta* **726**, 205–243.

Rice, C. W. and Hempfling, W. P. (1978). Oxygen-limited continuous culture and respiratory energy conservation in *Escherichia coli*. *Journal of Bacteriology* **134**, 115–124.

# Part VII
# Applications of Mass Spectrometry

# 17

# Mass Spectrometric Monitoring of Dissolved Gases

DAVID LLOYD AND ROBERT I. SCOTT[1]

*Department of Microbiology, University College, Cardiff, Wales, United Kingdom*

## Introduction

The measurement of gas uptake or output by organisms continues to play a key role in the investigation of biological activities. The more traditional and well-tested methods [including volumetric, manometric, potentiometric, isotopic and chromatographic techniques (Umbreit *et al.*, 1957)], are now complemented and extended by newly devised approaches. Although one is confronted with a wide choice of methods for $O_2$ determination (Degn *et al.*, 1976, 1980), most of these are most easily applied to the determination of partial pressures of $O_2$ in the gas phase; perhaps the most successful are paramagnetic oxygen analysers and solid-state oxygen sensors operating at high temperatures. These measurements are never as useful as direct liquid-phase measurements; it is the concentration of *dissolved* gas that determines biological activity. The development of membrane-covered electrodes (Gnaiger and Forstner, 1983) for the measurement of dis-solved oxygen has been an important step toward the better understanding of aerobic microbiological processes. Measurements made with this technique, however, may be subject to error as a result of compounds which are also reduced at the electrode surface and contribute to the electrode current which is detected (Moss *et al.*, 1981). The sensitivity of the electrode in liquids is limited by the unstirred layer (Nernst, 1904) on the surface of the electrode at about 0.1 $\mu M$ oxygen (Degn *et al.*, 1980). Even for measurements in the gas phase, the polarographic method becomes unreliable at low tensions, because (1) the oxy-gen-dependent current produced becomes small relative to the "residual cur-rent" (still observed at zero oxygen tension) and (2) the residual current is

---

[1]Present address: The Polytechnic of Central London, School of Engineering and Science, 115 New Cavendish Street, London W1M 8JS, United Kingdom.

239

MICROBIAL GAS METABOLISM:
MECHANISTIC, METABOLIC
AND BIOTECHNOLOGICAL ASPECTS

dependent on the recent history of the electrode with respect to exposure to oxygen (Degn et al., 1980; Lloyd et al., 1981a). Photometric indicators such as haemoglobin (Bârzu and Satre, 1970) and luminescent bacteria (Lloyd et al., 1981a, 1982b) are finding applications at oxygen concentrations lower (<0.1 $\mu M$) than the limit of sensitivity of membrane-covered electrodes.

A number of other gases and volatile substances are important, not only as reactants or products, but also as effectors, of microbiological growth and activity. There is increasing evidence to suggest that the level of carbon dioxide plays a major role in the control of many fermentations, both aerobic and anaerobic (Wimpenny, 1969; Jones and Greenfield, 1982). Although many attempts have been made to develop a sterilizable $CO_2$ electrode (for example, Severinghaus et al., 1978), unsolved problems still limit its usefulness for long-term measurements. Similar shortcomings also hinder the application of other electrochemical sensors to microbiological processes (Reuss et al., 1975).

Electrodes specific to other gases, including dissolved hydrogen, nitrous oxide and sulphur dioxide, have also been developed (Albery et al., 1979; Alefounder and Ferguson, 1982; Kell, 1980). For some other gases in solution, chemical estimations may be required (Orland, 1965). Thus improved methods will facilitate monitoring the effects of dissolved gases on microbiological processes.

Non-invasive continuous measurement of several gases and vapours simultaneously with a single probe may be carried out by using a membrane-inlet mass spectrometer. The technique may be used in both the gas and liquid phases. The sensitivity of the technique is limited, like that of all membrane-covered sensors (e.g. selective electrodes), by the unstirred layer at the liquid–membrane interface (Nernst, 1904). In the case of oxygen, errors caused by consumption in this layer become significant at about 0.1 $\mu M$ $O_2$ (Lundsgaard et al., 1978). This limit cannot be decreased, even by stirring the liquid more rapidly, as it is not possible to reduce the thickness of the stationary layer next to the membrane to less than 30 $\mu m$. Mathematical treatment of mass spectrometric data for the analysis of complex gas mixtures may be necessary (Tunnicliff and Wadsworth, 1965; Ruth, 1968; Todd, 1981; Norris and Scrivens, 1981; Braithwaite and Henthorn, 1981; Crawford et al., 1981; Schorr et al., 1982). In this chapter, we outline the principles of membrane-inlet mass spectrometry and show how we (Lloyd and Scott, 1983; Lloyd et al., 1983c; Degn et al., 1985) and other groups (Bohátka et al., 1983; Heinzle et al., 1983) have applied the technique.

## Methodology

### Quadrupole Analysers

Before entering a quadrupole mass analyser, molecules are ionized in an ion source. The most common type of ion source uses a beam of electrons derived from a tungsten, rhenium or thoriated iridium filament; usually positive ions produced by electron detachment are analysed (Willard *et al.*, 1981). Ions are then focussed through an analyser (the quadrupole filter), which separates them according to values of mass-to-charge ratios ($m/z$), before detection as an ion current (Fig. 1a).

A quadrupole field is formed by four electrically conducting, precisely parallel molybdenum or steel rods. One (diagonally opposite) pair of rods is held at $+U$ volts and the other pair at $-U$ volts. A radio frequency (rf) oscillator supplies a

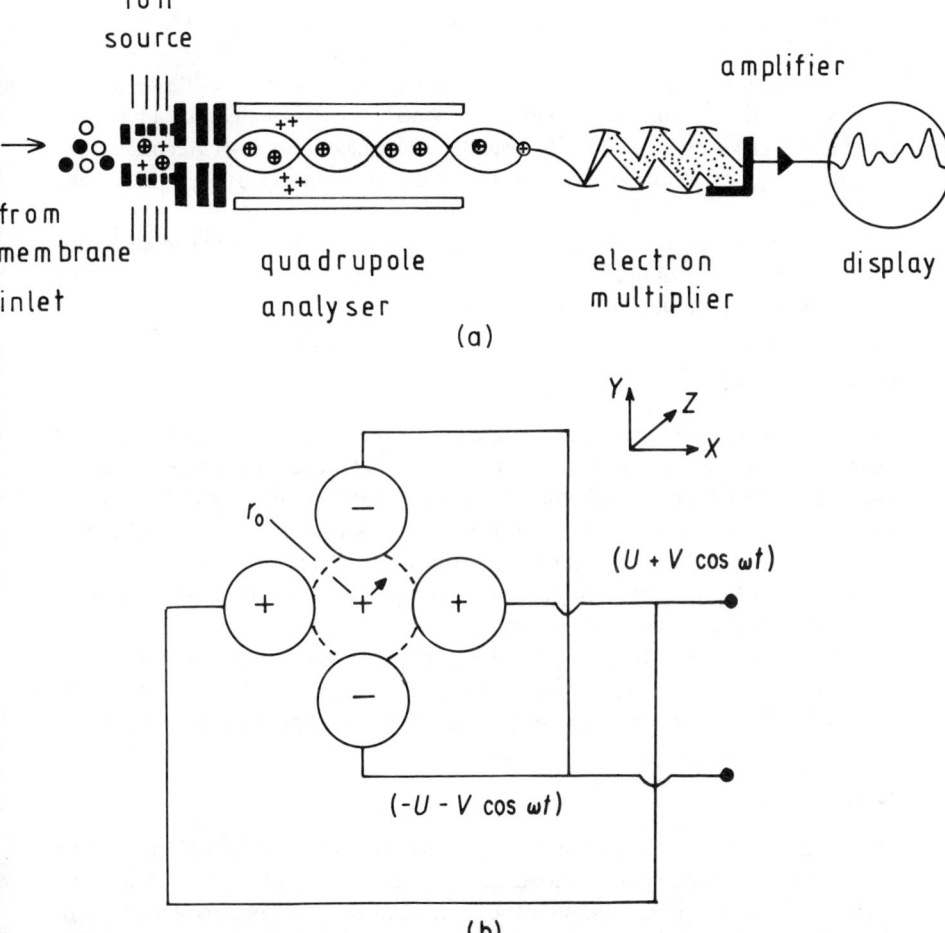

**Fig. 1.** Principles of operation of a quadrupole mass spectrometer, showing (a) the configuration of components and (b) a schematic diagram of the quadrupole filter principle. (From VG Gas Analysis product information.)

signal to the first pair of rods that is $+ V \cos \omega t$, and an rf signal retarded by $180°$ ($-V \cos \omega t$) to the second pair (Fig. 1b). An ion injected at one end of the array parallel to the $Z$ axis will be deflected into a complex trajectory by the rf and dc fields, which are perpendicular to the $Z$ axis. The ion may oscillate about the $Z$ axis and eventually emerge from the opposite end of the filter, or it may ultimately become neutralized on striking an electrode. Its fate depends on the parameters $U$ and $V$, the frequency of the rf voltage, the radius $r_0$ (Fig. 1b) and its value of $m/z$. By predetermined selection of suitable values, ions within a given range of $m/z$ will be transmitted through the filter, whilst others will strike an electrode. At any point in the scan, only one $m/z$ ratio can pass through the quadrupole mass analyser and be detected for a given combination of rf potential and frequency. An entire spectrum can be produced by varying the rf frequency while the rf and dc potentials remain constant, or by varying the rf potential (and the dc potential simultaneously so that these ratios of potentials remain fixed) while the rf frequency is held constant. As operation is dependent only on electric fields, very rapid mass scans can be achieved, typically 1000 mass units $s^{-1}$. Accurate mass measurement is possible because of the linear relationship between mass and voltage $V$.

Ion detection can be either on a simple plate detector (Faraday plate detector), with an electron multiplier having a multistage dynode assembly, or in a channeltron with a continuous dynode structure.

## Vacuum Systems

Quadrupole analysers detect gases in the partial pressure range from about $10^{-5}$ to $10^{-13}$ mbar. Various vacuum systems have been used; oil diffusion pumps give different pumping speeds for different gases (approximately related to molecular weights), but are preferable in some cases to ion pumps, which pump inert gases (e.g. helium or argon) very slowly by comparison with other gases. Either type of high-vacuum pump requires prepumping, for example by a sorption pump, down to about $10^{-2}$ mbar. However, the most convenient vacuum system consists of a turbomolecular pump (pumping rate $50 \, l \, s^{-1}$) backed by a rotary pump. This has the advantage that a vacuum of $10^{-6}$ mbar can be obtained within a few minutes of switch-on; all gases are pumped at similar rates.

## Membrane-Inlet Systems

Mass spectrometers with membrane-inlet systems were initially used in medical research for measurement of blood gases (Woldring et al., 1966; Delpy and Parker, 1975). The inlet consists of a membrane-covered, porous glass disc or stainless steel grid (Reuss et al., 1975; Lundsgaard et al., 1976; Doerner et al., 1982). The inlet leads directly to the ion source of the mass spectrometer.

A number of different membranes have been used. Teflon FEP fluorocarbon film is chemically inert and solvent-resistant. It is continuously serviceable over the range $-240$ to $+200°C$; its melting range is 260–280°C. It has a high resistance to impact, abrasion and tearing. Its permeabilities [in $cm^3$ mm/($m^2$ h bar)] to $CO_2$, $H_2$, $N_2$ and $O_2$ are 27.0, 35.6, 5.2 and 12.1, respectively.

The permeability of silicone rubber membrane is not simply a function of diffusion; it depends on the solubility of gases or vapours in the membrane, as well as their molecular size, so that the membrane exerts considerable selectivity (Berezhkovskii et al., 1980; Greenwalt et al., 1983). The advantage of silicone rubber is that it is exceptionally highly permeable to some gases (Table 1) and to some organic compounds. It is, for example, approximately $1.5 \times 10^5$-fold more permeable than Teflon, and 25-fold more permeable than natural rubber, to $O_2$. Its operating temperature range is from $-60$ to $+200°C$ and it will stand repeated steam or dry-heat sterilization. Doerner et al. (1982) found that in a typical inlet configuration, a 50.8-μm-thick silicone membrane gave the following response times (in s) in the gas phase: oxygen, 25; carbon dioxide, 26; ethanol, 17; acetone, 8; n-butanol, 73; acetic acid, 280; butyric acid, 30. Response times for either gas- or liquid-phase analysis depend on the membrane polymer and its thickness, the temperature and the pressure difference across the membrane.

*Closed and Open Systems*

Consumption of a gas in a closed system is measured by its time-dependent decreasing concentration; production gives a continuous increase to the limit of

**Table 1.** *Dimethylsilicone rubber membrane: Permeability to gases and vapours[a]*

| Gas | Permeability[b] | Gas | Permeability[b] |
|-----|-----|-----|-----|
| $H_2$ | 55 | $CO_2$ | 270–320 |
| He | 30 | $N_2O$ | 365 |
| $NH_3$ | 500 | $NO_2$ | 635 |
| $H_2O$ | 3000 | $SO_2$ | 1250 |
| CO | 30 | $CS_2$ | 7500 |
| $N_2$ | 25 | $CH_4$ | 80 |
| NO | 50 | $C_2H_6$ | 210 |
| $O_2$ | 50–60 | $C_2H_4$ | 115 |
| $H_2S$ | 840 | $C_2H_2$ | 2200 |
| Air | 50 | $C_3H_8$ | 340 |

[a]Data from General Electric Permaselective Membrane Bulletin GEA-8685A.

[b]Units: $1 \times 10^{-9} \dfrac{cm^3 \text{ gas (RTP) cm thick}}{cm^2 \text{ s cmHg } \Delta P}$

solubility. Oxygen electrode systems as usually employed to monitor respiration rates are of the "closed" type. The rates of consumption or production of a gas are monitored directly; these rates do not vary over those gas concentrations where saturation of a gas-consuming or -producing reaction system occurs. At non-saturating gas concentrations, however, this approach is not satisfactory, especially as the time taken for the gas concentration to pass through a range of interest is not easily controlled (Degn et al., 1980). Also, prolonged measurements of gaseous exchange cannot be carried out in the closed system. The difficulty of studying the rate of gaseous exchange at different gas concentrations in the non-saturating range has meant that systematic steady-state kinetic observations have not been carried out.

In the open system, the surface of the rapidly stirred liquid is maintained in contact with a gas mixture of known composition. The rate of exchange of gases into solution is linearly dependent on the difference in partial pressures between gas and liquid. Measurement of the dissolved gas concentration and the gaseous exchange constant enables calculation of steady-state production or consumption rates to be carried out.

$$V_r = k(T_G - T_L) \tag{1}$$

where $V_r$ is the production or consumption rate ($\mu M$ min$^{-1}$)

$k$ is the gaseous exchange constant (min$^{-1}$)

$T_G$ is the equivalent gaseous concentration in the gas phase ($\mu M$)

$T_L$ is the concentration of dissolved gas ($\mu M$)

($k$ is $\log_e 2/t_{1/2}$ where $t_{1/2}$ is the half-time for equilibration between gas and liquid in the absence of biological material). The temperature, rate of stirring and surface-to-volume ratio of the liquid sample all affect the value of $k$ (Finn, 1954; Danckwerts, 1970). Determination of gas production or consumption rates ($V_r$) at a given $T_L$ can be made by setting $T_G$ to a constant value and waiting until $T_L$ has become constant. Resetting $T_G$ repeatedly and waiting for a new steady state to occur gives a series of values for steady-state gas consumption versus gas concentration.

*Apparatus*

The reaction vessel (Fig. 2b) consists of a thermostated cylindrical stainless steel vessel of total volume 7 ml (working volume 4.5 ml). A stirring shaft enters vertically through the lid (Degn et al., 1980). Accuracy of measurements is critically dependent on the constancy of surface area between gas and liquid phases; excellent reproducibility of results has been obtained with a cross-shaped stirrer driven by a constant-speed motor. The lid also carries gas inlet and outlet tubes; gas mixtures have been made by using a calibrated gas mixer (Degn and Lundsgaard, 1980). The membrane-covered inlet to the mass spectrometer is

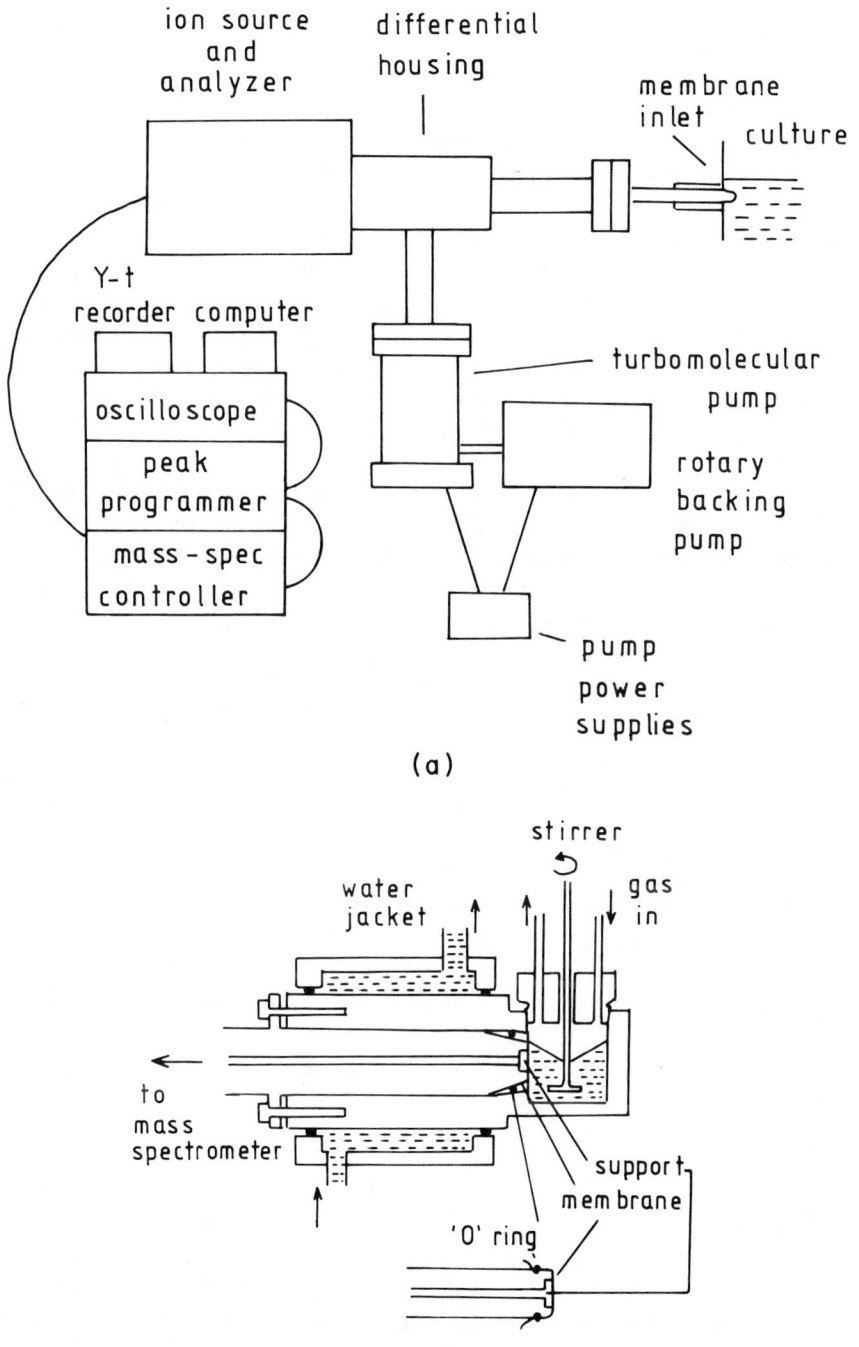

**Fig. 2.** Membrane-inlet mass spectrometer, showing (a) the main components and (b) the membrane inlet and reaction vessel.

sealed into a port lying below the level of the liquid. The membrane provides the selective barrier between the reaction mixture and the mass spectrometer. We have used a quadrupole mass spectrometer type SX200 (from VG Gas Analysis Ltd., Aston Way, Middlewich, Cheshire, U.K.) fitted with either a turbomolecular pump (backed by a rotary pump) or an ion pump (backed by a sorption pump). The mass spectrometer control unit enables a choice of filament to be made; the emission status of the filament is indicated together with the ability to degas the filament when necessary. The amplifier gain, the mass range monitored and the speed of scan through the mass spectrum may also be selected. Standard features included are visual indication of filament or emission failure, automatic selection of leak detection mode, digital display of mass number and a cursor for mass identification. The quadrupole analyser has both a Faraday plate and an electron multiplier as detectors.

A digital peak programmer (DPP 16) has 16 independently programmable channels. Each channel may be programmed independently for mass number (1 to 200 mass units), gain range ($10^{-5}$ to $10^{-13}$ mbar), detection system (Faraday plate or electron multiplier) and the accuracy level to which the measurements are required. Change in peak heights of programmed mass numbers may be monitored with time. Outputs provided are $X-Y$ output for oscilloscope display in a histogram form of the first 16 channels, eight analogue outputs for recorder display of channels 1–8 and an RS 232 serial interface for a host computer.

## Applications

The cracking patterns of some gases of microbiological interest are shown in Table 2. Judicious choice of mass numbers is required to separate components with the same molecular ion: measurement of one component at more than one mass number may be necessary. Portable mass spectrometers have recently been used as sensitive and versatile pollution monitors for organic vapours (Arnold and Robbiano, 1974; Meier, 1978; Ottley, 1981; Wilson and Ottley, 1981). The following microbiological processes have been studied by using quadrupole mass spectrometry.

### Respiration

The effect of varying $O_2$ and $NH_4^+$ concentrations on $CO_2$ production has been monitored in the yeasts *Saccharomyces uvarum*, *Candida utilis* and *Schizosaccharomyces pombe* (Lloyd *et al.*, 1983a,b). When suspensions were made anaerobic, the steady-state level of dissolved $CO_2$ was increased; reversal of this stimulation of glycolytic $CO_2$ production (the Pasteur effect) was observed upon aerobiosis. The Pasteur effect was abolished in the presence of the

**Table 2.** Mass spectra cracking patterns of some gases of microbiological interest[a]

| m/z | Hydrogen | Helium | Methane | Ammonia | Water | Acetylene | Ethylene | Carbon monoxide | Nitrogen | Ethane | Nitric oxide | Oxygen | Hydrogen sulphide | Argon | Carbon dioxide | Nitrous oxide | Nitrogen dioxide |
|-----|----------|--------|---------|---------|-------|-----------|----------|-----------------|----------|--------|--------------|--------|-------------------|-------|----------------|---------------|------------------|
| 2 | 100 | | 3 | | 1 | | | | | | | | | | | | |
| 4 | | 100 | | | | | | | | | | | | | | | |
| 12 | | | 2 | | | 3 | 3 | 5 | | | | | | | 6 | | |
| 13 | | | 8 | | | 6 | 4 | | | | | | | | | | |
| 14 | | | 19 | | | | | 1 | 7 | 3 | 8 | | | | | 13 | 10 |
| 15 | | | 36 | 7 | | | | | | 5 | 2 | | | | | | |
| 16 | | | 100 | 80 | 1 | | | 1 | | | 2 | 11 | | | 9 | 5 | 22 |
| 17 | | | | 100 | 23 | | | | | | | | | | | | |
| 18 | | | | | 100 | | | | | | | | | | | | |
| 20 | | | | | | | | | | | | | | 11 | | | |
| 22 | | | | | | | | | | | | | | | 1 | | |
| 24 | | | | | | 6 | 4 | | | | | | | | | | |
| 25 | | | | | | 20 | 12 | | | 4 | | | | | | | |
| 26 | | | | | | 100 | 62 | | | 23 | | | | | | | |
| 27 | | | | | | 3 | 65 | | | 33 | | | | | | | |
| 28 | | | | | | | 100 | 100 | 100 | 100 | | | | | 11 | 11 | |
| 29 | | | | | | | 2 | 1 | 1 | 22 | | | | | | | |
| 30 | | | | | | | | | | 26 | 100 | | | | | 31 | 100 |
| 32 | | | | | | | | | | | | 100 | 44 | | | | |
| 33 | | | | | | | | | | | | | 42 | | | | |
| 34 | | | | | | | | | | | | | 100 | | | | |
| 35 | | | | | | | | | | | | | 3 | | | | |
| 36 | | | | | | | | | | | | | 4 | | | | |
| 40 | | | | | | | | | | | | | | 100 | | | |
| 44 | | | | | | | | | | | | | | | 100 | 46 | |
| 45 | | | | | | | | | | | | | | | 1 | 1 | |
| 46 | | | | | | | | | | | | | | | | | 37 |

[a]Figures are expressed as a percentage of the maximum peak height detected.

uncoupler carbonyl cyanide $m$-chlorophenylhydrazone (CCCP). The $K_m$ for respiration of $S.$ $uvarum$ was 1.8 $\mu M$; the $K_m$ value for $O_2$ of the Pasteur effect was significantly higher (5 to 13 $\mu M$). Figure 3a shows one approach to study the Pasteur effect, that is, mass spectrometric monitoring of changes in dissolved $O_2$ and $CO_2$ levels upon illumination of a suspension of $S.$ $uvarum$, in which respiration had been partially inhibited by CO. That the dynamic response of the mass spectrometric system is adequate for the study of metabolic oscillations is indicated in Fig. 3b. In this experiment, damped respiratory oscillations with a period of the order of minutes follow anaerobic perturbation of the obligately aerobic yeast $Candida$ $utilis$ (Lloyd $et$ $al.$, 1983a). Other experimental systems for studying yeast cultures have recently been described (Weaver $et$ $al.$, 1980b; Heinzle $et$ $al.$, 1983).

Polarographic oxygen electrodes, a membrane inlet mass spectrometer and a redox electrode were used to continuously measure dissolved $H_2$, $O_2$ and $CO_2$ during chemolithotrophic growth of $Alcaligenes$ $eutrophus$ (Heinzle and Lafferty, 1980). The dynamic characteristics of the mass spectrometer were comparable with conventional steam-sterilized oxygen electrodes; comparison with a redox electrode for $H_2$ measurement led to the conclusion that the accuracy of the mass spectrometric measurement was much greater.

*Nitrogen Fixation*

The effect of the concentration of dissolved nitrogen on the rate of cyanobacterial nitrogen fixation has been studied by using membrane inlet mass spectrometry (Jensen $et$ $al.$, 1981; Jensen and Cox, 1983a,b). Jensen $et$ $al.$ (1981) showed that the $K_m$ for nitrogen of $Anabaena$ $flos$-$aquae$ was 165 $\mu M$ and the $V_{max}$ 3.6 $\mu$mole $N_2$ (mg chlorophyll)$^{-1}$ hr$^{-1}$. A $K_m$ value for $N_2$ of 65 $\mu M$ was reported by Jensen and Cox (1983a) for the same organism; the corresponding $K_m$ for acetylene reduction (a more common assay of nitrogenase activity) was 385 $\mu M$. Figure 4 shows simultaneous recording of dissolved $H_2$, $N_2$ and $O_2$ in a suspension of $Anabaena$. Illumination of the suspension resulted in a decrease in dissolved $N_2$ (due to increased $N_2$ fixation), an increase in the $O_2$ concentration (due to photosynthetic $O_2$ evolution) and transient increases in dissolved $H_2$. Jensen and Cox (1983a) suggested that the conversion factor between $C_2H_2$ and $N_2$ reduction should be 4 rather than 3. A conversion factor of 3 $C_2H_2$ reduced per $N_2$ reduced is often assumed, based on the fact that acetylene reduction requires two electrons and $N_2$ reduction requires six electrons. The results of Jensen and Cox (1983a) correlated well with the following mechanism for $N_2$ reduction:

$$N_2 + 8H^+ + 8e^- \rightarrow 2NH_3 + H_2 \qquad (2)$$

Nitrogenase and hydrogenase activities have been measured by a less direct method in $Azospirillum$ $brasilense$ (Berlier and Lespinat, 1980). Studies were

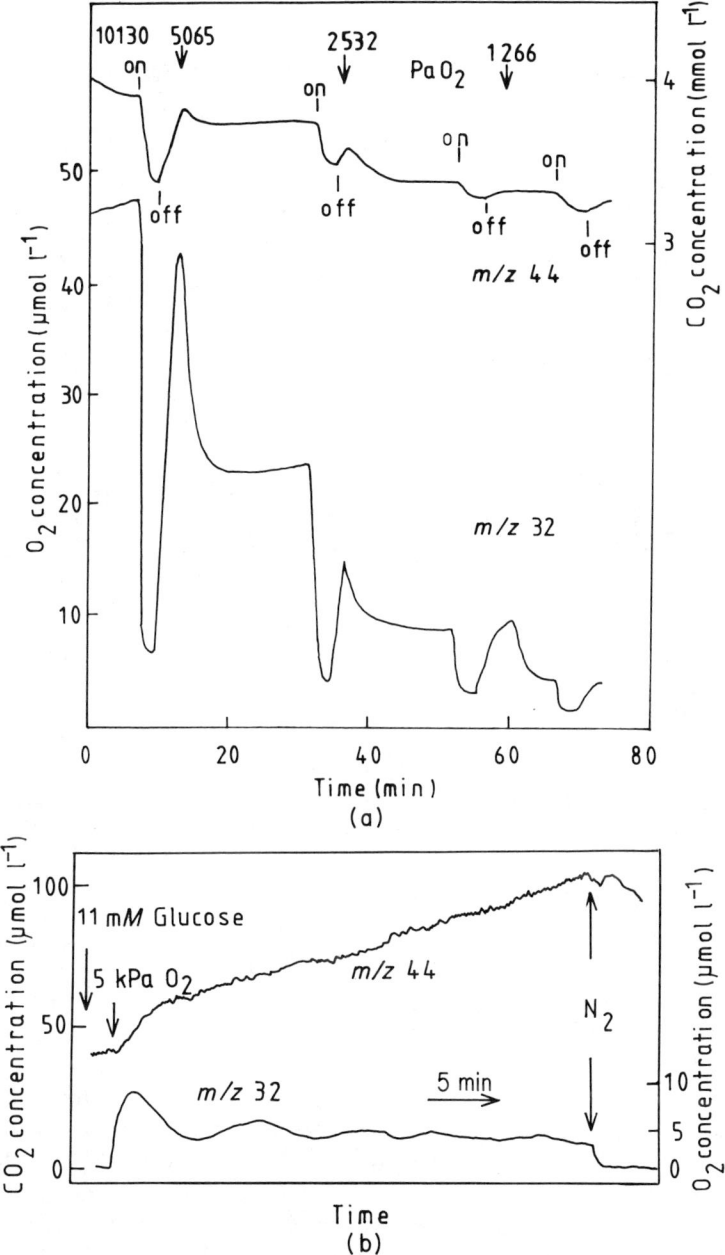

**Fig. 3.** Mass spectrometric measurement of yeast respiration. In (a) $O_2$ uptake and $CO_2$ production were measured in *Saccharomyces uvarum*. Cell respiration was partially inhibited by using a gas phase of a mixture of CO and $O_2$. The partial pressure of $O_2$ in the gas phase is shown. Reversal of the inhibition by CO was carried out by illuminating the cell suspension with light at 590 nm from a liquid dye laser. In (b), oscillations of respiration in *Candida utilis* are shown, following the addition of glucose and then $O_2$ to the anaerobic suspension.

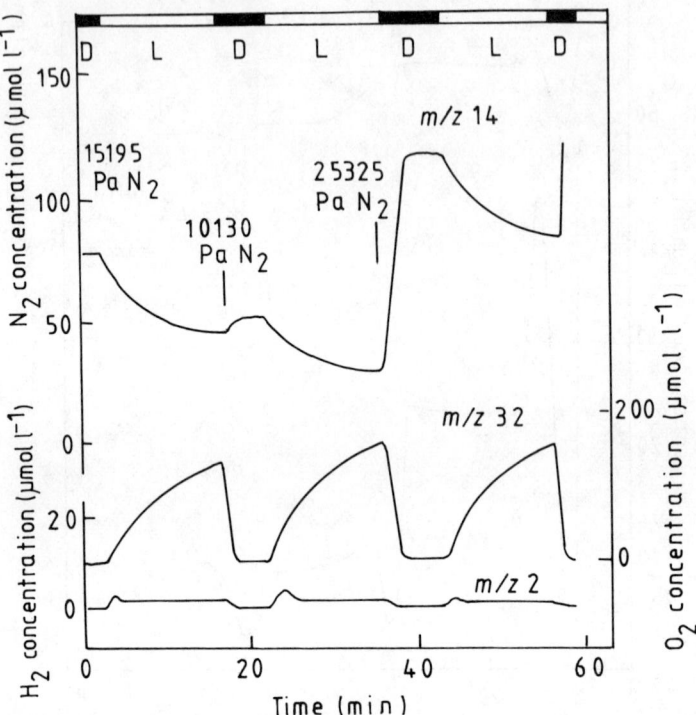

**Fig. 4.** Changes in dissolved $N_2$ ($m/z$ 14), $O_2$ ($m/z$ 32) and $H_2$ ($m/z$ 2) in a suspension of *Anabaena variabilis*. The gase phase contained argon and different concentrations of $N_2$ as indicated. Cells were illuminated (L) or kept dark (D) at the times shown. (Data from Jensen and Cox, 1983a.)

similar to those used to measure hydrogenase activity in *Rhodopseudomonas capsulata* (Jouanneau *et al.*, 1980) and *Azotobacter chroococcum* (Lespinat *et al.*, 1978) and employed measurements of the production of $H_2$ and HD (masses 2 and 3), uptake of deuterium (mass 4), and $O_2$ (mass 32) and production of $CO_2$ (mass 44). Reduction of acetylene to ethylene could be monitored by introducing deuterated acetylene ($C_2D_2$) and recording the evolution of deuterated ethylene ($C_2D_2H_2$) at mass 30. This procedure avoided mass 28, where the normal ethylene peak would have been masked by residual $N_2$ and by the gaseous inhibitor carbon monoxide. Hydrogen production by nitrogenase in *R. capsulata* was inhibited by acetylene and restored by carbon monoxide; hydrogen evolution coupled with HD formation and deuterium uptake (H–D exchange) was unaffected by acetylene and carbon monoxide. Cultures lacking nitrogenase activity also exhibited H–D exchange activity, which was catalysed by a membrane-bound hydrogenase present in the chromatophores of *R. capsulata*. Net hydrogen uptake, mediated by hydrogenase, was observed when electron acceptors such as ferricyanide, $CO_2$ or $O_2$ were present (Jouanneau et al., 1980). In A. brasilense, a microaerophilic nitrogen fixer, oxygen is required for nitrogenase function. Aerobically, an $H_2$-consuming hydrogenase recycles all the $H_2$ produced by nitrogenase; anaerobically, hydrogenase functions bidirectionally. The uptake hydrogenase may be involved in the respiratory protection of nitrogenase against oxygen inactivation and in the recycling of reducing power lost in hydrogen evolution (Berlier and Lespinat, 1980). Reversibility of the hydrogenase of *Paracoccus denitrificans* has also been shown by mass spectrometry (Vignais et al., 1982).

*Denitrification*

There have been no published accounts of attempts to measure the products of denitrification by membrane-inlet mass spectrometry. Dissimilatory denitrification by many microorganisms has been proposed to involve the following reactions: $NO_3^- \to NO_2^- \to NO \to N_2O \to N_2$. The final product, dinitrogen, is

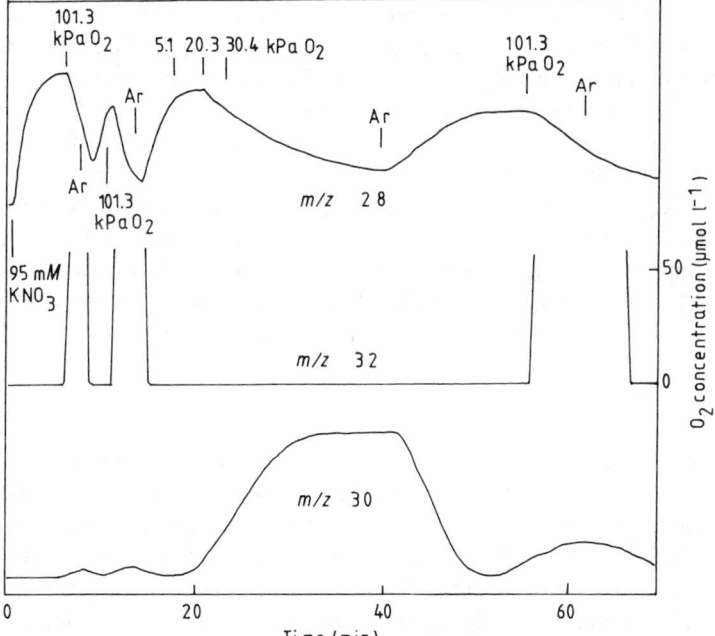

**Fig. 5.** Denitrification by *Paracoccus denitrificans*. The effects of various partial pressures of $O_2$ (*m/z* 32) on $N_2$ formation (*m/z* 28) and on *m/z* 30 are shown. The gas phase was changed to argon at the times labelled Ar; the partial pressure of $O_2$ in the gas phase is also indicated. Curves for *m/z* 28 and 30 are plotted on arbitrary scales.

evolved by the microorganism into the environment (Wharton and Weintraub, 1980). NO and $N_2O$ may also be evolved (Carlson and Ingraham, 1983); intermediates on the pathway of denitrification have been measured in various ways, including gas chromatography (Carlson and Ingraham, 1983), nitrous oxide electrodes (Alefounder and Ferguson, 1982) or gas chromatography/mass spectrometry (Wharton and Weintraub, 1980). Figure 5 shows the effects of $O_2$ on denitrification in *P. denitrificans* grown anaerobically with $NO_3{}^-$. Levels of $O_2$ >50 $\mu M$ cause rapid inhibition of $N_2$ production; this inhibition is rapidly reversed upon return to anaerobic conditions. The component at $m/z$ 30 may have contributions from $N_2O$, NO and $NO_2$ (see Table 2); monitoring additional mass numbers is required to elucidate these components. Exposure to lower levels of $O_2$ in the gas phase (5 to 30 kPa), giving undetectable $O_2$ in the liquid phase (<0.25 $\mu M$ $O_2$), again leads to inhibition of $N_2$ production and a much greater increase in the concentration of the $m/z$ 30 component.

*Sulphate Reduction*

Sulphate reduction and methanogenesis are microbial processes of particular importance in anaerobic environments as they are the major terminal oxidation steps in the flow of carbon and electrons (Senior *et al.*, 1982). Sulphate reduction has been measured with sulphide electrodes (Jørgensen and Revsbech, 1983), by trapping $H_2S$ as ZnS (Kristjansson *et al.*, 1982) or by using labelled $^{35}SO_4{}^{2-}$ (Lovley *et al.*, 1982). Figure 6 shows the effects of $O_2$ on $H_2S$ formation by *Desulfovibrio desulfuricans* measured by mass spectrometry. The addition of 12.2 $\mu M$ $SO_4{}^{2-}$ causes stimulation of $H_2S$ production; a partial pressure of 5065 Pa $O_2$ in the gas phase (insufficient to give detectable dissolved $O_2$) leads to a decrease in the $H_2S$ concentration of about 50%. Detectable levels of dissolved $O_2$ cause the $H_2S$ concentration to drop to zero; this inhibition is not reversed by anaerobiosis, or by the addition of a further 12.2 m$M$ $SO_4{}^{2-}$.

*Methanogenesis*

Membrane inlet mass spectrometry has been used to measure methanogenesis in two systems in our laboratory: (1) samples of rumen liquor and (2) samples from an anaerobic digestor. We have recently shown (unpublished results with L. Boddy and T. N. Williams) the feasibility of using a portable mass spectrometer with a flexible membrane inlet to measure dissolved gases *in situ* in the soil. This system may have great use in measuring methanogenesis *in situ* in, for example, anaerobic sediments, which are another important source of methane.

The rumen has often been assumed to be an anaerobic environment, although significant entry of $O_2$ must occur through the blood supply and with saliva and food (Czerkawski and Breckenridge, 1969). Oxygen has been detected in rumen

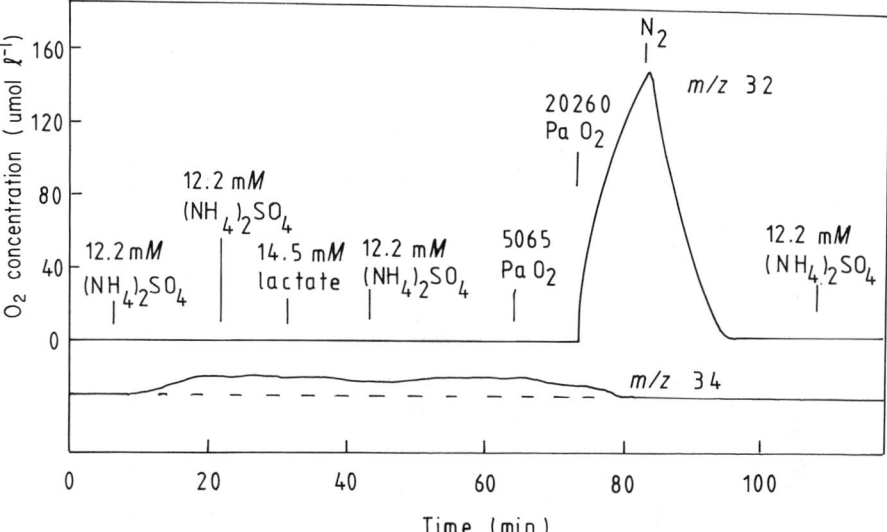

**Fig. 6.** Sulphate reduction by *Desulfovibrio desulfuricans*. The effect of various partial pressures of $O_2$ ($m/z$ 32) on $H_2S$ production ($m/z$ 34) is shown. $H_2S$ is on an arbitrary scale; zero $H_2S$ is shown by the dashed line. The gas phase was changed from 20,262 Pa $O_2$ to 101 kPa $N_2$ (100%) at the time labelled $N_2$.

gas (McArthur and Miltimore, 1961; Czerkawski and Clapperton, 1968) and we have recently shown its presence (at certain times after feeding) in the bulk liquid phase (Scott *et al.*, 1983b). These observations prompted our investigation, using membrane-inlet mass spectrometry, of the effect of $O_2$ on methanogenic systems.

Methanogenesis in samples of rumen liquor is characterized by being reversibly inhibited by $O_2$ (Scott *et al.*, 1983a,b). Low levels of $O_2$ in the gas phase (<1 kPa) lead to decreased methanogenesis and concomitant increased $H_2$ production. This effect is reversed on return to anaerobiosis. Figure 7 shows the effects of $O_2$ and glucose on $CH_4$ production in a sample of rumen liquor. Oxygen uptake by facultative organisms present can be increased by addition of glucose (analogous to the feeding of a ruminant). On return to anaerobiosis, methanogenesis recovers, even after this prolonged (25 min) exposure to detectable levels of dissolved $O_2$. Indeed, increased $O_2$ uptake by facultative organisms upon addition of glucose may itself lead to increased methanogenesis, even in the continued presence of some (2 kPa) $O_2$ in the gas phase (Scott *et al.*, 1983a). Membrane-inlet mass spectrometry has been used in conjunction with a *Photobacterium* probe for measurement of low levels of dissolved $O_2$ (Scott *et al.*, 1983a). Even at undetectable levels of dissolved $O_2$ (<30 n$M$), methanogenesis is reversibly inhibited. Hydrogen production in the rumen protozoon

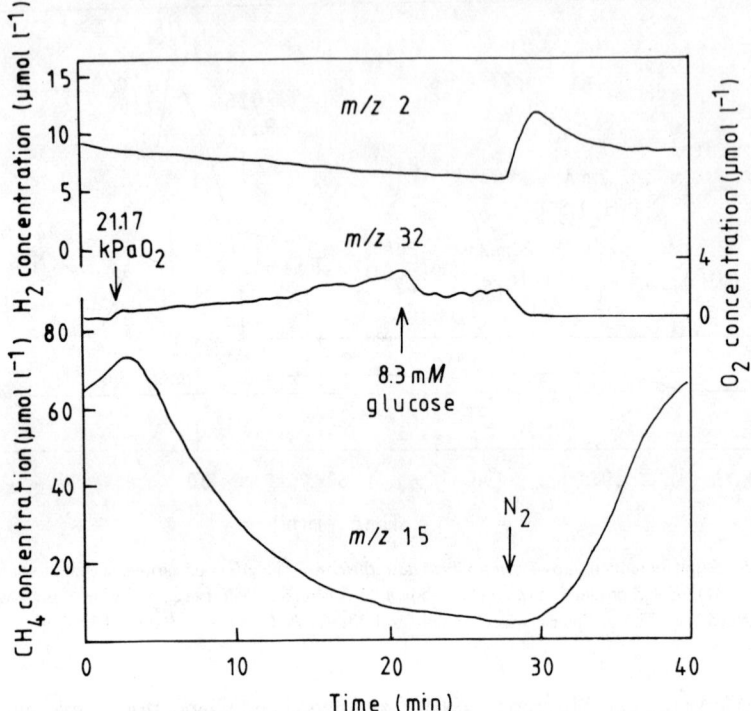

**Fig. 7.** Effects of $O_2$ on methanogenesis in a sample of bovine rumen liquor. $H_2$ was measured at $m/z$ 2, $O_2$ at $m/z$ 32 and $CH_4$ at $m/z$ 15. The gas phase was changed from 21.17 kPa $O_2$ to 101 kPa $N_2$ (100%) at the time labelled $N_2$.

*Dasytricha ruminantium* is reversibly inhibited at $O_2$ concentrations <2.8 $\mu M$, but irreversibly inactivated at levels of $O_2$ higher than this (Yarlett *et al.*, 1983).

Methanogenesis in anaerobic digestor samples is characteristically *irreversibly* inhibited by $O_2$, even if the concentration of dissolved $O_2$ is below 30 n$M$ (Scott *et al.*, 1983a). Figure 8a shows the effects of $O_2$ on methanogenesis in a sample from an anaerobic digester. Partial inhibition of $CH_4$ production occurs at gas-phase partial pressures of 0.2 kPa $O_2$. This inhibition is not reversed by re-exposure to $N_2$, that is, the rate of methanogenesis does not return to its original level. Exposure to even lower partial pressures of $O_2$ (0.025 kPa) for 9 hr leads to a reduction in the rate of $CH_4$ production by 75% (Fig. 8b). All of the data discussed so far have been obtained in a rapidly stirred, open reaction vessel. Addition of small amounts of $O_2$ to slowly stirred, closed samples results in rapid $O_2$ uptake and an increase in methanogenesis (Scott *et al.*, 1983c). Low levels of $O_2$ also stimulated methanogenesis (measured via gas chromatography) in the experiments of Pirt and Lee (1983). Mass spectrometry has also been used to measure methane and $O_2$ uptake in methanotrophic bacteria (Joergensen, 1983).

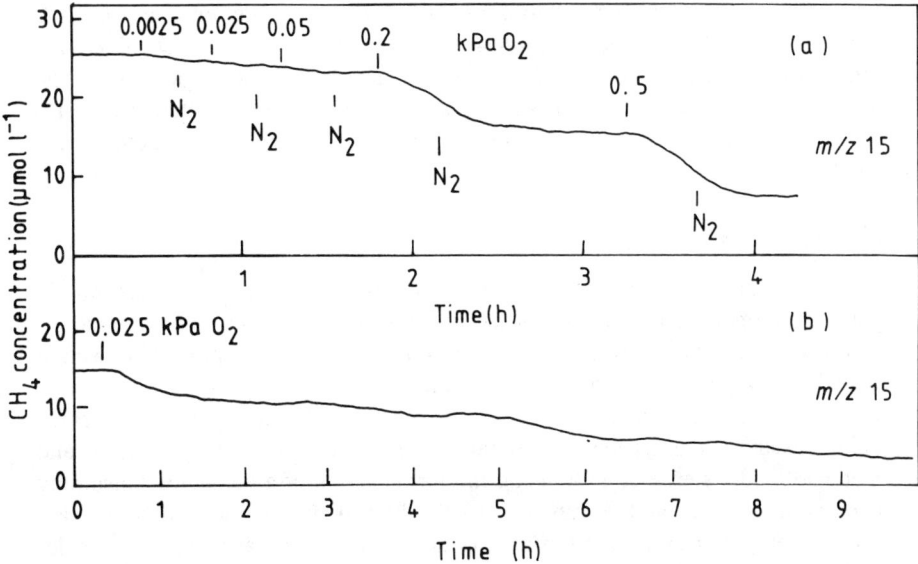

**Fig. 8.** Effect of $O_2$ on $CH_4$ formation in a sample from an anaerobic digester (domestic waste). (a) Effect of various partial pressures of $O_2$. (b) Effect of longer exposure to $O_2$. Dissolved $H_2$ and $O_2$ were both undetectable. The gas phase was changed from the partial pressure of $O_2$ indicated to 100% $N_2$ (101 kPa) at the times labelled $N_2$.

*Photosynthesis*

Mass spectrometry has been used for direct measurements of $O_2$ and $CO_2$ exchange during algal photosynthesis (Hoch and Kok, 1963; Stuart *et al.*, 1972; Radmer and Ollinger, 1980). Radmer and Ollinger (1981) extended the technique to study hydrazine photooxidation by photosystem II of isolated spinach chloroplasts.

*Gaseous Inhibitors*

Membrane-inlet mass spectrometry has been used to study the kinetics of inhibition of various gaseous inhibitors of respiration, for example, CO, HCN and $H_2S$. In many microorganisms terminal oxidases alternative to the main respiratory chain terminal oxidase may be responsible for a large proportion of total cell respiration (Degn *et al.*, 1978). The presence of these alternative oxidases may be shown by studying the effects of combinations of different respiratory inhibitors on cell respiration. In *Acanthamoeba castellanii*, an oxidase that was not inhibited by azide, CO or salicylhydroxamic acid but was inhibited by 1 mM cyanide has been shown to be present by mass spectrometry (Lloyd *et al.*, 1982a). The use of a mass spectrometer in studying the effects of $H_2S$ on respiratory inhibition is invaluable, as this inhibitor interferes with polarographic oxygen measurements. Sulphide-insensitive pathways of electron transport have been demonstrated in the protozoon *Tetrahymena pyriformis* (Lloyd *et al.*, 1980). *Acanthamoeba castellanii* has two means of protection against $H_2S$: the ability to oxidize it at low concentrations, and the presence of alternative pathways of electron transport which are not blocked by the inhibitor (Lloyd *et al.*, 1981b).

*Fermentation Monitoring*

The applications discussed so far have all involved monitoring of gases, and in general are on a small laboratory scale. Mass spectrometry may also be used for measuring dissolved gases and volatile components of low molecular weight directly in larger scale fermentation liquids (Reuss *et al.*, 1975; Weaver and Abrahams, 1979; Doerner *et al.*, 1982; Bohátka *et al.*, 1980, 1983). Reuss *et al.* (1975) used a silicone rubber membrane-covered inlet to measure $CO_2$, $O_2$ and methanol under sterile conditions during fermentation of a methanol-consuming bacterium. Weaver and Abrahams (1979) showed that a low-resolution mass spectrometer could measure through a dialysis membrane ammonia, formaldehyde, methanol, ethanol, acetaldehyde, $CO_2$ and formic, acetic and pyruvic acids. The measurement of dissolved volatile metabolites (ethanol or $CO_2$) produced by immobilized *Saccharomyces cerevisiae* enabled an indirect assay of sugars to be made by Weaver *et al.* (1980a); glucose could be measured at a concentration as low as 0.17 m$M$ in this way. A silicone membrane interface was used by Doerner *et al.* (1982) for an investigation of the acetone-butanol fermentation by *Clostridium acetobutylicum*. Differences in the molar fractions of components on the two sides of the membrane due to non-Fickian mass transfer made direct quantitation impossible. Calibration functions were obtained from gas chromatography data, enabling on-line determinations of butanol, acetone, ethanol, acetic and butyric acids to be made.

Membrane-inlet mass spectrometry has been used to continuously monitor $N_2$, $O_2$, $CO_2$ and ethanol in the outlet gas of a yeast batch fermentation (Pungor *et al.*, 1980). A computer was used to control the mass spectrometer, perform correction calculations and account for interferences between some ion peaks. The ability to scan continuously multiple peaks and to use $N_2$ as an internal standard provided rapid and accurate monitoring of the fermentation performance. Calibration curves obtained were linear for $CO_2$ (0 to 6.2%), $O_2$ (15 to 20%) and ethanol (0 to 15 g/l).

**Conclusions**

In this chapter, we have illustrated the usefulness of a quadrupole mass spectrometer with a membrane inlet for monitoring biological gaseous exchange reactions. The advantages of the technique may be summarized as follows:

1. Determinations are non-invasive.
2. Any number of gases or vapours can be monitored simultaneously.
3. Measurements can be made in the gaseous or liquid phase.
4. A continuous readout of partial pressures of gases and vapours is provided.
5. Long-term stability is such that calibration is required infrequently.
6. Response times in the liquid phase are on the order of minutes.
7. Interference between channels can be eliminated by judicious choice of mass to charge ratios, based on known cracking patterns for each gas.
8. The device has a wide dynamic range (single-figure ppm levels to tens of percents).
9. The lower limit of sensitivity provides information not previously attainable. For example, $0.25$ $\mu M$ $O_2$ is detectable and is below the reliable range of many oxygen electrodes.
10. Equipment can be made portable.
11. Although the cost of mass spectrometers is high, versatility more than compensates.

The technique that we have described is likely to be further developed in the near future. This is likely to involve development of multiparameter monitoring systems, for example with the *Photobacterium* probe (Lloyd *et al.*, 1981a; Scott *et al.*, 1983a), ion-selective electrodes for nonvolatile compounds (Kell, 1980) or spectrophotometric or fluorometric measurements. Further development of new methods of estimating non-volatile components in solution by using immobilized organisms (Weaver *et al.*, 1980a) is required. The ease of interfacing with computers (Bohátka *et al.*, 1983) makes process control (Taillez and Hume, 1981) and optimization of product formation by feedback of mass spectrometer signals an attractive development. Finally, application to systems that have been little studied (e.g. methanotrophic, denitrifying or sulphate-reducing organisms) and using the technique for non-invasive monitoring of dissolved gases in natural ecosystems hold great promise.

## References

Albery, W. J., Brooks, W. N., Gibson, S. P., Heslop, M. W. and Hahn, W. E. (1979). An electroanalytical method for the determination of $N_2O$. *Electrochimica Acta* **24,** 107–108.

Alefounder, P. R. and Ferguson, S. J. (1982). Electron transport-linked nitrous oxide synthesis and reduction by *Paracoccus denitrificans* monitored with an electrode. *Biochemical and Biophysical Research Communications* **104,** 1149–1155.

Arnold, J. T. and Robbiano, P. (1974). Portable vapor surveillance system. *United States Army Technical Report* **LWL-CR-03 P68C.**

Bârzu, O. and Satre, M. (1970). Determination of oxygen affinity of respiratory systems using oxyhaemoglobin as oxygen donor. *Analytical Biochemistry* **36,** 428–433.

Berezhkovskii, M. A., Korolinskii, M. I., Ozerov, L. N., Pushkin, M. A. and Slutskii, M. E. (1980). One-step membrane admission system for a mass spectrometer. *Pribory i Tekhnika Eksperimenta* **4**, 164–165.

Berlier, Y. M. and Lespinat, P. A. (1980). Mass spectrophotometric kinetic studies of the nitrogenase and hydrogenase activities in *in vivo* cultures of *Azospirillum brasilense* sp. 7. *Archives of Microbiology* **125**, 67–72.

Bohátka, S., Berecz, I. and Langer, G. (1980). Contamination measurements with quadrupole mass spectrometer. *Vide, les Couches Minces* **201**, Supplement, 243–246.

Bohátka, S., Langer, G., Szilágyi, J. and Berecz, I. (1983). Gas concentration determination in fermentors with quadrupole mass spectrometer. *International Journal of Mass Spectrometry and Ion Physics* **48**, 277–280.

Braithwaite, A. and Henthorn, K. (1981). Interfacing a microprocessor to a quadrupole mass spectrometer. In "Dynamic Mass Spectrometry" Vol. 6 (Eds. D. Price and J. F. J. Todd), pp. 141–154. Heyden, London.

Carlson, C. A. and Ingraham, J. L. (1983). Comparison of denitrification by *Pseudomonas stutzeri*, *Pseudomonas aeruginosa* and *Paracoccus denitrificans*. *Applied and Environmental Microbiology* **45**, 1247–1253.

Crawford, R. W., Bedford, R. G., Wong, C. M., Brand, H. R. and Kishiyama, K. I. (1981). Use of an automated mass spectrometer for underground coal gassification field test. In "Dynamic Mass Spectrometry" Vol. 6 (Eds. D. Price and J. F. J. Todd), pp. 195–211. Heyden, London.

Czerkawski, J. W. and Breckenridge, G. (1969). The effect of oxygen on fermentation of sucrose by rumen micro-organisms *in vitro*. *British Journal of Nutrition* **23**, 67–80.

Czerkawski, J. W. and Clapperton, J. L. (1968). Analysis of gases produced by metabolism of micro-organisms. *Laboratory Practice* **17**, 994–996, 1012.

Danckwerts, P. V. (1970). "Gas-Liquid Reactions." McGraw-Hill, London.

Degn, H. and Lundsgaard, J. W. (1980). Dynamic gas mixing techniques. *Journal of Biochemical and Biophysical Methods* **3**, 233–242.

Degn, H., Balslev, I. and Brook, R. (Eds.) (1976). "Measurement of Oxygen." Elsevier, Amsterdam.

Degn, H., Lloyd, D. and Hill, G. C. (Eds.) (1978). "Functions of Alternative Terminal Oxidases." Pergamon Press, Oxford.

Degn, H., Lundsgaard, J. S. and Petersen, L. C. (1980). Polarographic measurement of steady state kinetics of oxygen uptake by biochemical samples. *Methods of Biochemical Analysis* **26**, 47–77.

Degn, H., Cox, R. and Lloyd, D. (1985). Direct continuous measurements of dissolved gases in biological systems by membrane inlet mass spectrometry. *Methods of Biochemical Analysis* (in press).

Delpy, D. and Parker, D. (1975). Transcutaneous measurement of arterial blood gas tensions by mass spectrometry. *Lancet* **1**, 1016.

Doerner, P., Lehmann, J., Piehl, H. and Megnet, R. (1982). Process analysis of the acetone–butanol fermentation by quadrupole mass spectrometry. *Biotechnology Letters* **4**, 557–562.

Finn, R. K. (1954). Agitation–aeration in the laboratory and in industry. *Bacteriological Reviews* **18**, 254–274.

Gnaiger, E. and Forstner, H. (Eds.) (1983). "Polarographic Oxygen Sensors." Springer-Verlag, Berlin and New York.

Greenwalt, C. C., Voorhees, K. J. and Futrell, J. H. (1983). Transmission of organic molecules by a silicone membrane gas chromatograph/mass spectrometer interface. *Analytical Chemistry* **55**, 468–472.

Heinzle, E. and Lafferty, R. M. (1980). Continuous mass spectrometric measurement of dissolved $H_2$, $O_2$ and $CO_2$ during chemolithotrophic growth of *Alcaligenes eutrophus* strain H16. *European Journal of Applied Microbiology and Biotechnology* **11**, 17–22.

Heinzle, E., Furukawa, K., Dunn, I. J. and Bourne, J. R. (1983). Experimental methods for on-line mass spectrometry in fermentation technology. *Biotechnology* **1**, 181–188.

Hoch, G. and Kok, B. (1963). A mass spectrometer inlet system for sampling gases dissolved in liquid phases. *Archives of Biochemistry and Biophysics* **101**, 160–170.

Jensen, B. B. and Cox, R. P. (1983a). Direct measurements of steady-state kinetics of cyanobacterial $N_2$ uptake by membrane-leak mass spectrometry and comparisons between nitrogen fixation and acetylene reduction. *Applied and Environmental Microbiology* **45**, 1331–1337.

Jensen, B. B. and Cox, R. P. (1983b). Cyanobacterial $N_2$-uptake measurements using membrane-inlet mass spectrometry. *Society for General Microbiology, Quarterly* **10**, Part 4, M9.

Jensen, B. B., Cox, R. P. and Degn, H. (1981). Mass spectrometric measurements of steady state kinetics of cyanobacterial nitrogen fixation by monitoring dissolved $N_2$ in an open system. *FEMS Microbiology Letters* **12**, 37–40.

Joergensen, L. (1983). Methane oxidation measured by membrane-inlet mass spectrometry. *Society for General Microbiology, Quarterly* **10**, Part 4, M9.

Jones, R. P. and Greenfield, P. F. (1982). Effect of carbon dioxide on yeast growth and fermentation. *Enzyme and Microbial Technology* **4**, 210–223.

Jørgensen, B. B. and Revsbech, N. P. (1983). Colorless sulphur bacteria, *Beggiatoa* spp. and *Thiovulum* spp., in $O_2$ and $H_2S$ microgradients. *Applied and Environmental Microbiology* **45**, 1261–1270.

Jouanneau, Y., Kelley, B. C., Berlier, Y., Lespinat, P. A. and Vignais, P. M. (1980). Continuous monitoring by mass spectrometry of $H_2$ production and recycling in *Rhodopseudomonas capsulata*. *Journal of Bacteriology* **143**, 620–636.

Kell, D. B. (1980). The role of ion-selective electrodes in improving fermentation yields. *Process Biochemistry* **15**, No. 1, 18–23, 29.

Kristjansson, J. K., Schönheit, P. and Thauer, K. (1982). Different $K_s$ values for hydrogen of methanogenic bacteria and sulphate reducing bacteria: An explanation for the apparent inhibitions of methanogenesis by sulphate. *Archives of Microbiology* **131**, 278–282.

Lespinat, P. A., Gerster, R. and Berlier, Y. (1978). Direct mass spectrometric determination of the relationship between respiration hydrogenase and nitrogenase activities in *Azotobacter chroococcum*. *Biochimie* **60**, 339–341.

Lloyd, D. and Scott, R. I. (1983). Direct measurement of dissolved gases in microbiological systems using membrane inlet mass spectrometry. *Journal of Microbiological Methods* **1**, 313–320.

Lloyd, D., Kristensen, B. and Degn, H. (1980). The effects of inhibitors on the oxygen kinetics of terminal oxidases of *Tetrahymena pyriformis* ST. *Journal of General Microbiology* **121**, 117–125.

Lloyd, D., James, K., Williams, J. and Williams, N. (1981a). A membrane-covered photobacterium probe for oxygen measurements in the nanomolar range. *Analytical Biochemistry* **116**, 17–21.

Lloyd, D., Kristensen, B. and Degn, H. (1981b). Oxidative detoxification of hydrogen sulphide detected by mass spectrometry in the soil amoeba *Acanthamoeba castellanii*. *Journal of General Microbiology* **126**, 167–170.

Lloyd, D., Kristensen, B. and Degn, H. (1982a). The effects of cyanide, azide, carbon monoxide and salicylhydroxamic acid on whole cell respiration of *Acanthamoeba castellanii*. *Journal of General Microbiology* **128**, 185–188.

Lloyd, D., Williams, J., Yarlett, N. and Williams, A. G. (1982b). Oxygen affinities of the hydrogenosome-containing protozoa *Tritrichomonas foetus* and *Dasytricha ruminantium* and two aerobic protozoa, determined by bacterial bioluminescence. *Journal of General Microbiology* **128**, 1019–1022.

Lloyd, D., Kristensen, B. and Degn, H. (1983a). Glycolysis and respiration in yeasts: The Pasteur effect studied by mass spectrometry. *Biochemical Journal* **212**, 749–754.

Lloyd, D., Kristensen, B. and Degn, H. (1983b). Glycolysis and respiration in yeasts: The effects of ammonium ions studied by mass spectrometry. *Journal of General Microbiology* **129**, 2125–2127.

Lloyd, D., Scott, R. I. and Williams, T. N. (1983c). Membrane inlet mass spectrometry—Measurement of dissolved gases in fermentation liquids. *Trends in Biotechnology* **1**, 60–63.

Lovley, D. R., Dwyer, D. F. and Klug, M. J. (1982). Kinetic analysis of competition between sulphate reducers and methanogens for hydrogen in sediments. *Applied and Environmental Microbiology* **43**, 1373–1379.

Lundsgaard, J. S., Petersen, L. C. and Degn, H. (1976). Mass spectrophotometric determination of oxygen kinetics. *In* "Measurements of Oxygen" (Eds. H. Degn, I Balslev and R. Brook), pp. 168–183. Elsevier, Amsterdam.

Lundsgaard, J. S., Grønlund, J. and Degn, H. (1978). Error in oxygen measurements in open system due to oxygen consumption in unstirred layer. *Biotechnology and Bioengineering* **20**, 809–819.

McArthur, J. M. and Miltimore, J. E. (1961). Rumen gas analysis by gas solid chromatography. *Canadian Journal of Animal Science* **41**, 187–196.

Meier, R. W. (1978). A field portable mass spectrometer for monitoring organic vapours. *American Industrial Hygiene Association Journal* **39**, 233–239.

Moss, A. J., Jr., Nagle, W. A. and Baker, M. L. (1981). A chemiluminescent assay for oxygen in aqueous solution. *Photochemistry and Photobiophysics* **2**, 271–277.

Nernst, W. (1904). Theorie der Reaktionsgeschwindigkeit in heterogenen systemen. *Zeitschrift für Physikalische Chemie* **47**, 52–55.

Norris, P. E. and Scrivens, J. H. (1981). On-line mass spectrometry evaluation and implementation. *In* "Dynamic Mass Spectrometry" Vol. 6 (Eds. D. Price and J. F. J. Todd), pp. 156–166. Heyden, London.

Orland, H. P. (Ed.) (1965). "Standard Methods for the Examination of Water and Waste Water," 12th ed. American Public Health Association, New York.

Ottley, T. W. (1981). A quadrupole system for atmospheric pollution monitoring. *In* "Dynamic Mass Spectrometry" Vol. 6 (Eds. D. Price and J. F. J. Todd), pp. 212–219. Heyden, London.

Pirt, S. J. and Lee, Y. K. (1983). Enhancement of methanogenesis by traces of oxygen in bacterial digestion of biomass. *FEMS Microbiology Letters* **18**, 61–63.

Pungor, E., Jr., Perley, C. R., Cooney, C. L. and Weaver, J. C. (1980). Continuous monitoring of fermentation outlet gas using a computer coupled MS. *Biotechnology Letters* **2**, 409–419.

Radmer, R. and Ollinger, O. (1980). Light driven uptake of oxygen, carbon dioxide and bicarbonate by the green alga *Scenedesmus*. *Plant Physiology* **65**, 723–729.

Radmer, R. and Ollinger, O. (1981). Mass spectrometric studies of hydrazine photo-oxidation by illuminated chloroplasts. *Biochimica et Biophysica Acta* **637**, 80–87.

Reuss, M., Piehl, H. and Wagner, F. (1975). Applications of mass spectrometry to the measurement of dissolved gases and volatile substances in fermentations. *European Journal of Applied Microbiology* **1**, 323–325.

Ruth, J. M. (1968). Mathematical solution of the mass spectrophotometric standard mixture problem. *Analytical Chemistry* **40**, 747–750.

Schorr, W. K., Duschner, H. and Stark, K. (1982). Mathematical treatment of mass spectrophotometric data for analysis of gas mixtures. *Analytical Chemistry* **54**, 671–674.

Scott, R. I., Williams, T. N. and Lloyd, D. (1983a). Oxygen sensitivity of methanogenesis in rumen and anaerobic digester populations using mass spectrometry. *Biotechnology Letters* **5**, 375–380.

Scott, R. I., Yarlett, N., Hillman, K., Williams, T. N., Williams, A. G. and Lloyd, D. (1983b). The presence of oxygen in rumen liquor and its effects on methanogenesis. *Journal of Applied Bacteriology* **55**, 143–149.

Scott, R. I., Williams, T. N., Whitmore, T. N. and Lloyd, D. (1983c). Direct measurement of methanogenesis in anaerobic digestion by membrane inlet mass spectrometry. *European Journal of Applied Microbiology and Biotechnology* **18**, 236–241.

Senior, E., Lindström, E. B., Banat, I. M. and Nedwell, D. B. (1982). Sulphate reduction and methanogenesis in the sediment of a salt-march on the east coast of the United Kingdom. *Applied and Environmental Microbiology* **43**, 987–996.

Severinghaus, J. W., Stafford, M. and Bradley, A. F. (1978). $P_{CO_2}$ electrode design, calibration and temperature gradient problems. *Acta Anaesthesiologica Scandinavica, Supplementum* **68**, 118–122.

Stuart, T. S., Herold, E. W., Jr. and Gaffron, H. (1972). A simple combination mass spectrometer inlet and oxygen electrode chamber for sampling gases dissolved in liquids. *Analytical Biochemistry* **46**, 91–100.

Taillez, B. Y. and Hume, S. H. (1981). The use of time-of-flight mass spectrometry (TOFMS) in process monitoring and control. *In* "Dynamic Mass Spectrometry" Vol. 6 (Eds. D. Price and J. F. J. Todd), pp. 181–192. Heyden, London.

Todd, J. F. J. (1981). A survey of the current state of quadrupole mass spectrometry. *In* "Dynamic Mass Spectrometry" Vol. 6 (Eds. D. Price and J. F. J. Todd), pp. 3–13. Heyden, London.

Tunnicliff, D. D. and Wadsworth, P. A. (1965). A stepwise regression program for quantitative interpretation of mass spectra. *Analytical Chemistry* **37**, 1082–1085.

Umbreit, W. W., Burris, R. H. and Stauffer, J. R. (1957). "Manometric Techniques." Burgess, Minneapolis, Minnesota.

Vignais, P. M., Henry, M. F., Berlier, Y. and Lespinat, P. A. (1982). Effect of pH on H–$^2$H exchange, $H_2$ production and $H_2$ uptake catalysed by the membrane-bound hydrogenase of *Paracoccus denitrificans*. *Biochimica et Biophysica Acta* **681**, 519–529.

Weaver, J. C. and Abrahams, J. H. (1979). Use of variable pH interface to a mass spectrometer for the measurement of dissolved volatile compounds. *Review of Scientific Instruments* **50**, 478–481.

Weaver, J. C., Perley, C. R., Reames, F. M. and Cooney, C. L. (1980a). Temporarily immobilized micro-organisms: Rapid measurements using a mass spectrometer. *Biotechnology Letters* **2**, 133–137.

Weaver, J. C., Perley, C. R. and Cooney, C. L. (1980b). Mass spectrometer monitoring of a yeast fermentation. *Enzyme Engineering* **5**, 85–88.

Wharton, D. C. and Weintraub, S. T. (1980). Identification of nitric oxide and nitrous oxide as products of nitrite reduction by *Pseudomonas* cytochrome oxidase (nitrite reductase). *Biochemical and Biophysical Research Communications* **97**, 236–242.

Willard, H. H., Merrit, L. L., Jr., Dean, J. A. and Settle, F. A., Jr. (1981). "Instrumental Methods of Analysis," 6th ed. Wadsworth Publishing Co., Belmont, California.

Wilson, H. K. and Ottley, T. W. (1981). The use of a transportable mass spectrometer for the direct measurement of industrial solvents in breath. *Biomedical Mass Spectrometry* **8**, 606–610.

Wimpenny, J. W. T. (1969). Oxygen and carbon dioxide as regulators of microbial growth and metabolism. *Symposia of the Society for General Microbiology* **19**, 161–197.

Woldring, S., Owens, G. and Woodford, D. C. (1966). Blood gases: Continuous *in vivo* recording of partial pressures by mass spectrography. *Science* **153**, 885–887.

Yarlett, N., Scott, R. I., Williams, A. G. and Lloyd, D. (1983). Effects of oxygen on hydrogen production by the rumen protozoa *Dasytricha ruminantium* Schuberg. *Journal of Applied Bacteriology* **55**, 359–361.

# Discussion

*D. Kell:* What are the problems with the membrane inlet?

*D. Lloyd:* None; the membrane is supported on a grid or sintered disc and in the case of Teflon lasts as long as a year. The membrane (two layers of 25-μm-thick Teflon) is remarkably resistant.

*Kell:* Is the limit of detection a function of mass number?

*Lloyd:* No, not at these small mass numbers, but more than one gas may be detected at some values of *m*/*z*; you have to be aware of the cracking patterns.

*O. Meyer:* How much does it cost?

*Lloyd:*   About £10,000.

*R. K. Thauer:*   I gather that CO and $N_2$ give the same mass peaks in the mass spectrometer. Is it therefore right that CO in the presence of $N_2$ cannot be determined by mass spectrometry?

*Lloyd:*   I didn't have time to talk about the problems! Both $N_2$ and CO give the largest peak at $m/z$ 28. We have measured CO as a respiration inhibitor, but you have to use argon and exclude $N_2$.

*R. K. Poole:*   What influence, if any, does the differential permeability of the membrane to various gases have in multiparameter recordings?

*Lloyd:*   This consideration is important only for gas-phase measurements. In the dissolved gas determinations described here, in an open system, the rate-limiting step is diffusion at the interface between the stirred liquid and the gas.

*C. Anthony:*   So you could have a manifold with different kinds of membranes?

*Lloyd:*   Yes, we have considered this possibility.

*A. P. F. Turner:*   How long may the apparatus you have described be continuously operating?

*Lloyd:*   Using the turbopump rotary pump system, an oil change is needed every 5000 hr. With an ion pump–sorption pump, even longer periods of continuous use have been employed (e.g. in our laboratory more than 2 years).

# 18

# Mass Spectrometric Determinations of the Effect of Oxygen on Methanogenesis: Inhibition or Stimulation?

ROBERT I. SCOTT,[1] T. NORMAN WILLIAMS, TIMOTHY N.
WHITMORE AND DAVID LLOYD

*Department of Microbiology, University College, Cardiff, Wales, United
Kingdom*

## Introduction

Methanogenic bacteria grow, and produce methane, only when cultivated under strictly anaerobic conditions in media with a redox potential below $-330$ mv (Smith and Hungate, 1958; Zehnder, 1978). This has given rise to the belief, held by most microbiologists, that this group of archaebacteria are extremely susceptible to oxygen. Recently, however, qualitative and quantitative studies have led to a reappraisal of the possible effects of oxygen on methanogenic systems.

Bacterial methanogenesis is a ubiquitous process in most anaerobic environments. The association of this event with anaerobic decomposition of organic matter in microbial habitats such as sewage sludge digestors, the rumen and intestinal tract of animals, and in sediments and muds of various aquatic habitats has been recognized and documented for more than a century (Zeikus, 1977). The degree of anaerobiosis of some of these habitats has, however, been questioned: oxygen has been found in the gas phase of the rumen (Czerkawski and Breckenridge, 1971; Barry *et al.*, 1977; McArthur and Miltimore, 1961) and also in the liquid phase (Scott *et al.*, 1983a). In a review of the composition of gases leaving 48 anaerobic digestors, Wheatley (1980) reported that in 20 of these oxygen was detectable; the effect of this oxygen on methanogenesis was not clear.

---

[1]Present address: The Polytechnic of Central London, School of Engineering and Science, 115 New Cavendish Street, London W1M 8JS, United Kingdom.

MICROBIAL GAS METABOLISM:
MECHANISTIC, METABOLIC
AND BIOTECHNOLOGICAL ASPECTS

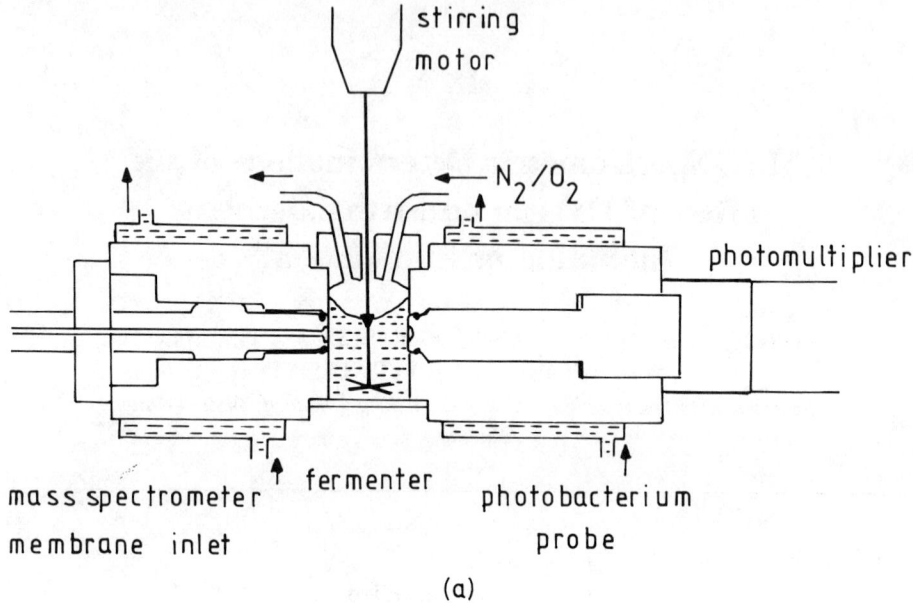

(a)

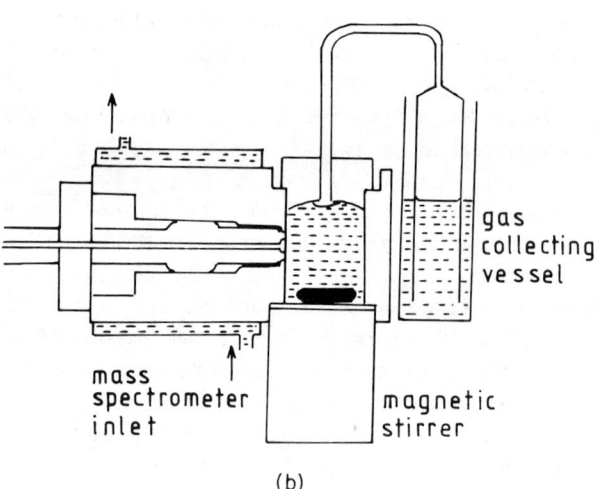

(b)

**Fig. 1.** (a) The mass spectrometer membrane inlet and photobacterium probe. The open reaction vessel contained 4.5 ml of either rumen liquor or anaerobic digestor contents, and was maintained at 28°C. The speed of the motor-driven, cross-shaped stirrer ranged from 600 to 1000 rev min$^{-1}$. An $N_2/O_2$ mixture was passed over the surface of the stirred sample at 200 ml min$^{-1}$. (b) The closed reaction vessel. The digestor contents (8 ml) were slowly stirred (100 rev min$^{-1}$) by a magnetic stirrer bar.

There have been few quantitative studies of the effect of oxygen on methanogens; these studies have been concerned with effects on pure (for example, Kiener and Leisinger, 1982, 1983) and mixed (Scott *et al.*, 1983a, b, c: Pirt and Lee, 1983) cultures. Kiener and Leisinger (1982, 1983) conducted a study of the oxygen tolerance of five different species of methanogenic bacteria. *Methanobacterium thermoautotrophicum, Methanobrevibacter arboriphilus* and *Methanosarcina barkeri* were able to survive after exposure to air for up to 30 hr. In the case of *M. barkeri,* protection against oxygen occurred at the level of cell aggregates. The arrangement of cells in packets probably led to protection of the cells in the interior and thereby secured the survival of a cell packet as a colony-forming unit during extended periods of oxygen stress.

Membrane-inlet mass spectrometry has been used to measure the concentrations of dissolved $H_2$, $CH_4$ and $O_2$ in samples of rumen liquor and anaerobic digestor contents (Scott *et al.*, 1983a, b, c,). Methanogenesis in samples of rumen liquor was characteristically *reversibly* inhibited after exposure to oxygen. In anaerobic digestor samples methane formation was *irreversibly* inhibited by oxygen. These effects were observed at undetectable levels of dissolved oxygen (that is $<0.25$ $\mu M$ $O_2$). Clearly, increased sensitivity of detection of $O_2$ is necessary to elucidate the effect of $O_2$ on these methanogenic systems.

In the present paper we describe the use of a photobacterium probe, in conjunction with a membrane-inlet mass spectrometer, to measure lower levels of dissolved $O_2$ (down to 30 n$M$ $O_2$) in methanogenic systems.

## Methods

Bovine rumen liquor samples were obtained as described by Scott *et al.* (1983a), anaerobic digestors were maintained and sampled as described by Scott *et al.* (1983b, c). The rapidly stirred open reaction vessel with membrane-covered mass spectrometer inlet and membrane-covered photobacterium probe (Lloyd *et al.*, 1981) is shown in Fig. 1a. The theory of the open system, and details of the mass spectrometer, are presented in an accompanying article (Lloyd and Scott, Chapter 17, this volume). In the open system the steady-state concentration of an evolved gas is directly proportional to its rate of formation.

The slowly stirred closed vessel is shown in Fig. 1b. Stirring was by a magnetic stirrer bar at about 100 rev min$^{-1}$.

## Results and Discussion

Figure 2a shows the concentrations of $O_2$ and $CH_4$ in a sample from an anaerobic digestor in the rapidly stirred open system. Exposure to a gas phase containing

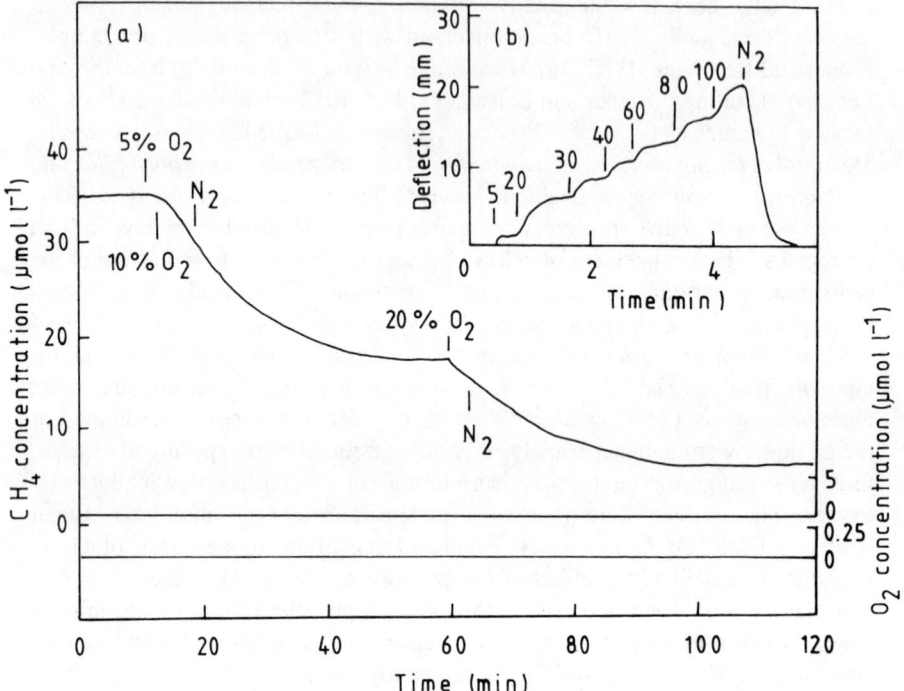

**Fig. 2.** (a) The effect of $O_2$ on $CH_4$ production in a sample from an anaerobic digestor in the rapidly stirred open system. Oxygen was measured by using the mass spectrometer (0–5 µM) and the photobacterium probe (0–0.25 µM). (b) Calibration of the photobacterium probe after 5 hr exposure to the digestor sample. Numbers refer to percentages (v/v) of a gas phase containing 0.05% $O_2$ in $N_2$. A gas phase of 0.05% $O_2$ is equivalent to 0.56 µM $O_2$.

10% (v/v) $O_2$ for 6 min led to a decrease in the steady-state level of methane from 35 to 18 µM. Oxygen was undetectable by either the mass spectrometer (that is <0.25 µM $O_2$) or the photobacterium probe (that is <30 nM $O_2$). Exposure to 20% (v/v) $O_2$ for 5 min caused a further 34% reduction in the level of methane. Again $O_2$ was undetectable. That the photobacterium probe was still functioning even after exposure to the anaerobic digestor contents is shown in Fig. 2b. A gas phase containing 0.0025% $O_2$ (equivalent to 31 nM $O_2$) is clearly detectable.

The addition of various amounts of $O_2$ to digestor contents in the slowly stirred closed vessel is shown in Fig. 3. At (a) 0.2 ml of $O_2$-saturated water was added; $O_2$ was undetectable (presumably being utilized rapidly by the facultative organisms present) and the methane concentration *increased* from 370 to 390 µM within 1 hr. The addition of 0.4 ml $O_2$-saturated water (b) led to an increase in the level of dissolved methane from 390 to 420 µM; $O_2$ became transiently

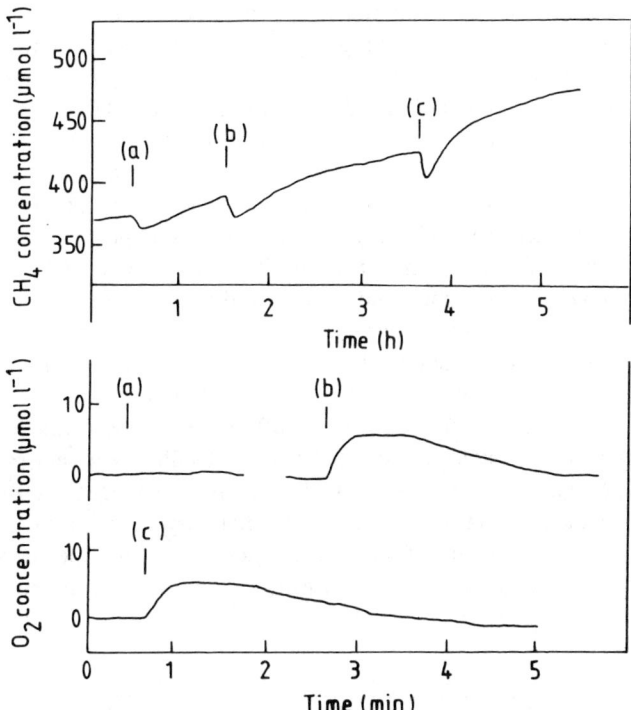

**Fig. 3.** The addition of $O_2$ to samples of digestor contents in the slowly stirred closed system. At (a), (b) and (c), 0.2, 0.4 and 0.6 ml, respectively, of $O_2$-saturated water were added to 8 ml of digestor sludge. The top trace shows the concentration of dissolved methane and the bottom trace dissolved oxygen.

detectable. Similar results were obtained on addition of 0.6 ml $O_2$-saturated water (c). Low levels of oxygen also led to increased $CH_4$ production from the digestion of algal biomass (measured by gas chromatography) in the experiments of Pirt and Lee (1983). Oxygen introduced into the gas phase by Pirt and Lee (1983) was found to be consumed within 3 days; no measurements of dissolved gases were made, however.

This qualitative difference in the response of the rapidly and slowly stirred anaerobic digestor samples may be due to breaking up cell aggregates (flocs) in the rapidly stirred system and thus exposing the methanogens to toxic levels of $O_2$. Resistance of *M. barkeri* to $O_2$ in pure culture is at least partly due to cell aggregates (Kiener and Leisinger, 1983). Increased methanogensis in slowly stirred samples (presumably with undisturbed cell aggregates) may arise from stimulation by $O_2$ of facultative organisms present which may release methanogenic precursors.

The effect of various concentrations of air on levels of dissolved $H_2$ and $CH_4$ in a sample of rumen liquor in the rapidly stirred open vessel is shown in Fig. 4. Exposure to 100% air [20.9% (v/v) $O_2$] for 3 min led to inhibition of methanogensis and $H_2$ production; dissolved $O_2$ was undetectable with the mass spectrometer but reached 0.25 $\mu M$ as indicated by the photobacterium probe. Changing the gas phase back to $N_2$ enabled methanogensis to recover completely within 20 min. Exposure to 35% (v/v) air also caused inhibition of $CH_4$ and $H_2$ production. Oxygen was undetectable by mass spectrometry or by using the photobacterium probe. Even greater sensitivity of detection is required to clarify the effect of $O_2$ on the rumen and anaerobic digestor methanogenic systems.

Figure 5 shows the effect of prolonged exposure to air on a rapidly stirred sample of rumen liquor. The methane concentration fell from a maximum of 73 $\mu M$ to 3 $\mu M$; $O_2$ was detectable with the mass spectrometer. The addition of glucose (analogous to feeding of a ruminant) did not stimulate methanogenesis but led to increased $O_2$ uptake (presumably by the facultative organisms present). Returning the gas phase to $N_2$ led to increased methanogenesis and a rapid overshoot in the $H_2$ concentration. The reversibility of methanogenesis after exposure to $O_2$ in the rapidly stirred rumen liquor samples may be due to a higher proportion of methanogens that are intrinsically more $O_2$-tolerant or possibly to

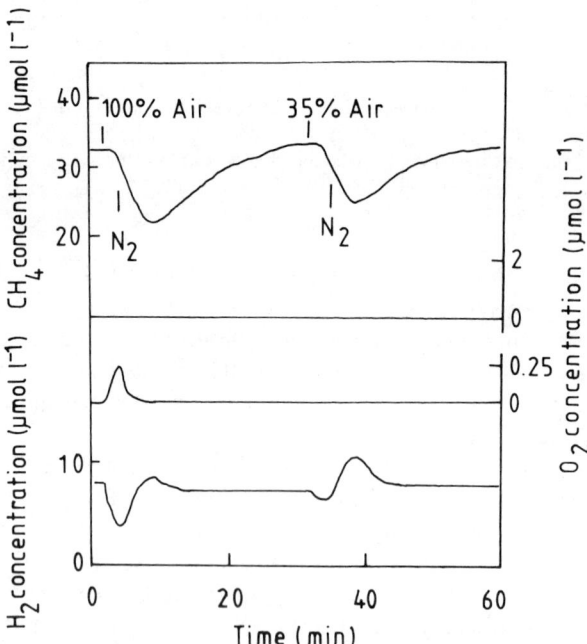

**Fig. 4.** The effect of short-term exposure to $O_2$ on methanogenesis in a rapidly stirred rumen liquor sample. Oxygen was measured with the mass spectrometer (0–2 $\mu M$) and the photobacterium probe (0–0.25 $\mu M$).

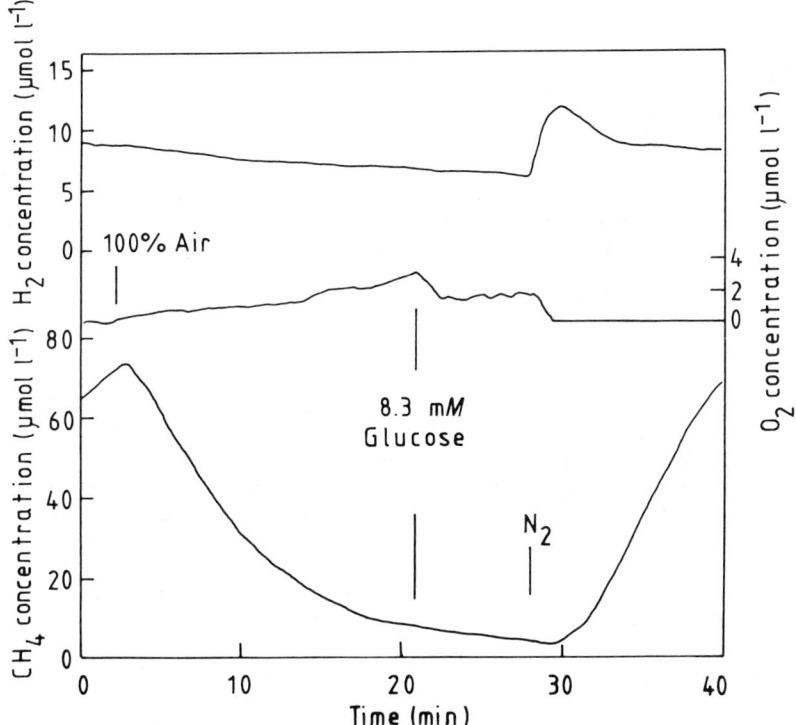

**Fig. 5.** The effect of longer term exposure of rumen contents to $O_2$. Oxygen was measured with the mass spectrometer.

more stable cell aggregates, which are not broken up by stirring and which protect highly $O_2$-sensitive organisms. Further work is required to differentiate between these possibilities and to study the effect of $O_2$ on methanogenesis of pure cultures by mass spectrometry.

## Acknowledgment

This work was supported by the Science and Engineering Research Council.

## References

Barry, T. N., Thompson, A. and Armstrong, D. G. (1977). Rumen fermentation studies on two contrasting diets. 1. Some characteristics of the *in vivo* fermentation, with special reference to the composition of the gas phase, oxidation/reduction state and volatile fatty acid proportions. *Journal of Agricultural Science* **89,** 183–195.

Czerkawski, J. W. and Breckenridge, G. (1971). Determination of concentration of hydrogen and some other gases dissolved in biological fluids. *Laboratory Practice* **20**, 403–405, 413.

Kiener, A. and Leisinger, T. (1982). Plating efficiency of methanogenic bacteria. *Experientia* **38**, 1379.

Kiener, A. and Leisinger, T. (1983). Oxygen sensitivity of methanogenic bacteria. *Systematic and Applied Microbiology* **4**, 305–312.

Lloyd, D., James, K., Williams, J. and Williams, N. (1981). A membrane-covered photobacterium probe for oxygen measurements in the nanomolar range. *Analytical Biochemistry* **116**, 17–21.

McArthur, J. M. and Miltimore, J. E. (1961). Rumen gas analysis by gas solid chromatography. *Canadian Journal of Animal Science* **41**, 187–196.

Pirt, S. J. and Lee, Y. K. (1983). Enhancement of methanogenesis by traces of oxygen in bacterial digestion of biomass. *FEMS Microbiology Letters* **18**, 61–63.

Scott, R. I., Yarlett, N., Hillman, K., Williams, T. N., Williams, A. G. and Lloyd, D. (1983a). The presence of oxygen in rumen liquor and its effects on methanogenesis. *Journal of Applied Bacteriology* **55**, 143–149.

Scott, R. I., Williams, T. N., Whitmore, T. N. and Lloyd, D. (1983b). Direct measurement of methanogenesis in anaerobic digestors by membrane inlet mass spectrometry. *European Journal of Applied Microbiology and Biotechnology* **18**, 236–241.

Scott, R. I., Williams, T. N. and Lloyd, D. (1983c). Oxygen sensitivity of methanogenesis in rumen and anaerobic digestor populations using mass spectrometry. *Biotechnology Letters* **5**, 375–380.

Smith, P. H. and Hungate, R. E. (1958). Isolation and characterization of *Methanobacterium ruminantium* n. sp. *Journal of Bacteriology* **75**, 713–718.

Wheatley, B. I. (1980). The gaseous products of anaerobic digestion—biogas. *In* "Anaerobic Digestion" (Eds. D. A. Stafford, B. I. Wheatley and D. E. Hughes), pp. 415–428. Applied Science Publishers, London.

Zehnder, A. J. B. (1978). Ecology of methane formation. *In* "Water Pollution Microbiology" Vol. 2 (Ed. R. Mitchell), pp. 349–376. Wiley, New York.

Zeikus, J. G. (1977). The biology of methanogenic bacteria. *Bacteriological Reviews* **41**, 514–541.

# 19

# The Effects of Oxygen on Hydrogen Production by Rumen Holotrich Protozoa, as Determined by Membrane-Inlet Mass Spectrometry

K. HILLMAN, DAVID LLOYD, ROBERT I. SCOTT[1] AND
A. G. WILLIAMS*

*Department of Microbiology, University College, Cardiff, Wales, United Kingdom and *Department of Animal Nutrition and Production, Hannah Research Institute, Ayr, Scotland, United Kingdom*

Oxygen may enter the rumen by various means; it is ingested with food, it is present in high concentrations in saliva (which is constantly swallowed) and it may enter through the rumen wall from the bloodstream (Czerkawski, 1969).

In order that those species assumed to be strictly anaerobic, for example the methanogens, can survive this input of oxygen, the rumen contents must possess some ability to dispose of it, either chemically (the rumen liquor is known to be highly reducing) or by some organisms that possess a facility for rapid oxygen consumption.

The facultatively anaerobic species present in the rumen rapidly utilise oxygen when the animal is fed, although this activity is likely to decrease as the readily available nutrients are removed and metabolic activity slows down. In fact, the dissolved oxygen in rumen liquor was observed to drop to undetectable levels immediately after feeding (despite the added intake of oxygen with food and saliva) and to increase again approximately 3 hr after feeding (Scott *et al.,* 1983a). Therefore, in an animal fed twice daily, detectable levels of oxygen are present in the rumen liquor for about 18 hr out of 24, so the anaerobic organisms must have some form of protection from this oxygen.

Experiments with a Clark-type oxygen electrode demonstrated the presence *in situ* in fistulated sheep, goats and cows of up to 1.6 $\mu M$ oxygen in rumen liquor

---

[1]Present address: The Polytechnic of Central London, School of Engineering and Science, 115 New Cavendish Street, London W1M 8JS, United Kingdom.

MICROBIAL GAS METABOLISM:
MECHANISTIC, METABOLIC
AND BIOTECHNOLOGICAL ASPECTS

(Scott *et al.*, 1983a). This observation led to investigations into the oxygen sensitivity of the activities and viability of the mixed population of organisms from anaerobic digestors (Scott *et al.*, 1983b,c) as well as of individual bacterial methanogens, often assumed to be strictly anaerobic (Kiener and Leisinger, 1983). In this report we confirm the high affinity for oxygen of rumen holotrich protozoa (Yarlett *et al.*, 1982; Lloyd *et al.*, 1982) and quantitate the inhibitory effect of oxygen on hydrogen generation by these genera (Yarlett *et al.*, 1983a).

A quadrupole mass spectrometer with a 50-μm Teflon membrane inlet was used to determine the effect of oxygen on isolates of the two genera of holotrich protozoa present in the rumen: *Isotricha* and *Dasytricha*. These two genera comprise the following three species in the rumen: *Isotricha intestinalis, Isotricha prostoma* and *Dasytricha ruminantium*.

Organisms were separated from rumen liquor and freed from contaminating bacteria by filtration through a series of micropore cloths (Williams and Yarlett, 1982). The holotrich protozoa (and also some rumen entodiniomorphs) possess a special organelle, the hydrogenosome, which has been implicated in both hydrogen generation and oxygen utilisation (Yarlett *et al.*, 1981, 1983a,b, 1984).

The membrane-inlet mass spectrometer (Lloyd *et al.*, 1983; Lloyd and Scott, 1983; Lloyd and Scott, Chapter 17, this volume; Degn *et al.*, 1985) was fitted with a stirred reaction vessel of working volume 6 ml. Calibrations were $O_2/N_2$ mixtures from a digital gas mixer (Lundsgaard and Degn, 1973). Half-times for solution of the gases in a buffer consisting of (in mg ml$^{-1}$): $KH_2PO_4$, 0.55; $K_2HPO_4$, 10.8; NaCl, 0.65; $NaHCO_3$, 9.0; $MgSO_4 \cdot 7H_2O$, 0.09; and $CaCl_2$, 0.01, and with stirring at 1100 rev min$^{-1}$, were 2.5 and 2.1 min for hydrogen and oxygen, respectively.

Cell suspensions were of the order of $10^4$–$10^6$ cells ml$^{-1}$. The lower limit of sensitivity for oxygen of the mass spectrometer was 0.25 μM.

Double-reciprocal plots for determination of $K_m$ $O_2$ values show that both *D. ruminantium* and mixed *Isotricha* suspension exhibited a high affinity for oxygen (Figs. 1a and 2a). Figure 3 shows a typical trace from the mass spectrometer demonstrating the decrease in steady-state levels of dissolved hydrogen that corresponded with stepwise increases in dissolved oxygen, produced by step changes of gas phase. Figures 1b and 2b show almost linear inverse dependence of hydrogen production on dissolved oxygen, until values of around 1.5 μM are exceeded; rapid inhibition of hydrogen evolution occurred at higher oxygen concentrations. These effects were fully reversible on switching back to low oxygen. Table 1 summarises these results.

The results lead to the following conclusions:

1. The rumen holotrich protozoa (*Isotricha* spp. and *D. ruminantium*) are capable of scavenging oxygen at the low levels found *in situ* in the rumen; the lower value of $K_m$ for oxygen in the latter species makes it more efficient in this respect.

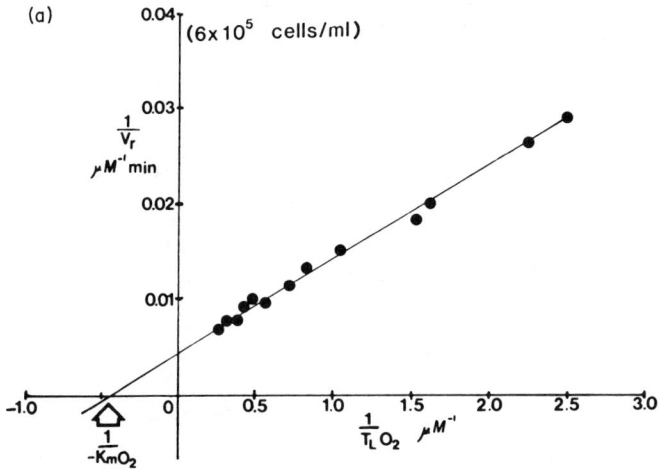

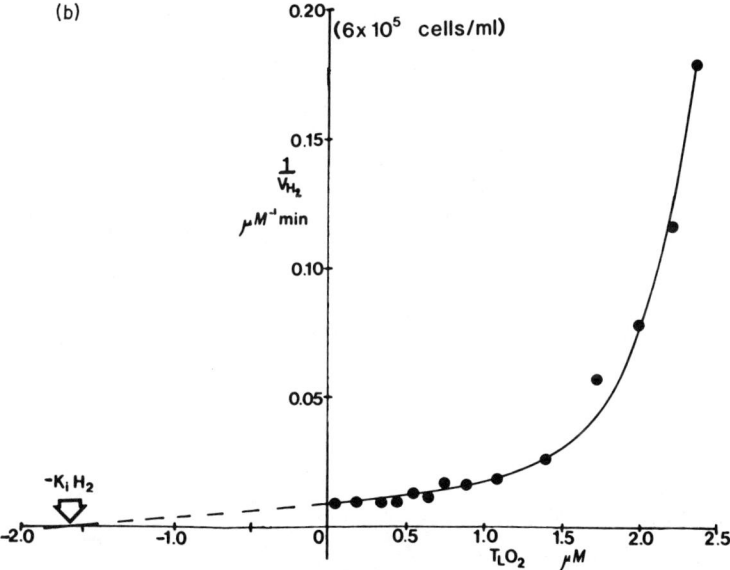

**Fig. 1.** Respiration and hydrogen production in *Isotricha* spp. (a) A double-reciprocal plot and the calculation of $K_m$ $O_2$; (b) the inhibitory effect of oxygen on hydrogen production.

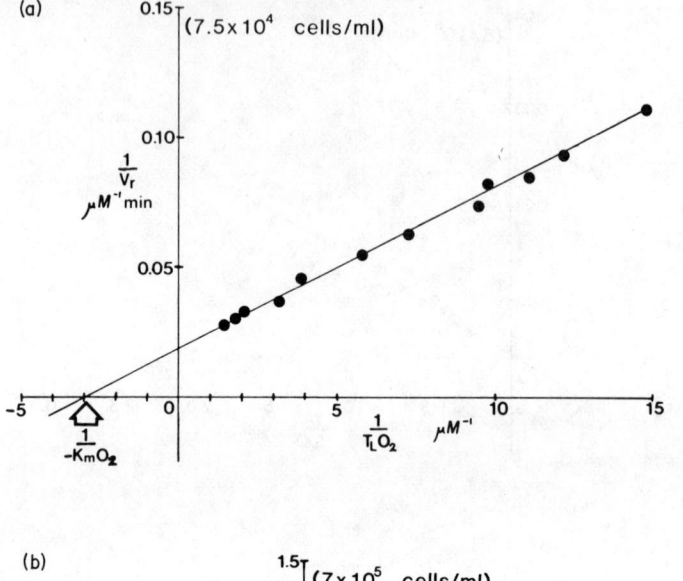

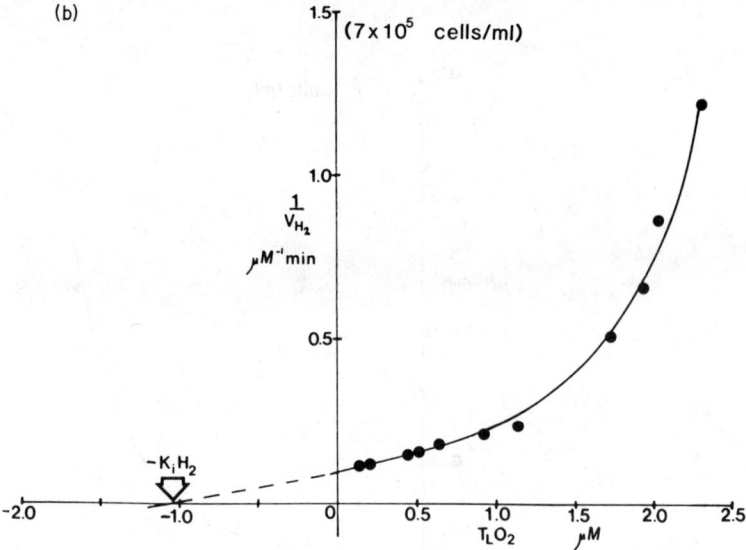

**Fig. 2.** Respiration and hydrogen production in *D. ruminantium*. (a) A double-reciprocal plot and the calculation of $K_m$ $O_2$; (b) the inhibitory effect of oxygen on hydrogen production.

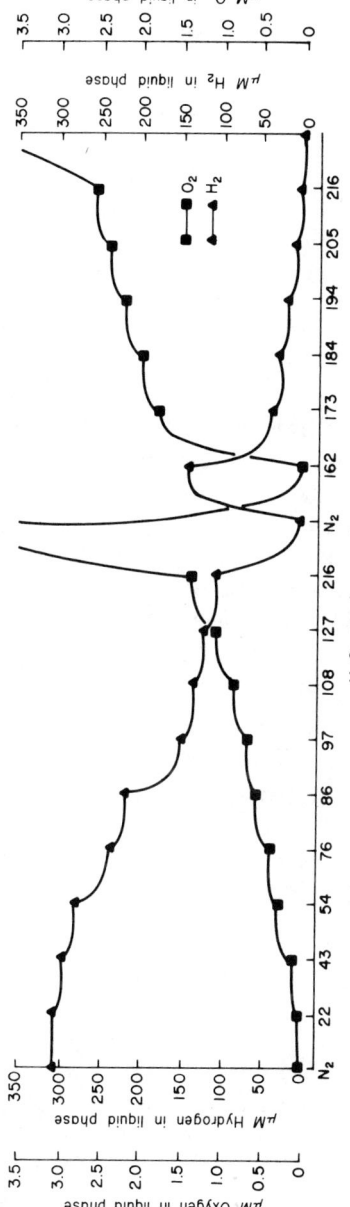

**Fig. 3.** A typical trace from the mass spectrometer measuring hydrogen production and oxygen in solution during an experiment with *Isotricha* spp.

**Table 1.**    *Oxygen utilisation and hydrogen production in
Isotricha spp. and D. ruminantium[a]*

| Value calculated | $O_2$ concentration ($\mu M$) |
|---|---|
| *Isotricha* spp. | |
| Apparent $K_m$ for $O_2$ utilisation | 2.333 ± 0.480 (3) |
| Apparent $K_i$ for $H_2$ production | 1.648 ± 0.467 (4) |
| *Dasytricha ruminantium* | |
| Apparent $K_m$ for $O_2$ utilisation | 0.336 ± 0.203 (3) |
| Apparent $K_i$ for $H_2$ production | 1.106 ± 0.901 (3) |

[a]Results are means ± standard deviations for the number of
experiments (at 39°C) shown in parentheses.

2. Reversible inhibition of hydrogen production increases almost linearly with
   increasing oxygen concentrations up to the maximum concentrations found
   *in situ* in the rumen; higher levels of oxygen have a drastic inhibitory effect
   on hydrogen production.
3. Hydrogenosome-containing protozoa may play an important role in the
   balance between hydrogenogens and hydrogen-utilising organisms, and
   ambient levels of dissolved oxygen exert an important regulatory effect.
4. Membrane-inlet mass spectrometry provides a powerful technique for the
   investigation of these relationships.

## Acknowledgments

K. Hillman held a SERC CASE studentship and D. Lloyd and R. I. Scott a
SERC research grant (GR/B 60569) during the course of this work.

## References

Czerkawski, J. W. (1969). Methane production in ruminants and its significance. *World Review of
Nutrition and Dietetics* **11**, 240–282.
Degn, H., Cox, R. and Lloyd, D. (1985). Direct measurement of gaseous exchange in biological
systems by membrane inlet mass spectrometry. *Methods of Biochemical Analysis* (in press).
Kiener, A. and Leisinger, T. (1983). Oxygen sensitivity of methanogenic bacteria. *Systematic and
Applied Microbiology* **4**, 305–312.
Lloyd, D. and Scott, R. I. (1983). Direct measurement of dissolved gases in microbiological systems
using membrane-inlet mass spectrometry. *Journal of Microbiological Methods* **1**, 313–320.
Lloyd, D., Williams, J. L., Yarlett, N. and Williams, A. G. (1982). Oxygen affinities of the
hydrogenosome-containing protozoa *Tritrichomonas foetus* and *Dasytricha ruminantium* and two

aerobic protozoa determined by bacterial bioluminescence. *Journal of General Microbiology* **128**, 1019–1022.

Lloyd, D., Scott, R. I. and Williams, T. N. (1983). Membrane inlet mass spectrometry—measurement of dissolved gases in fermentation liquids. *Trends in Biotechnology* **1**, 60–63.

Lundsgaard, J. S. and Degn, H. (1973). Digital regulation of gas flow rates and composition of gas mixtures. *IEEE Transactions on Biomedical Engineering* **BME=20**, 384–387.

Scott, R. I., Yarlett, N., Hillman, K., Williams, T. N., Williams, A. G. and Lloyd, D. (1983a). The presence of oxygen in rumen liquor and its effects on methanogenesis. *Journal of Applied Bacteriology* **55**, 143–149.

Scott, R. I., Williams, T. N., Whitmore, T. N. and Lloyd, D. (1983b). Direct measurement of methanogenesis in anaerobic digestors by membrane inlet mass spectrometry. *Applied Biotechnology* **18**, 236–241.

Scott, R. I., Williams, T. N. and Lloyd, D. (1983c). Oxygen sensitivity of methanogenesis in rumen and anaerobic digestor populations using mass spectrometry. *Biotechnology Letters* **5**, 375–380.

Williams, A. G. and Yarlett, N. (1982). An improved technique for the isolation of holotrich protozoa from rumen contents by differential filtration with defined aperture textiles. *Journal of Applied Bacteriology* **52**, 267–270.

Yarlett, N., Hann, A. C., Lloyd, D. and Williams, A. G. (1981). Hydrogenosomes in the rumen protozoon *Dasytricha ruminantium* Schuberg. *Biochemical Journal* **200**, 365–372.

Yarlett, N., Lloyd, D. and Williams, A. G. (1982). Respiration of the rumen ciliate *Dasytricha ruminantium* Schuberg. *Biochemical Journal* **206**, 259–266.

Yarlett, N., Scott, R. I., Williams, A. G. and Lloyd, D. (1983a). A note on the effects of oxygen on hydrogen production by the rumen protozoon *Dasytricha ruminantium* Schuberg. *Journal of Applied Bacteriology* **55**, 359–361.

Yarlett, N., Hann, A. C., Lloyd, D. and Williams, A. G. (1983b). Hydrogenosomes in a mixed isolate of *Isotricha prostoma* and *Isotricha intestinalis* from ovine rumen contents. *Comparative Biochemistry and Physiology B* **74B**, 357–364.

Yarlett, N., Coleman, G. S., Williams, A. G. and Lloyd, D. (1984). Hydrogenosomes in known species of rumen entodiniomorphid protozoa. *FEMS Microbiology Letters* **21**, 15–19.

# 20

# Cyanobacterial Dinitrogen Uptake Measurements Using Membrane-Inlet Mass Spectrometry

B. B. JENSEN AND R. P. COX

*Institute of Biochemistry, Odense University, Odense, Denmark*

## Methods for Measuring Nitrogen Fixation

Nitrogen fixation involves the consumption of $N_2$, and of the various microbial processes involving dissolved gases, $N_2$ uptake is arguably the most difficult to measure. This is because the substrate is relatively unreactive so that no polarographic method is available, whilst the low affinity of the nitrogen-fixing apparatus for $N_2$ and the low solubility of the gas effectively preclude measurements of uptake by following changes in the composition of the gas phase with gas chromatography, although this can be used where production of $N_2$ is being measured, as in studies of denitrification.

Two approaches to the measurement of nitrogen fixation have thus become established amongst workers in the field. The more popular by far takes advantage of the lack of specificity of nitrogenase by measuring the conversion of acetylene to ethylene. In this case a gaseous product is obtained which can be readily measured with high sensitivity by gas chromatography. The introduction of the acetylene reduction assay revolutionised research on nitrogen fixation, and as a semi-quantitative assay of potential nitrogenase activity in field and laboratory conditions it has many advantages. However, uncertainty about the conversion factors between $N_2$ and $C_2H_2$ has led to some unease amongst workers in the field, and it is frequently argued that the assay needs to be calibrated against nitrogen measurements. This is usually done by using the stable isotope [15]N, which can be detected in the organic nitrogen fraction after incubation and suitable treatment by either mass spectrometry or atomic emission spectroscopy. The high cost of [15]N and the relative complexity of the measurements, which involve destruction of the sample, have limited the use of this approach. This is

279

MICROBIAL GAS METABOLISM:
MECHANISTIC, METABOLIC
AND BIOTECHNOLOGICAL ASPECTS

in spite of the fact that many physiological investigations require use of the natural substrate rather than an artificial replacement such as acetylene.

We have developed a method for measuring nitrogen fixation which involves following changes in the concentration of dissolved $N_2$ in a bacterial suspension by using a mass spectrometer with a membrane-covered inlet (Jensen et al., 1981; Jensen and Cox, 1983). In this case (atmospheric) $^{14}N_2$ is used, the measurements are continuous and they do not involve destruction of the sample. This approach is in fact not new; mass spectrometers with membrane-covered inlets have been used to monitor $O_2$ and $CO_2$ in suspensions of photosynthetic cells for many years, and an isolated report of measurements of nitrogen uptake by Rhodospirillum rubrum appeared in 1974 (Paschinger, 1974). More recently, Berlier and Lespinat (1980) used a similar apparatus in several investigations in which $H_2$ and acetylene metabolism by nitrogen-fixing bacteria were measured. Two factors have probably prevented mass spectrometry being more widely used for directly measuring $N_2$ uptake. Firstly, the older generation of magnetic mass spectrometers were more complex and expensive than the quadrupole mass analysers which we use for our experiments. Secondly, successful measurement of dissolved $N_2$ in aqueous solution requires the use of a membrane with a high selective permeability to non-polar gases compared with water and a pumping system capable of maintaining a high vacuum, even when the gas phase comprises a high proportion of argon, which is not removed effectively by ion pumps. This means that the requirements for measurement of $N_2$ are somewhat more demanding than those for the monitoring of other gases.

## The Mass Spectrometric Measuring System

The measuring apparatus is shown in Fig. 1. It consists of a cylindrical sample chamber made from stainless steel with a volume of 7.0 ml and containing 4.0 ml of reaction medium. It has a perspex bottom to allow illumination. A vertically mounted motor drives a cross-shaped stirrer fixed on a spindle. This provides a rapidly stirred solution and a stable interface. The sample chamber is closed by a circular lid which has two peripheral holes, one for the entry and one for the exit of the gas phase, and a central hole for the spindle. The temperature of the sample chamber is controlled by circulating water from a thermostat-controlled water bath.

The apparatus is based on the open-system approach developed by Degn and Wohlrab (1971). The basic idea behind the open-system measurements of nitrogen fixation is that the rate of diffusion ($V$) of $N_2$ across a stable interface between a rapidly stirred liquid phase and a mobile gas phase containing $N_2$ is

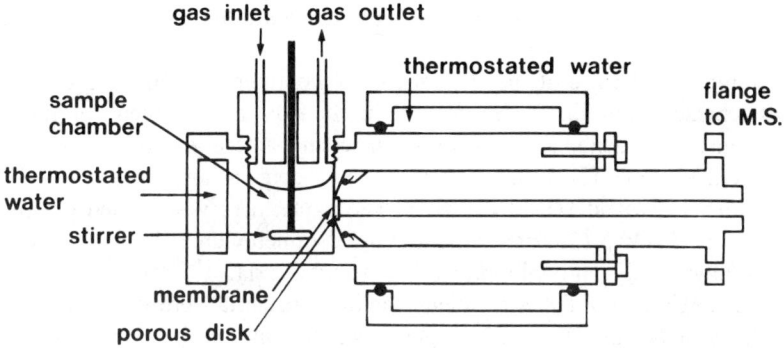

**Fig. 1.** Schematic diagram of the sample chamber. (M.S. = mass spectrometer.)

directly proportional to the difference between the nitrogen tension of the gas $(T_G)$ and the nitrogen tension in the liquid $(T_L)$. Under steady-state conditions the rate of diffusion of $N_2$ from the gas phase to the liquid phase is equal to the $N_2$ consumption by the cyanobacteria, and so the rate of nitrogen fixation can be calculated from

$$V = K(T_G - T_L) \tag{1}$$

$T_L$ is measured with a mass spectrometer with an inlet covered by a 12.5-μm Teflon membrane through which gases diffuse. We use a quadropole mass spectrometer (type Q8 from VG-Micromass Ltd., Winsford, U.K.) fitted with a turbopump (type TPH 100 from A. Pfeiffer Vakuumteknik Wetzlar G.m.b.H., Asslar, F.R.G.). The $N_2$ concentration was measured by selected ion monitoring (SIM) of the peak corresponding to mass-to-charge ratio $m/z = 14$. This peak was selected because $CO_2$ affects the $m/z = 28$ signal. The instrument allows a rapid scan of up to four different $m/z$ values, allowing simultaneous measurements of four gases.

The value of $T_G$ is given by the signal from the mass spectrometer when the liquid and gas phases are in equilibrium. The dissolved-gas concentration corresponding to a particular partial pressure is determined from published values of its solubility. The transfer constant $K$ for a particular combination of experimental conditions can be determined from the time course of the change in the concentration of dissolved gas following a sudden change in the composition of the gas phase. These transients show an exponential decay to the new equilibrium value with a rate constant equal to $K$. Degn et al. (1980) have provided an extensive discussion of the theoretical background for open-system measurements.

## Some Examples of Applications

An example of the use of the system for the continuous monitoring of several gases is presented in Fig. 2. This shows simultaneous measurements of nitrogen, hydrogen and oxygen in a suspension of the cyanobacterium *Anabaena variabilis* (strain CCAP 1403/13a = ATCC 29413). Nitrogen was measured at $m/z = 14$, $O_2$ at $m/z = 32$ and $H_2$ at $m/z = 2$. The experiment was carried out in the presence of 5 $\mu M$ 3-(3,4-dichlorophenyl)-1,1-dimethylurea (DCMU) to inhibit photosynthetic oxygen evolution by the cyanobacteria.

At the beginning of the experiment, the cyanobacteria were in the dark with a gas phase of $N_2$ and Ar. No $N_2$ uptake was possible in the absence of either respiration or light-induced electron transport. Addition of $O_2$ to the gas phase caused a transient. When the steady state was reached, the concentrations of dissolved $O_2$ and $N_2$ were both lower than the equilibrium values as the result of respiration and nitrogen fixation. Switching on the light caused another transient. At the new steady state, there was a much greater rate of $N_2$ uptake and an inhibition of respiration. Some respiration still occurred, since $H_2$ produced by nitrogenase was scarcely detectable in the steady state. Removal of $O_2$ from the gas phase caused a third transient; note the rapid depletion of $O_2$ as the result of respiration. The $H_2$ could not be reoxidised in the absence of $O_2$; higher concentrations of $H_2$ were present in the new steady state with a consequent inhibition of the rate of $N_2$ uptake.

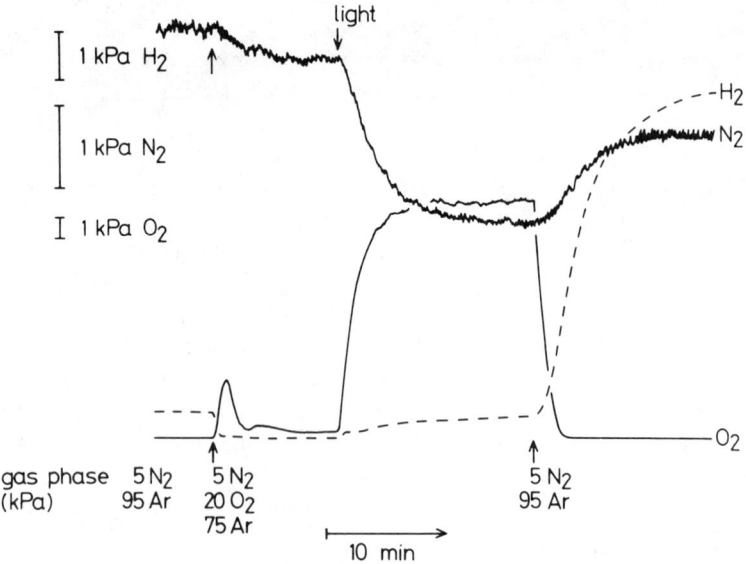

**Fig. 2.** Changes in dissolved-gas concentrations in a suspension of *Anabaena variabilis*.

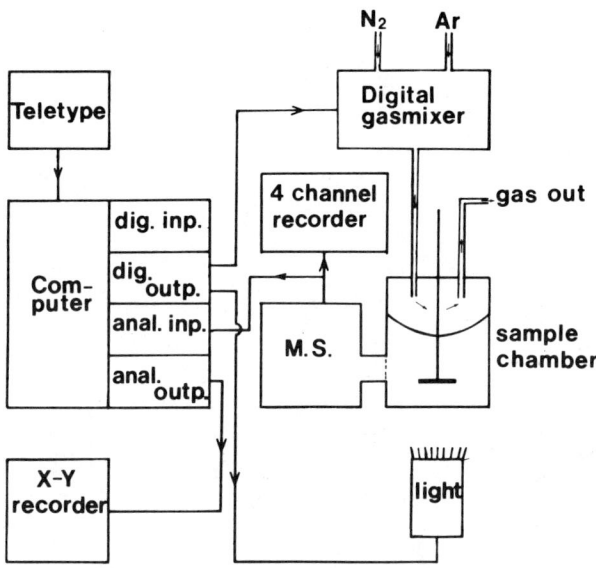

**Fig. 3.** Schematic diagram of the computer-controlled mass spectrometer with membrane inlet for on-line monitoring and control of dissolved gases. (M.S. = mass spectrometer.)

As shown in Fig. 3, we have further developed the system by connecting it to a small laboratory computer. The signal from the mass spectrometer was monitored by the computer, allowing the experiment to be performed automatically. The dissolved-$N_2$ concentration could be changed by a routine controlling a digital gas mixer (Degn *et al.*, 1980) through feedback from the mass spectrometer signal. The computer also controlled a shutter to allow illumination of the sample and collected readings from the mass spectrometer. Figure 4 shows a typical series of traces from the mass spectrometer under computer control.

At the start of the experiment the cyanobacteria were in the dark in the absence of $O_2$; the $N_2$ trace was the same as that observed in the absence of cyanobacteria and corresponded to the concentration in equilibrium with the gas phase. Illumination caused the $N_2$ concentration to move to a new steady-state level. The difference between the light and dark steady-state levels can be multiplied by the gas transfer constant $K$ to give the rate of light-induced nitrogen fixation as described in Eq. (1). The shutter was then closed and the $N_2$ concentration changed to a new value by the computer, using the feedback routine. The cyanobacteria were illuminated again when the dissolved-$N_2$ concentration had stabilised at a new value. The computer allowed rapid stabilisation of the dissolved-gas concentrations at the new level between measurements, allowed measurements of several substrate concentrations and allowed $V_{max}$ and $K_m$ for $N_2$ fixation to be determined within 90 min (Jensen and Cox, 1983).

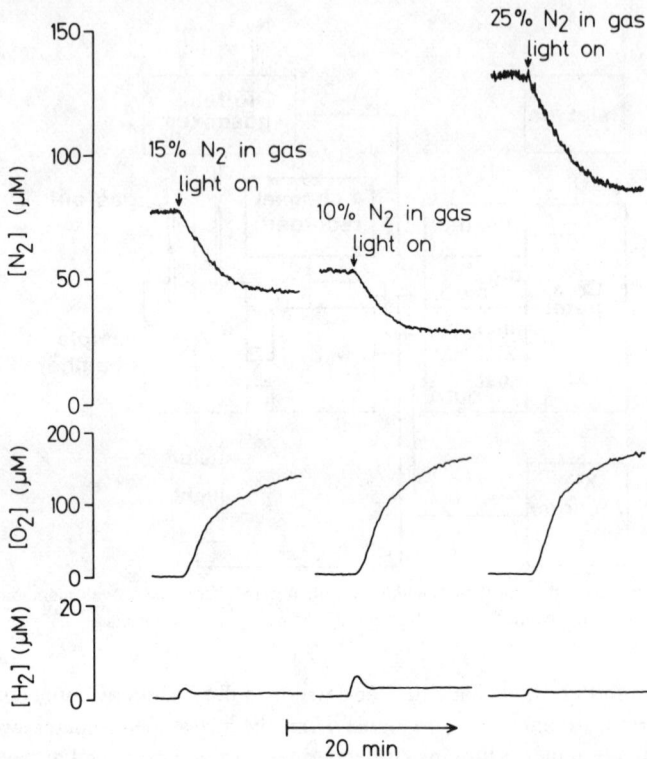

**Fig. 4.** Typical series of traces from the mass spectrometer during a computer-controlled experiment with *Anabaena variabilis*.

Figure 4 also shows measurements of $O_2$ ($m/z = 32$) and $N_2$ ($m/z = 2$). In the dark, the concentrations of $O_2$ and $H_2$ were, as expected, indistinguishable from zero. At the start of the illumination, the $H_2$ trace increased as the result of light-induced $H_2$ production by the cyanobacteria, but as soon as the $O_2$ concentration in the medium increased as a consequence of photosynthetic oxygen evolution, the $H_2$ trace fell to almost zero and stayed there throughout the illumination period. This was because, in the presence of $O_2$, the cyanobacteria reoxidised the $H_2$ produced by nitrogenase. As shown in Fig. 2, the $H_2$ trace attained a steady-state level indicating a constant rate of $H_2$ production if DCMU was added to inhibit photosynthetic oxygen evolution.

## Future Potential

A quadrupole mass spectrometer with a membrane-covered inlet provides a unique measuring apparatus for experiments in microbial physiology where dissolved gases are produced or consumed. When applied to studies of nitrogen fixation it allows $N_2$ uptake to be followed in a way which is conceptually identical with the use of a polarographic oxygen electrode to monitor respiratory $O_2$ uptake. In addition, an apparatus capable of scanning allows simultaneous measurements of $H_2$ and $O_2$, both gases of particular significance in studies of nitrogen fixation, as well as others such as $CO_2$, $CH_4$ or $H_2S$, which may be of interest in some experiments. Application of this technique should thus allow detailed information to be easily obtained about the interactions of these gases in cyanobacteria and other diazotrophic systems.

# References

Berlier, Y. M. and Lespinat, P. A. (1980). Mass spectrometric kinetic studies of the nitrogenase and hydrogenase activities in in vivo cultures of Azospirillum brasilense Sp. 7. Archives of Microbiology 125, 67–72.

Degn, H. and Wohlrab, H. (1971). Measurement of steady-state values of respiration rate and oxidation levels of respiratory pigments at low oxygen tensions. A new technique. Biochimica et Biophysica Acta 245, 347–355.

Degn, H., Lundsgaard, J. S., Petersen, L. C. and Ormicki, A. (1980). Polarographic measurement of steady-state kinetics of oxygen uptake by biochemical samples. Methods of Biochemical Analysis 26, 47–77.

Jensen, B. B. and Cox, R. P. (1983). Direct measurements of steady-state kinetics of cyanobacterial $N_2$ uptake by membrane-leak mass spectrometry and comparisons between nitrogen fixation and acetylene reduction. Applied and Enviromental Microbiology 45, 1331–1337.

Jensen, B. B., Cox, R. P. and Degn, H. (1981). Mass spectrometric measurements of steady-state kinetics of cyanobacterial nitrogen fixation by monitoring dissolved $N_2$ in an open system. FEMS Microbiology Letters 12, 37–40.

Paschinger, H. (1974). A changed nitrogenase activity in Rhodospirillum rubrum after substitution of tungsten for molybdenum. Archives of Microbiology 101, 379–389.

# 21

# Methane Oxidation by *Methylosinus trichosporium* Measured by Membrane-Inlet Mass Spectrometry

LARS JOERGENSEN

*Institute of Biochemistry, Odense University, Odense, Denmark*

## Introduction

The obligate methanotrophic bacteria are dependent on methane or other $C_1$ compounds and oxygen for cell growth. They oxygenate methane to methanol by the enzyme methane monooxygenase, which requires molecular oxygen and NAD(P)H as cosubstrates. Methanol is oxidized to formaldehyde and assimilated as cell material or further oxidized to formate and carbon dioxide. Oxygen is used both as a substrate for the initial oxygenation of methane and as the final electron acceptor by the respiratory chain. Many methanotrophs need no other organic growth factors than methane. Thus their growth kinetics can be investigated by following their methane and oxygen consumption. Both methane and oxygen in solution can be measured directly by the membrane-inlet mass spectrometry technique (Joergensen and Degn, 1983), where a quadrupole mass spectrometer fitted with a membrane-covered inlet is connected to an open-system cuvette. The open system consists of a rapidly stirred sample in contact with a continually renewed and controlled gas phase (Degn *et al.*, 1980). The methane utilization rate by the sample is calculated from

$$V = K(T_G - T_L) - dT_L/dt \qquad (1)$$

where $K$ is a diffusion constant, $T_G$ is the methane tension in the gas phase and $T_L$ is the methane tension in the liquid phase.

We have used the membrane-inlet mass spectrometry technique to measure the kinetics of methane and oxygen utilization by the methanotrophic bacteria *Methylosinus trichosporium*.

MICROBIAL GAS METABOLISM:
MECHANISTIC, METABOLIC
AND BIOTECHNOLOGICAL ASPECTS

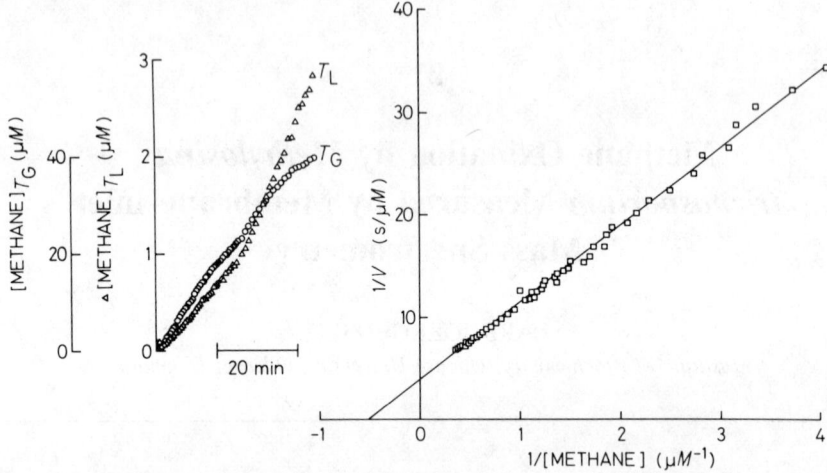

**Fig. 1.** Methane utilization by *M. trichosporium*. The methane tension of the gas phase (O) was continually adjusted by a computer-driven feedback regulation of a gas mixer to give the methane gradient in the sample (△). The gradient was initially 3 μM methane hr$^{-1}$ and changed to 6 μM methane hr$^{-1}$ after 20 min to reduce the experimental time. The oxygen tension of the gas phase was kept at 10% (v/v) during the experiment by another gas mixer. Argon was used as the carrier gas in both gas mixtures. The solubility of methane at 1 atm partial pressure, 25°C, was assumed to be 1354 μM (Barnes *et al.*, 1976) and used to calculate the methane concentration. The methane utilization rate was calculated from Eq. (1) and plotted in a double-reciprocal plot (□). *Methylosinus trichosporium* OB3b was kindly provided by Professor H. Dalton, University of Warwick, U.K. Cells were grown in batch in a 1.8-l fermentor and harvested at cell densities between 100 and 500 mg dry weight l$^{-1}$. They were centrifuged at 5000 *g* for 5 min and resuspended in 20 m*M* potassium phosphate buffer, ph 7.0, with 5 m*M* MgCl$_2$. The cell content of the sample was 600 mg dry weight l$^{-1}$. Methane was measured at *m/e* = 15.

## Methane and Oxygen Utilization

*Methane utilization.* Figure 1 shows a double-reciprocal plot of methane utilization by a sample of *M. trichosporium*. The apparent $K_m$ for methane was 2 μ*M* and the maximal respiration rate was 26 nmol CH$_4$ min$^{-1}$ (mg dry weight)$^{-1}$. The linearity of the double-reciprocal plot indicates that methane utilization followed Michaelis-Menten kinetics and that it was probably catalyzed by only one enzyme. If there were two enzymes with different $K_m$ values we would have expected to obtain a curved plot.

*Oxygen utilization.* Figure 2 shows a double-reciprocal plot of oxygen utilization by a sample of *M. trichosporium* oxidizing methane. In this experiment, we used a negative oxygen gradient to avoid an initially reduced respiration rate, which always followed anaerobiosis. The double-reciprocal plot deviated from a

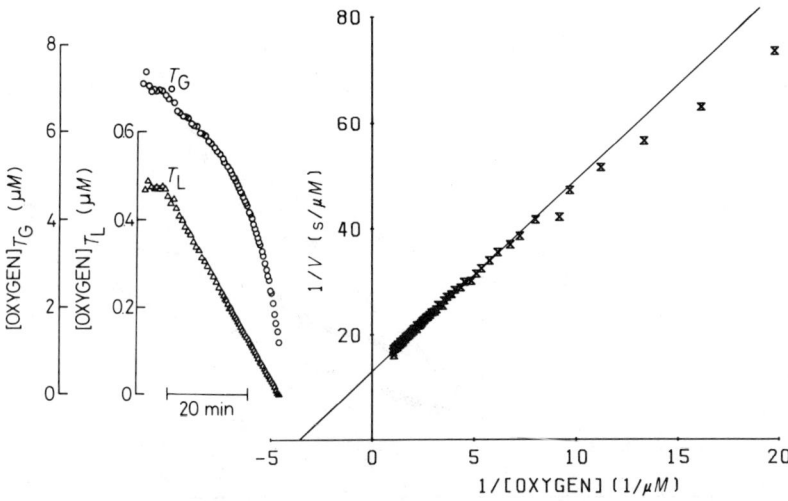

**Fig. 2.** Oxygen utilization by *M. trichosporium*. The oxygen tension of the gas phase (○) was continually regulated to give a linearly decreasing oxygen tension in the sample (△). The oxygen gradient was $-1$ $\mu M$ oxygen hr$^{-1}$. The methane tension of the gas phase was 5% (v/v) during the experiment. The solubility of oxygen at 1 atm partial pressure, 25°C, was assumed to be 1275 $\mu M$ (Barnes *et al.*, 1976) and used to calculate the oxygen concentration. The oxygen utilization rate was calculated from Eq. (1) and plotted in a double-reciprocal plot (x). The cell content was 100 mg dry weight l$^{-1}$. Oxygen was measured at *m/e* = 32. The other experimental conditions were as in Fig. 1.

straight line at low oxygen tension. This may indiate that two enzymes catalyze oxygen utilization. The apparent $K_m$ for oxygen was 0.3 $\mu M$ for the low-affinity enzyme and the maximal rate was 45 nmol O$_2$ min$^{-1}$ (mg dry weight)$^{-1}$.

The oxygen utilization by *M. trichosporium* was found to depend on the energy source. Figure 3 shows double-reciprocal plots of oxygen utilization when different C$_1$ compounds were oxidized. The plots were linear and parallel when the intermediates of methane oxidation, methanol, formaldehyde and formate, were used as the energy source. The apparent $K_m$ values and oxygen utilization rates are given in Table 1. The ratio of oxygen utilization rates for methanol, formaldehyde and formate oxidation was 3:1.95:1.3. The complete oxidation of methanol to CO$_2$ is catalyzed by two or three dehydrogenases and delivers six electrons to the respiratory chain. The first step in methanol oxidation is catalyzed by the NAD$^+$-independent methanol dehydrogenase, which has a pyrroloquinoline quinone as prosthetic group and probably passes the electrons on to the respiratory chain at the level of cytochrome *c* (Duine and Frank, 1981). The second step is the oxidation of formaldehyde by either the NAD$^+$-dependent formaldehyde dehydrogenase or the NAD$^+$-independent methanol dehydrogenase, and the final step is the oxidation of formate to CO$_2$, which may occur by an NAD$^+$-dependent formate dehydrogenase (Zatman, 1981). If the oxidation steps

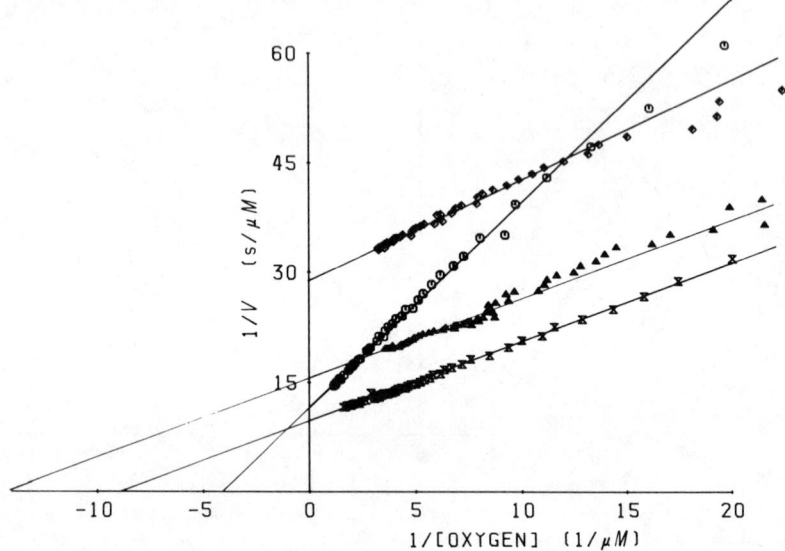

**Fig. 3.** Oxygen utilization by *M. trichosporium* oxidizing methane (○), methanol (x), formaldehyde (▲) or formate (♦). The experimental conditions were as in Fig. 2 except that no methane was added to the gas phase when other $C_1$ compounds were oxidized. The concentrations of methanol, formaldehyde and formate were 8, 2.5 and 10 m$M$, respectively.

have equal activities the ratio of oxygen utilization rates will be 3:2:1 for methanol, formaldehyde and formate oxidation. This is close to the observed ratio and shows that the turnover rate of the respiratory chain depends on the compound oxidized. The parallel plots may indicate a ping-pong reaction mechanism for the terminal oxidase and its reaction with the electron donor and oxygen. The terminal oxidase is probably cytochrome $aa_3$, as it has been found in *M. trichosporium* (Tonge *et al.*, 1974).

**Table 1.** *Oxygen utilization rates and apparent $K_m$ values for oxygen by M. trichosporium with various $C_1$ compounds as the energy source[a]*

| Energy source | $K_m$ (μ$M$) | $V_{max}$ [nmol $O_2$ min$^{-1}$ (mg dry wt.)$^{-1}$] |
| --- | --- | --- |
| Methane | 0.3 | 46 |
| Methanol | 0.1 | 60 |
| Formaldehyde | 0.07 | 39 |
| Formate | 0.04 | 26 |

[a]The data were obtained from Fig. 3.

The maximal rate of oxygen utilization for methane oxidation was between the values found for methanol and formaldehyde oxidation. If methane was oxidized to $CO_2$ and $H_2O$ at the same rate as the other $C_1$ compounds, the ratio of oxygen utilization rates would be 4:3:2:1 for methane, methanol, formaldehyde and formate oxidation. Thus the methane-dependent oxygen utilization rate should be about 80 nmol $O_2$ min$^{-1}$ (mg dry weight)$^{-1}$. The observed rate is 46 nmol $O_2$ min$^{-1}$ (mg dry weight)$^{-1}$. A likely explanation for this discrepancy is that the first enzyme in methane oxidation, the methane monooxygenase, was rate-limiting. The observed methane-dependent oxygen utilization kinetics can then be ascribed to the methane monooxygenase.

*Cyanide inhibition.* Figure 4 shows double-reciprocal plots of methane utilization by *M. trichosporium* at varying concentrations of hydrogen cyanide. The plots were found to intersect in the third quadrant, resulting in an inhibition pattern characteristic of mixed-type inhibition. The secondary plots of intercept and slope versus inhibitor concentration were straight lines. The $K_{ii}$ found from the intercept replot was 1 $\mu M$ HCN and the $K_{is}$ found from the slope replot was 2.6 $\mu M$ HCN.

Figure 5 shows the effect of varying concentrations of hydrogen cyanide on methane-dependent oxygen utilization by *M. trichosporium*. The double-reciprocal plots intersect in the second quadrant and show a mixed-type inhibition

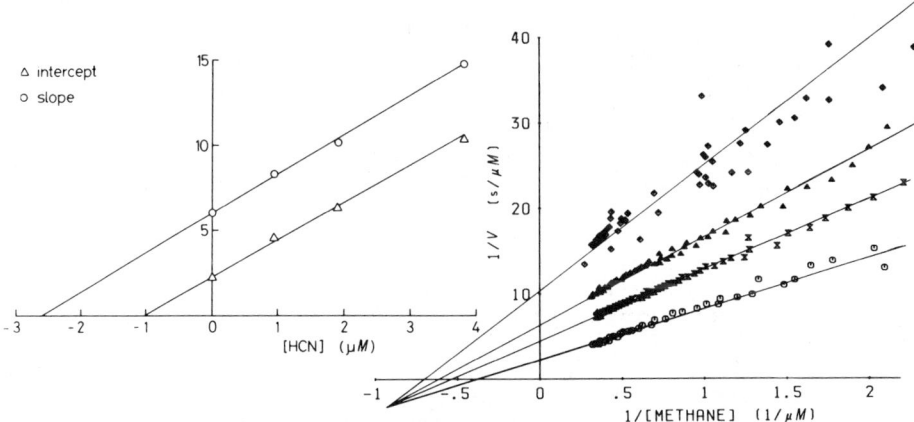

**Fig. 4.** Effect of hydrogen cyanide on methane utilization by *M. trichosporium*. Double-reciprocal plots and secondary plots of methane utilization for varying concentrations of hydrogen cyanide. The experimental procedure was as described in Fig. 1 except that an HCN/$N_2$ gas mixture was added to the gas phase as previously described (Degn and Kristensen, 1981). The cyanide concentrations were 0 (o), 0.95 (x), 1.9 (▲) and 3.8 (♦) $\mu M$. The cell content of the sample was 1000 mg dry weight l$^{-1}$.

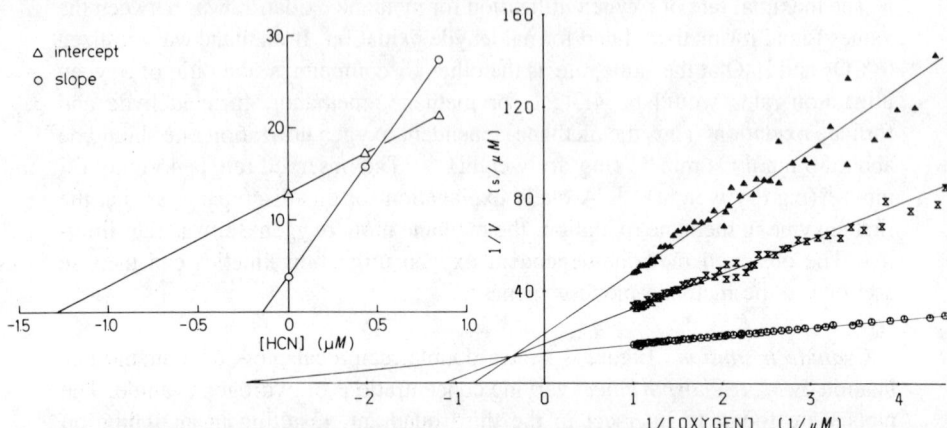

**Fig. 5.** Effect of hydrogen cyanide on methane-dependent oxygen utilization by *M. trichosporium*. Double-reciprocal plots and secondary plots of oxygen utilization for varying concentrations of hydrogen cyanide. The experimental procedure was as described in Fig. 2 except that an HCN/N₂ gas mixture was added to the gas phase. The cyanide concentrations were 0 (●), 0.43 (x) and 0.85 (▲) μ*M*. The cell content of the sample was 100 mg dry weight l⁻¹.

pattern. The slope and intercept replots show straight lines. The $K_{ii}$ was 1.27 μ*M* HCN and the $K_{is}$ was 0.15 μ*M* HCN.

The inhibition pattern found for hydrogen cyanide can be explained by a random bireactant system where cyanide is a non-exclusive inhibitor (Segel, 1975). According to the nomenclature given by Segel, the interaction between cyanide and methane binding can be expressed by the interaction factor γ and the interaction between cyanide and oxygen binding can be expressed by the interaction factor β. The relationships between the inhibition constants and the interaction factors are given in Table 2. According to the model, the βγ$K_i$ value for cyanide inhibition of methane utilization at saturating oxygen concentrations

**Table 2.** *Kinetic constants (Segel, 1975) for cyanide inhibition of oxygen and methane utilization by M. trichosporium*[a,b]

| Replot type | Intercept, $K_{ii}$ βγ$K_i$ | Slope, $K_{is}$ γ$K_i$ | β$K_i$ |
|---|---|---|---|
| Methane utilization | 1.0 μ*M* | | 2.6 μ*M* |
| Oxygen utilization | 1.27 μ*M* | 0.15 μ*M* | |

[a]The data were obtained from Figs. 4 and 5.
[b]$K_{ii}$ is the $K_i$ determined from the intercept and $K_{is}$ is the $K_i$ determined from the slope of the line.

must be equal to the $\beta\gamma K_i$ value for cyanide inhibition of oxygen utilization at saturating methane concentrations. This requirement was fulfilled within the accuracy of the measurements. The dissociation constant $K_i$ for cyanide binding to the methane monooxygenase can be calculated from these data as 0.34 $\mu M$.

## Conclusion

We used the membrane-inlet mass spectrometry method to study *in vivo* metabolism by *M. trichosporium*. Our results show that methane utilization follows Michaelis-Menten kinetics. The apparent $K_m$ for methane was 2 $\mu M$ and the maximal rate was 26 nmol $CH_4$ min$^{-1}$ (mg dry weight)$^{-1}$. The methane-dependent oxygen utilization is catalyzed by two enzymes, the methane monooxygenase and the terminal oxidase, which was probably cytochrome $aa_3$. The methane monooxygenase enzyme was found to be rate-limiting and had an apparent $K_m$ for oxygen of 0.3 $\mu M$. The apparent $K_m$ for oxygen by the terminal oxidase varied with the compound oxidized and was 0.1, 0.07 or 0.04 $\mu M$ when methanol, formaldehyde or formate was oxidized. This variation in apparent $K_m$ was probably due to the different turnover rates of the respiratory chain when these three components were oxidized, the terminal oxidase exhibiting a ping-pong reaction mechanism. The effect of hydrogen cyanide on methane and methane-dependent oxygen utilization showed that the reaction mechanism for methane monooxygenase could be described as a random bireactant system with cyanide as a nonexclusive inhibitor. The inhibitor constant $K_i$ was 0.34 $\mu M$ HCN.

## Acknowledgment

I would like to thank Dr. Hans Degn for his constant advice and encouragement. This work was supported by the Danish Science Research Council, grant 113495.

## References

Barnes, L. J., Drozd, J. W., Harrison, D. E. F. and Hamer, G. (1976). Process considerations and techniques specific to protein production from natural gas. *In* "Microbial Production and Utilization of Gases (H$_2$,CH$_4$,CO)" (Eds. H. G. Schlegel, G. Gottschalk and N. Pfennig), pp. 389–402. Akademie der Wissenschaflen, Göttingen.

Degn, H. and Kristensen, B. (1981). Low sensitivity of *Tubifex* sp. respiration to hydrogen sulfide and other inhibitors. *Comparative Biochemistry and Physiology B* **69B**, 809–817.

Degn, H., Lundsgaard, J. S., Petersen, L. C. and Ormicki, A. (1980). Polarographic measurement of steady state kinetics of oxygen uptake by biochemical samples. *Methods of Biochemical Analysis* **26**, 47–77.

Duine, J. A. and Frank, J. (1981). Methanol dehydrogenase: A quinoprotein. *In* "Microbial Growth on $C_1$ Compounds" (Ed. H. Dalton), pp. 31–41. Heyden, London.

Joergensen, L. and Degn, H. (1983). Mass spectrometric measurements of methane and oxygen utilization by methanotrophic bacteria. *FEMS Microbiology Letters* **20**, 331–335.

Segel, I. H. (1975). "Enzyme Kinetics," pp. 293–297. Wiley, New York.

Tonge, G. M., Knowles, C. J., Harrison, D. E. F. and Higgins, I. J. (1974). Metabolism of one carbon compounds: Cytochromes of methane- and methanol-utilizing bacteria. *FEBS Letters* **44**, 106–110.

Zatman, L. J. (1981). A search for patterns in methylotrophic pathways. *In* "Microbial Growth on $C_1$ Compounds" (Ed. H. Dalton), pp. 42–54. Heyden, London.

# Index

295